Surface Well Testing

With easy oil extraction becoming a thing of the past, new technologies and processes of discovery have been introduced into the exploration of oil and gas. These advancements rely on precise and accurate data, in many cases live during operations. Surface well testing operations acquire the necessary data during exploration, production, and development, and clean data is essential and heavily relied upon. *Surface Well Testing: A Practical Guide* guides readers on the fundamentals and techniques of surface well testing operations and data acquisition to ensure proper operational procedures and standards.

- Explains actual operations, equipment, and data acquisition and quality
- Introduces readers to the processes and techniques of surface well testing, the required measurements and readings, and how to get the right data to perform accurate reservoir and petroleum engineering calculations
- Bridges the gap between practical field operations and simulated engineering and mathematical models

This book supports readers and organisations in the oil and gas industry as an operations reference and training manual to ensure standardisation of operating procedures and accuracy of results.

Surface Well Testing
A Practical Guide

Paul Budworth and Abdullah Tanira

CRC Press
Taylor & Francis Group
Boca Raton London New York

CRC Press is an imprint of the
Taylor & Francis Group, an **informa** business

First edition published 2024
by CRC Press
2385 NW Executive Center Drive, Suite 320, Boca Raton FL 33431

and by CRC Press
4 Park Square, Milton Park, Abingdon, Oxon, OX14 4RN

CRC Press is an imprint of Taylor & Francis Group, LLC

© 2024 Paul Budworth and Abdullah Tanira

ISBN: 978-1-032-62364-1 (hbk)
ISBN: 978-1-032-62365-8 (pbk)
ISBN: 978-1-032-62368-9 (ebk)

DOI: 10.1201/9781032623689

Typeset in Times
by Apex CoVantage, LLC

The American Petroleum Institute in their first Glossary of Oilfield Production Technology (GOT) defines a well test as:

"The measurement of any factor, or factors relating to production or injection of oil, water or gas from, or into, a well in a given length of time for a given or established set of conditions to assist in prediction of production or injection capability."

Contents

Authors

Paul Budworth has been working with oilfields for many decades, including as an Operations Training Instructor at Gulf Drilling and Maintenance Co., Kuwait and Well Test and QA Coordinator. He served as Technical Director at Fardux, a major UK-based oil and gas data acquisition organisation, for 24 years.

After completing a four-year apprenticeship with Erie Electronics in Great Yarmouth, Paul followed in his father's footsteps and entered the oil field. He joined Expro (North Sea) in 1976 as a Trainee Well Tester and left in 1989 as a Senior Engineer. The highlight of his Expro career was his oversight of the company's data acquisition and surface readout departments. His innovative contributions earned him multiple promotions and recognitions as a pioneer in the industry.

Paul propagated his career with Production Testers from 1989 as a Field Services Supervisor to update and enhance their data acquisition and memory gauge systems. After this he was transferred to Dimas Utama in Singapore as Senior Well Test Engineer to set up the training of local personnel in well testing and data acquisition. This was followed by temporary assignments to PT Atlas in Bohai in China to both supervise and train local personnel.

From 1990 to 2014, Paul worked at Fardux as Technical Director, responsible for the development of new downhole gauges, initially invented by him and his younger brother.

In addition to his role at Fardux, Paul accepted a 1992 secondment at Geoservices Production, based in Stavanger and Paris. There he designed and developed new data acquisition programmable logic controller (PLC) systems. He later moved into writing Geoservices' data acquisition manuals, composing and presenting courses in both well testing and data acquisition. He was assigned to BP Columbia project until 1996 to improve Geoservices' operations, services, methods, and reports.

Returning from his secondment period to Fardux, Paul developed the data acquisition system side of the business. After constructing Fardux system courses, he then taught them throughout the world at different levels with different operating and service companies. This led to the early development of a wireless system as well as involvement with competency systems and online learning. He then wrote and implemented the PLC with automatic shutdown and control systems in underbalanced drilling.

From 2014 to 2015, Paul was involved with Fahud Oilfield Services in Oman. There he wrote new well testing and data acquisition systems courses and competency assessments for all levels of field personnel.

From 2015 to 2019, Paul worked with Gulf Drilling and Maintenance (GDMC), based in Kuwait, to provide technical support for improvements with their well-testing department operations and reporting. Paul assumed the role of Technical

Training Officer within the well testing, coil tubing, and slickline department. Kuwait Oilfield Company (KOC) adopted his courses, including high pressure–high temperature well testing. Paul also managed the integration of the multi-phase metering systems to the GDMC data acquisition system.

From 2019 to 2020, Paul served as a Technical Support Manager on a consultancy basis for GOFSCO and Eastern Well Test, in Kuwait, to propagate their existing well-testing operations into the HPHT market. He managed to successfully record all their procedures and methodologies.

Since 2020, Paul has invested his time in co-authoring this book with his vast and in-depth experience gathered over the years, while still assisting many around the globe in the industry regarding well testing and its operations.

Abdullah Tanira earned an MSc in petroleum production engineering at Robert Gordon University. He has been involved in petroleum engineering consultancy with projects in secondary and enhanced oil recovery, onshore concessions open for E&P investments, oil and gas production services, and waste management.

As an electrical engineering graduate, Abdullah embarked on a multifaceted career in the oil and gas industry. His journey began in 2007 as a trainee in well services at MBPS, when he delved into the intricacies of well testing and data acquisition, well intervention, downhole gauges, and instrumentation.

Abdullah rapidly progressed through various positions within well testing, slickline, and coil tubing departments across the Middle East and Africa. His expertise in well-testing data acquisition made him a subject matter expert, and he acted as the focal point for clients and operations. This led him to play a pivotal role in developing standard operating procedures, elevating quality assurance and control standards, in well testing, well service data acquisition, metering, and instrumentation.

In collaboration with Paul Budworth, during his Fardux position, they set up and commissioned the groundbreaking system integration of permanent downhole pressure/temperature sensors with well testing and hydraulic fracturing operations into a single live on-site data stream. This was an essential technical strategy for BP Oman to accomplish their first in-country hydraulic fracturing campaign.

Abdullah's unwavering dedication to innovation and excellence led him to redesign standard well-testing equipment, which revolutionised metering, data acquisition, and control systems. His expertise as an Enhanced Technical Services Coordinator and Lead played a pivotal role in pushing boundaries and driving progress within the industry.

In 2011, Abdullah took on the role of Technical Services and Development Manager at FOS Energy in the Sultanate of Oman. In this capacity, he managed specialised departments, implemented standard operating procedures for data acquisition, and spearheaded a comprehensive training program for senior well testers. His expertise in well-testing job designs and operational contracts brought a new level of efficiency to operations. He executed a reorganisation of the downhole production logging department and managed operations, successfully established and propagated the multi-phase flow meter department as an agent for MPFM, and introduced the Pason Rig Data Acquisition system to the market.

Driven by a thirst for knowledge, Abdullah pursued a master's degree in petroleum production engineering at Robert Gordon University. Leveraging his extensive production and well services experience, he focused his dissertation on software simulation that assessed drawdown and choke schedules for optimal post-fracture clean-up in horizontal/vertical wells.

He continued his involvement with oil and gas production services as well as petroleum engineering, specialised production chemicals, projects in secondary and enhanced oil recovery, onshore concessions open for E&P investments, and waste management.

Drawing upon the wealth of knowledge he amassed throughout his career, Abdullah decided to share his insights through the power of the written word. Abdullah's transition from an accomplished oil and gas professional to a promising author showcases his versatility and passion for sharing knowledge.

MEETING

Paul and Abdullah first met on a data acquisition (DAS) training course and subsequent advanced courses afterwards.

Over the years they worked together on several operations involving data acquisition and real-time reporting, merging data from several well test companies into a single database, non-intrusive sand detection on sand clean-ups, 360 degree real-time erosion monitoring, and general well testing activities.

They remain close friends, hoping their collaboration on this book would be considered an investment of knowledge and experience in fellow professionals and to the benefit of the industry.

THANKS

We would like to thank the myriad friends and colleagues involved in the oilfield companies that we have worked for and worked with, from Company Operational Personnel, Well Services, Reservoir Engineering, Well Test Operations & Management, as well as multiple Well Testers from Senior Supervisors to Junior Operators for all their suggestions and guidance who have added to our knowledge and experience over the active years of working within all aspects of the well testing sphere of influence in compiling this document.

DISCLAIMER

This document is written in good faith and we have tried to be as accurate and precise as possible. Any errors, omissions, and faults are not intentional.

List of Figures

1 Introduction

Well testing, often now referred to as surface well testing, has been practiced, mainly by specialised service companies, for many years with each generation adding its own innovation and technique.

What has to be realised is that data is the primary objective of well testing. This data is used in deriving reservoir fluid properties, characteristics and productivity by the use of a series of complicated mathematical models.

Avoidable errors in the actual well test will, and do, result in errors in the final reservoir calculations and ultimately to the financial decisions that can lead to the loss of millions of dollars.

A simple error on well testing, prior to the standard use of Data Acquisition Systems (DAS), is inputting the wrong orifice plate in the gas calculations. This drastically reduced the calculated flow rate which was expected to be in excess of 100 MMSCF/D to below 70 MMSCF/D.

The reservoir engineers, given this data, thought that there was either a fault or the well was at the edge of the reservoir. It was only at the end of the flow period when the orifice plate was raised that the error was discovered. All the data had to be manually recalculated and resubmitted. The operating company were understandably discontent with having to re-evaluate all their reservoir models. This error should have been "picked up" in the field.

It was made clear from reading posts on LinkedIn, Facebook, and other social media sites that have a specific well testing content that the junior and new hire personnel are not familiar with, trained for, or safety conscious on the basics of well testing. Additionally, years of experience with teaching well testing – and dealing with operating companies' inexperience – served as a catalyst for writing this book.

The purpose of this book is to ensure that all personnel involved in a well test understand how, where, and why the well testing parameters are derived from and the procedures involved, so that simple mistakes are minimised and more accurate data is reported whilst making operations safer.

Most operating companies and safety personnel do not have the formal training and exposure to practical surface well testing, even though they are expected to oversee the operations. Basic sediment and water (BS&W) errors, water cut, account for the majority of the erroneous reported flow rates solely because of malpractices and inadequate training of well testing crews.

This book intends to assist in comprehending surface well testing operations in connection with its appropriate data objectives, to overcome the current shortcomings. Also, by detailing the metrology and procedures involved, this book enhances safety awareness among the company personnel and well test safety officers (HSE).

This book will allow ALL involved personnel access to the fundamental techniques and methods when majority of the relevant documentation is copyrighted, proprietary to the major well test companies and absent from the public domain.

DOI: 10.1201/9781032623689-1

Several of our friends and ex-colleagues, in supervisory, safety, and management positions in different service companies, had asked us for help to improve their operator's basic well testing performance. This started with explaining BS&W and safety and has expanded over time to include several of the other aspects within this book.

1.1 WHY THIS BOOK

With the easy oil becoming a thing of the past, new technologies and processes of discovery are being introduced into the exploration of oil and gas. These advancements are relatively more expensive relying on precise and accurate data – in many cases live during operations.

The well testing companies' manuals, by large, do not detail the basic measurements, sampling methods, and respective techniques. It often just refers to a sample, measurement, or reading.

The fundamentals of surface well testing are referenced for the benefit of many.

   i. Service companies and personnel, by describing the operating of equipment, the processes, actual sampling procedures, essential measurements, readings, and calculations, as well as the reasons and precautions to ensure that the client receives the correct data and calculations

   ii. The operating company representatives and personnel that oversee operations (company employees): This manual will aid them in verifying that the operations are being carried out to a standard. The company representatives, often, have a diverse technical background with little well testing expertise, leading them to unquestionably accepting data handed over by the service company. With this book as a guide, operating company personnel would be able to confidently witness recorded readings and taken measurements

   iii. Reservoir engineers would find this book useful in understanding the origin of their input data, ascertaining the source of any errors when compared to other measurements such as downhole logging tools, and better designing their well testing campaigns with the understanding of the limitations at hand. Specialised field engineers coordinating a multi-discipline operation under their speciality often need to understand and confirm data received from well testing operations that complement their overall campaign

   iv. Private investors in exploration and production of hydrocarbons or land owners with land and mineral rights are often unaware of the source of data any development is built on. This book can be used to give these stakeholders an indication of how the testing of the wells should be carried out

There is nothing new or fundamentally different in the proposed methods within this book, rather, it contains the basics that have "fallen by the wayside" which have been established over years of operations.

The book lays out some of the recommendations, standard methods, practices, and techniques in carrying out a surface well test so that the petroleum and well test companies can be guaranteed a report that contains reliable, accurate and repeatable data, bringing quality and consistency back to the results and reports.

Working around the world in various countries and companies it has become prevailing that the overriding consideration for a well test supervisor is to have a degree, often in petroleum engineering or similar. Yet, surface well testing operations are not taught in depth in these curricula, and consequently the supervisor learns in the field from his crew. Often, this magnifies the inherited errors and malpractices without proper training, allowing them to flourish and eventually dominate the operations.

A reference is rarely found among company personnel, rig crew, field engineers, and HSE staff, leaving understanding of proper well test procedures in consideration of acquired data to the dominant opinion. This book states the overwhelmingly correct methods involved in carrying out a well test. A manual that supervisors and engineers can use to update their knowledge with and use to train crews is lacking. This book is a source to which operating company personnel who, if inexperienced in well testing, can now refer on how data is obtained rather than accepting any results.

It has always been – and should continue to be – a mandatory practice for the operating company dedicated well testing personnel on site to witness and check the unit's performance, in terms of methods, techniques, and procedures, evaluating the individuals on the job against their position and experience, guiding them when necessary. They are also often the last line of revision and approving authority on the data supplied to the engineers at the headquarters, submitted by them on a frequent basis.

The rise of enhanced oil recovery and advanced technologies/processes may influence the well testing process as well, and without the fundamental understanding it becomes difficult to adapt to the purposes and objectives of well testing. An extreme example is extra heavy oil (< API 10), which is produced by the use of chemicals, when on the surface the oil is so dense that water floats on top of the oil, as opposed to the natural order, which in turn flips the separator's function and all data measurements.

What has also happened is that with the advances and introduction of less human-involving new technologies, instrumentation, and computers, some of the basic and fundamental practices are no longer regularly or properly practiced. These flawed practices that are routinely employed by the well testers can fundamentally affect the following:

i. Flow rates
ii. Calculations
iii. Reporting
iv. Sensor correlation

It is often not the unit operators' fault but rather a symptom of improper training and a lack of understanding in well testing procedures, how the equipment functions, and how to properly operate it.

In times of austerity, which the oil industry regularly goes through, training is the first to suffer cutbacks and closures. This, with the retirement of experienced staff that are capable and equipped to train and guide the younger generation of professionals in the field, widens the gaps in the practice and theory of surface well testing.

What has to be remembered is that before the first slump of the 80s, well test service companies used to send their employees for training in periods of up to three months at a time at regional training centres. These days the equipment used in well testing is more complex yet the training periods are shorter and diminished.

Detrimental circumstances to results were presented to the client, and what is the more worrying is that many of the supervisors and operators are unaware of the shortcomings. In some cases, the policy of some of the service companies, "we always do it that way," prevails and little heed is given to the correct procedures.

During the operational phases of data acquisition over 90% of the issues are from manual readings being incorrect. This may not seem concerning, but it actually is; because if the manual readings are recorded incorrectly, the report is erroneous and costly efforts and time have to be spent in correcting the errors – if they were ever discovered. Otherwise, undiscovered errors are delivered to the client, used as input into their models and simulations and then used to set plans of large investments based on the results with amplified mathematical errors.

Detecting errors after the event is demanding and costly, to reputation at least. Once identified, it must be proved why and how and that only the errors identified are incorrect and nothing else. They then must be corrected, if mathematically possible or at the cost of repeating well test operations. Once data validity is suspect, the well test supervisor (occasionally also the operating company representative) is responsible for sufficient proof that the readings/data are correct; resetting the client's confidence can be very difficult.

Within this book, manual readings sheets are mentioned repeatedly, which is a handwritten document that contains all the activities of the crew throughout the test periods, including all valve changes, manual readings, observations, flare activities, etc. The manual readings sheets are mandatory and should be the primary reference in the event of any incident or accident by HSE or other investigating personnel. Non-compliance should be considered a disciplinary offence.

Unfortunately, manual readings sheets are often neglected by the crew, even though their importance cannot be emphasised enough and their content of all data and sequence of events are essential records. Having them updated frequently by the crew manning equipment allows them to detect any changes in the flow conditions, or even errors before they are prolonged, especially when they cannot see the data acquisition screen within the site office.

For example, if the BS&W readings are constantly at 3% and one is recorded as 9%, this should be instantly picked up by the crew, prompting a repeated sample reading. If the sample is correct the change should be reported to the well test supervisor for further investigation.

Many well testing crews find completing manual readings sheets a tedious task or refuse to comply with the task, and what is more concerning is that the well testing companies fail to impose their use. It will go a long way to improve tests, predicting possible HSE concerns, noticing and then proactively dealing with unexpected changes and most of all

Redundancy, a second set of recorded data in parallel to the data acquisition system:

   i. To compare with the electronic data acquisition, with independent measuring instruments
  ii. A back up of the well test data in case of any unexpected failures in the electronic data acquisition system due to power/system failures or loss of data

Multiple excuses exist for not filling in the manual readings sheets, from "data acquisition records it" to "it's difficult to read and write at the equipment." Not filling out the sheets is unacceptable practice and often a contractual obligation, unknown to field personnel until it's too late. More importantly it's the first line of defence for potential hazard and operating failures at no cost, and it re-enforces recognised practices.

On several occasions the service company have been forced to repeat a whole well test, even long-duration ones – or pay for a second service company to do it in their stead as their data was that questionable.

To stress the importance of proper manual readings it is best portrayed when something goes wrong and the well test supervisor is directly responsible for:

   i. Identifying the errors:
     By going through the data individually or using graphs to actually find where the event took place and trying to find the times. The manual readings sheet would ease this process, if it was used
  ii. Finding the correct reading:
     This involves trying to find the manual readings sheet or (individual crew members) tally books to find what the recorded reading was, and if there is any error the personnel involved have to be tracked and accounts verified. If A unified manual readings sheet was used all this would be avoided
 iii. Checking manual readings sheets:
     This is often quite challenging when over 70% of the job's sheets are not completed or the manual reading sheets are illegible from lack of care and scribbles
 iv. Checking all manual recording devices:
     Check Barton and Foxboro type recorders, if used. When left on the separator during unit mobilisation, the vibrations affect their internal mechanisms, frequently resulting in loss of calibration
  v. The calibration of a sensor was the cause of error, an argument seldomly accepted by client without substantial proof. Similarly, leaving the sensors on a separator during transport can and will cause calibration errors

Once the data error(s) are identified and corrected then the report has to be regenerated from the start of the test, while discarding all previous reports. It involves a lot of time and effort from the data acquisition engineer and supervisor(s) to ensure proper sequence of events and relative data. The well testers now should also provide sufficient proof that this time around the readings/data are correct and re-establishing the client's confidence will prove very difficult. Records of manual readings sheets are persuasive and a great assistance if not the only proof in tandem with the electronic data and charts.

If the data is being transmitted in real time then trying to update already transmitted data and files becomes very difficult. This is particularly troublesome with integrated services and live data is widely relied upon for on-the-spot decisions. Explaining wrong data has been passed on would certainly be met with extreme disappointment.

If the well test operators had done their jobs correctly then the onus on the data acquisition engineer to vet the data would be easier, with higher confidence and reports produced faster with more accurate results.

The current vogue is to utilise online competency training to compensate for the lack of true and proper operational training. Furthermore, what is of concern is the fact that a lot of the well testing competency courses have more safety sections than practical well test and data acquisition practices. This of course does not imply that safety is unimportant, but surface well testing should be treated as a separate subject that must be passed for competency, and courses can be expanded to include the topic in greater depth with more accurate assessments of individuals.

Even data acquisition courses are now focused more on the interaction with the human interface software than they are on what data is, how it is measured/acquired, and calculated results.

After interacting with staff responsible for writing and running in-house competency courses, the feel is the courses are designed to deliver material with intentions of students/operators to pass on completion rather than to truly assess the individual's competency. The course generally entails a series of witnessed tasks the individual has to complete and they have to be verified as having been performed correctly. Although these tasks are required during operations, neither the aspect of timing nor application are considered.

Concerning the question "have they performed it correctly?" – more importantly how "competent" is the witness to assess them? What training and qualifications does the witness have on the subject matter and delivery?

The tasks and questions should be designed, set, and marked by a qualified/experienced well tester, rather than a qualified assessor. An oilfield accreditation is normally sought, such as API, AGA, or SPE equivalent, to ensure the competency courses are recognised and substantial. Although a local accreditation may be accepted within local authorities and companies, they may not be recognised elsewhere even if conducted by colleges or universities.

These institutes set the standards upon which work methodologies and testing are based for their operations. Any competency course should align with these standards and receive equivalent recognition.

Service companies may provide online training courses, which transfer the required knowledge, but the completion certificates and grades should be issued based on a classroom and workshop set of exams. Otherwise, the system could be abused and assessments not truly reflective of a trainee's true capability. More decisive is to also have practical assessments on site to ensure that theory can be put into practice.

The approach and implementation of the competency courses is to demonstrate the ability of an individual. This should not be designed to compensate for the lack of true and fully competent well test operators or for insurance/indemnity type purposes.

It would be challenging to undertake well testing if there were no qualified crew members present. Since most well testing jobs are performed at the end of well planning, it often results in a rushed execution, managed by operating company representatives with little or no formal experience in surface well testing.

The rapid call outs, poorly prepared locations or rig, and above all the rush to get "rigged up" within a timeframe are impractical, in some cases dangerous and in breach of health and safety regulations. To compensate for the risks involved with surface well testing, the more modern practice of test the well on paper (TWOP) offers protection, but in many areas these pre-planning sessions are not adopted.

Well tests without proper planning and preparations often result in an unclear objective or structure. In some cases, the wrong type of test is carried out due to a lack of understanding by the on-site company representatives or improper equipment on site.

With the shortage of experienced work force, early individual promotions, and cross-department employee transfers, a lot of the operating company representatives assigned to well testing operations lack the proper information, knowledge, and operational awareness, relying heavily on the trust built between them and the service provider. This sort of reliance is not always deserving or warranted and it should not be unchecked. Auditing and accountability are for the greater good and both the service provider and operating company rely on the integrity and quality of data from the field, for safe operations and exploration/production decisions or activities. A lot of people's lives are at stake and expensive investments are at risk; unfortunately, there have been multiple instances with breaches in trust.

A standard book serving as a reference and manual allows company personnel, engineers, and even safety officers enough understanding to assess and ensure safe and true acquired data on behalf of their employer.

Moreover, the results and reports from the well test jobs are eventually used by the reservoir engineers in complex computer simulations to determine the properties of the reservoir. If the submitted well test results are not based on a standardised or documented procedure, then the reservoir engineer's derived results can be incorrect due to a simple manual measurement error.

This applies mainly to the submitted flow rates where multiple manual inputs are involved.

Besides the big international well test service companies, smaller more localised ones exist that only work in a specific area or country. These smaller companies rarely have full-time training and/or quality assurance engineers within their well testing departments. Well test professionals are employed with their relative experience and their practices are inherited and adopted from other companies. Seldom are these experiences checked and may even be conflicting between well test personnel from different schools or experiences.

This book does not intrude into the company safety systems, procedures, codes, and PPE. These are normally well laid out, policed, and enforced. Yet many of the well test crews only pay what is referred to as "lip service" to these procedures when they can use a shortcut. The HSE enforcement should be part of the supervisor's responsibility and regularly verified by company personnel.

For instance, assuming pressure of 2,000 psi when it is actually 4,000 or 6,000 psi could result in catastrophic well test equipment failures and flows to lower pressure rated vessels unable to handle higher flow rates.

We have to consider that the specified HSE is in place, but helmets and glasses are not going to prevent accidents. Having personnel sufficiently trained in the basic operations and limits or constraints of the well test are the company's preventive measures.

It is this whole area of misunderstandings, failings, and ignorance in well testing that this book hopes to address by bringing back the original practices developed over many years by many competent operators in many different service companies.

This document is not intended to be a definitive working procedure but a guide. It is not intended to be taken as an original piece of work but a collation of methods, techniques, and practices that intends to support the results of the technical and analytical techniques employed in the analysis of the well. Simple manual errors can have serious errors regarding the predicted results from incorrect procedures.

This document is not intended to be read or used sequentially and is written in a modular manner.

To explain if the well in question is a dry gas producing well, then the sections referring to oil and liquid measurement are not used. However, in the case of an oil producing well, then the sections on oil, gas, and liquids become relevant.

All descriptions and advice are based on generic equipment throughout this document and should not replace company instructions and training.

2 Well Testing Preparations

The level of well testing personnel assigned to any job varies, mostly subject to the following reasons.

i. Operations – A well test can be a complex or simple operation depending upon the design and amount of work that has to be carried out
ii. Company Policy – Often a service or operating company will contractually specify the number of people employed on location
iii. Space – On an offshore location space is often a premium and as such numbers are limited
iv. Cost – The one that nobody wants to talk about

2.1 OPERATIONS

A well test can vary from a simple clean-up/flowback to a full-blown test involving multiple service companies and operations.

A valuable process, currently exercised, is to test the well on paper (TWOP). Before the test plan is finalised or unit mobilised, the operating company conducts a series of meetings with all the involved service companies, HSE, and representatives of the rig/offshore platform. The service company's personnel should also include the proposed job supervisors where their operational requirements and safety concerns are explained, discussed, and rectified. This exercise should be carried out and run by the well test engineer or proposed employee who will oversee the test.

Through this process the well test can be properly planned for; designed with understood objectives, proper data, and reporting; and then the potential operational and safety issues can be identified and planned for.

At the end of the TWOP the operating company would publish the outcome of the process as the official proposed procedure, which would be signed off by all involved companies.

2.2 TWOP (TEST THE WELL ON PAPER)

Test the well on paper (TWOP) is an essential exercise often undertaken to effectively and safely review a well test program, the procedures and planned sequence of events, whether onshore or offshore. The TWOP process is similar to the drill the well on paper (DWOP) which helps identify potential drilling issues.

A TWOP exercise should analyse each individual operational step of the well testing process and events. Ensuring the well testing unit is equipped for the purpose, personnel are competently aware of the program, while taking into account hazard identification and analysis that are capable of causing either harm or damage. In addition to hazard identification (HAZID) and hazard and operability studies (HAZOP),

DOI: 10.1201/9781032623689-2

TWOP may include or lead to other processes such as layers of protection analysis (LOPA) that further HSE and operational efficiency.

In the case of repeated gas oil rate (GOR)/pressure gas oil rate (PGOR) testing within the same reservoir or concession area, involving a well testing campaign, there should be an adhered to initial or master procedure based on TWOP. If there is any significant change to the standard test then the initial TWOP should be reviewed.

This will act as a catalyst for improving and amending:

- Safety/HSE
- Operations
- Communication
- Performance
- Efficiency
- Costs

The TWOP requires active participation of all companies involved in the operations and procedures. Field personnel with diverse expertise would provide greater value when attending TWOP.

2.2.1   TWOP ATTENDEES

Company representatives and attendees involved with the test planning should include but not be limited to:

   i. Well Test Engineer – TWOP Coordinator
  ii. Company Petroleum Engineers
 iii. Company HSE Engineers
  iv. Well Site Manager
   v. Company Man/Rig Company Man/Company Representative
  vi. Rig Tool pusher or representative
 vii. Representatives from all service companies
      a. Operations Manager
      b. Job Supervisor
      c. Coordinator/Base Supervisor
      d. HSE Manager
      e. Tools Specialists

While well test services may vary from a simple clean-up to a complex well test, often multiple services are involved around or during well testing operations. Having representation from all the service companies, support divisions, or suppliers at the TWOP allows for a smoother execution and initiates rapport between them.

Attendance of field personnel allows for practical operational participations; also the discussed points and requirements are noted by them directly, making it more effective when cascaded down to their teams in context of the major plan and specifically to their responsibilities.

Ultimately TWOP exercises a planning for appropriate test design, proper equipment, necessary resources, reduced rapid call outs, ensured location or rig readiness, operational issues, safety issues, and above all practical timing of sequence of events to avoid any urgency or standbys that increase losses and/or incidents. The company representative, being the focal point, orchestrates the job on site, which makes their attendance critical.

The TWOP will be exercised in chronological order, starting from when the well is considered completed and ready for testing. The sequence of events discussed depend on the type of well test conducted, but in general the following events are covered:

- Positioning of equipment/loading operations
- Supply of air, water, and electricity for WT equipment
- Completion operations
- Bottom hole instruments
- Safety meetings
- Environmental concerns and requirements
- Pressure testing
- Perforation sequence
- Initial flow
- Burner operations
- Stimulation
- Slickline operations
- Main flow periods
- Sampling
- Production logging, PLT, etc.
- Build-Ups
- Rig down and abandonment

Each of these would typically have its standard operating procedure (SOP) created by the service provider, as well as HAZOP and HAZID analysis that should be revised during TWOP to ensure they reflect the design and purpose of the test.

Finally reporting and data corresponding with the well test objectives are extensively discussed during TWOP. The details vary from the units used to the way data is acquired, transmission of data, if necessary, daily reporting requirements, and the format for final report submission.

2.2.2 Pre-Test Data

Before the mobile well testing unit and its crew are mobilised, the operating company are required to provide data to the well test company if available and applicable, to assist in proper test planning, operations, and safety. This information may be based on historical records, engineering designs, deductions, or estimates. General information needed:

    i. Details of other wells within the same formation that are relevant
   ii. Well depth data; this will include:
      a. Total depth of drilling
      b. Formation depth details
      c. Number of zones
      d. Formation lithology
  iii. Projected depth of formation perforation(s); this will include:
      a. Projected perforation depth From – To
      b. Expected perforation type
   iv. Type(s) of formation – any sand or particulates expected
    v. Type of fluids expected; this will include:
      a. Oil gravity, GOR
      b. Gas gravity, H_2S, CO_2, N_2, other contaminants
      c. Water salinity, pH
   vi. Expected production of all phases, including:
      a. Maximum flow rates
      b. Maximum flow pressures
      c. Maximum flow temperatures
      d. Maximum expected choke
  vii. Expected shut in pressure at surface
 viii. Completion diagram – may be updated on location
   ix. Tubing dimensions – internal diameter for volume calculations and any potential flow restrictions
    x. Restrictions in tubing and depths:
      a. Gravel packing
      b. Smallest nipple
      c. Safety valve
      d. Downhole chokes
      e. Injection points
   xi. Fluids lost during drilling operations; this will include:
      a. Lost fluid(s) details
      b. Lost fluid(s) properties
      c. Volume of lost fluid(s)
      d. Type of lost fluid(s)
      e. Any completion issues
  xii. Stimulation information; this should include:
      a. HSE details of all materials
      b. Any fracturing operations
      c. Other stimulation operations
      d. Fluid lift operations
      e. Potential emulsion issues

xiii. Surface completion details; this should include:
    a. Wellhead details
    b. Wellhead pressure ratings
    c. Wellhead flanges' details
    d. Expected height from floor base

All of this data helps the well testers and reservoir engineers to understand expected performance, flow, and potential issues.

The information gathered influences the data acquired by the well test and typically is included in the final well test reports to provide the right context and circumstances for any future references. Without this detailed information, there would be many different possible interpretations of the data and inferences that could be deduced from the results.

Without such information as reference, the well test design may not be adequate or the well test results, over the life of the well, would likely be interpreted or evaluated incorrectly. The following examples demonstrate how information on the well and ongoing and previous well operations may influence well testing data:

- If a smallbore nipple is installed in the well completion throughout the lifetime of a well, post well test, it would function as a choke and restrict flow; it will act as a choke and restrict fluid flow. A well test design intended with high flow rates may be prevented from producing them by the nipple's flow restriction
- Tubing size is a major factor with high gas flow rates, since the flow friction caused may limit the well's flow potential. If this happens, the well test flow data will not be representative and the test's intended goals unachieved
- Fluids and chemicals used for drilling or stimulation operations will surface. For the sake of HSE and sampling, the foreign substances must be taken into consideration; H_2S if Scavengers are used, pH if acids haven't been neutralised, water specific gravity (SG) of injected fluid may be measured instead of formation water, proppant production may damage equipment, etc

Although an operating company would be most qualified to examine any unexpected results, given their background knowledge on all the operations carried out, well testers will always face questions regarding the performance of wells. An experienced well tester or data acquisition engineer would be able to better anticipate or explain the well test, any unexpected well performance, or acquired data while taking into account any influences, if they had a thorough understanding of variables, whether from concurrent or past well operations.

It is crucial to have the daily and final well test reports include any anomalies and deviations from expected results or unexpected events, together with descriptions of their causes and consequential effects. Engineers in the present and distant future will be able to understand the accurate history of the well performance and relate it to the operations. It is seldom simple to compose cause and effect with complete certainty, and it requires collaboration between different subject matter experts and focal points in the operating company, service companies, and the surface well testing service company.

Furthermore, within the well testing package and operations it is vital that the service company personnel are aware of variables that may also affect and misrepresent data. These are covered in further detail in later chapters, but as an introductory example here, the equipment moved from one well to another may still have residual fluids or flushing fluid left in them and would misrepresent data measured on the next job.

At the end of the TWOP, the operating company should then publish the final proposed procedure according to the exercise, which would be signed off and adopted by all the involved companies.

2.2.3 TEST TYPES

There are several different types of well tests depending upon the purpose, the reservoir, and the information the reservoir engineers require.

The classification of test types can be confusing. Well testers have different terminologies depending on geography, duration, sequence of events, equipment used, etc. such as flow after flow, pressure build-up, clean up, and multi-well/single well, but these common names have little to do with operating companies' types of tests.

Subsurface/reservoir tests are mainly classified into:

  i. Exploration/appraisal well testing
 ii. Development well testing
iii. Production facility well testing which is used to cross check facility equipment or to cover up for downtime with service and maintenance

Each test within these categories is to gain specific measurements or estimate parameters/values, either for reservoir evaluation, reservoir properties, or reservoir management. These are defined with numerous configurations of equipment, operations, and tasks.

It is critical to realise that test names or operations vary, and blindly repeating them in a manner of "Copy and Paste" would not suit the purpose. One must understand the core elements of well testing and apply them with confidence and assurance as and when required, regardless of used terminologies or changes in the nature of tests.

Throughout the history of what we now refer to as surface well testing, it was commonly and simply known as "well testing," which adequately described the operation. However, in technical terms well test refers to a reservoir engineering process, not the collection of equipment.

Well testing is used for acquiring data that is analysed for the purpose of understanding hydrocarbon reservoirs and their fluid properties. Depending on whether it is for exploration or development phases of a field or concession, the well testing, sequence, length of time, and number of variables tested vary. The types of reservoir well tests are generally comprised of:

• Radial flow analysis
• Pressure response
• Pressure drawdown
• Pressure build-up
• Type curve analysis

- Gas well testing
- Deliverability testing
- Drill stem testing
- Reservoir and production maintenance

In common terms of surface well testing from the service provider's perspective these can be broadly described in brief within the following sections as:

2.2.3.1 Simple Clean Up

Sometimes referred to as a flowback.

It is commonly stated that this is a one-person operation, from its simplicity, without taking into account all the tasks that the well testers are expected to complete.

The unit is comprised of the following major pieces of equipment:

  i. Choke manifold
 ii. Laboratory
iii. Pipework
 iv. This is often expanded to include burners or tanks

There is no such thing as a simple clean-up/flowback; there is always the unexpected event which cannot be covered with a minimum crew. Following are the main activities involved:

Rigging Up

This is never a one-person operation, rather a three person or more, depending on the equipment, size of the layout, and time constraints.

Also, rig or production personnel may be assigned to assist the well testers in completing the rig up and testing it.

Reporting

The operating company requires a report on measurements at regular intervals, with an operator stationed at the testing equipment at all times; this then sets the minimum level of the personnel required as two per shift.

The number of total personnel on shift may be larger, to include a supervisor, dedicated data acquisition engineers, senior operators, etc., relatively defined by onshore or offshore operations, contractual and local/company standards.

Safety

A standard shift for a well tester is normally 12 hours, excluding any travelling time. No one can be expected to man the equipment without a break; therefore, if the well tester leaves the equipment unmanned for any reason there is a potential for an accident.

At least two well testers should be assigned per shift, with one senior operator and the other a junior.

2.2.3.2   Exploration/Appraisal and Development Well Test

A basic well test is usually defined as an operation where the well is flowed for predetermined intervals. The flow rates, pressures, and other parameters are monitored and recorded in order to evaluate the well's performance.

This is employed with a new discovery, a work-over, a GOR test, or a routine check of the well's performance; once more the TWOP will aid in identifying potential issues. Regardless of the type of test conducted, the risk of an accident or explosion is greater than normal rig operations, and planning for potential accidents must be considered.

The following major pieces of equipment may be included in the set up:

      i. Flowhead
     ii. Flexible high-pressure pipe (Coflexip)
    iii. Surface safety valve
     iv. Choke manifold
      v. Heater/steam exchanger
     vi. Separator
    vii. Laboratory
   viii. Tanks
     ix. Various manifolds
      x. Pipework
     xi. Data acquisition and gauges
    xii. General sub surface equipment
   xiii. Sampling

The test can involve burners and burner booms but this may require the involvement of a specialised company.

2.2.3.3   Production Monitoring Well Test

A production monitoring well test, sometimes referred to as a gas oil rate (GOR)/pressure gas oil ratio (PGOR) test, is usually employed on established fields to monitor the wells' ongoing performance. In such tests, the well is typically flowed through a mobile separator package to ascertain the flowing parameters.

In order to observe the reservoir or concession historical performance and confirm deliverability, production monitoring testing is often carried out over short periods of time and carried out over multiple wells.

Typically, being a trailer-mounted package for quick rig up and mobilisation, the amount of rigging up is limited to the connection to the wellhead (or production manifold) and exporting to the gathering/production station.

The trailer mounted unit's main pieces of equipment are as follows:

      i. Flow lines and associated pipework
     ii. Surface safety valve
    iii. Choke manifold
     iv. Separator
      v. Laboratory
     vi. Tanks

2.3 RIGGING UP

With the number of equipment required, rigging up well testing equipment and other services is never a simple task. The length of time and amount of space required for a well test package are regularly underestimated by operating companies and the rig operators.

This would be a detail raised at TWOP and rectified before equipment is mobilised.

The whole well test crew are in charge of making sure that all equipment are certified and rated for operations. Usually, the principal equipment have the necessary certification but the slapdash use of uncertified or inappropriate fittings sometimes goes unnoticed. Especially with the excessive usage of PTFE (polytetrafluoroethylene) tape.

While acknowledging the necessity to optimise costs and maximise operations, rigging up must be given a fair time allotment without company staff constantly looking for rig ups to be hastened.

To reduce the time taken for rigging up the operating company should prioritise:

i. Space

There must be a designated space for well testing equipment. The service providers supply a detailed design on equipment positioning (commonly known as spotting).

If the well test package has been mobilised (or before) then designated space should be completely cleared on the location or rig. The natural tendency of other companies is to use this available space, even temporarily; if emptied in advance rig up times are shortened.

ii. Facilities

The service companies can provide the required facilities (power, air, water, etc.) they need with the layout. When applicable, in order to save time, the rig should have these dedicated outlets set up in the area before the equipment arrives. This is typically a prerequisite for all well test companies.

These outlets should be dedicated solely to well test equipment – safety or otherwise – and not shared for any other purpose.

The appropriate rig specialist (electrician, mechanic, etc.), who for no fault of their own cannot just leave their routine duties to suit the service providers' requirements causing further delays with equipment rig up.

iii. Cranes

Because of restricted crane capacity, there is always a delay with it. The accountable co-ordinator should ensure that enough cranes and drivers on location match the loading/offloading and positioning equipment as needed.

The crane operator's knowledge of where the equipment is to be spotted/positioned and their dedicated space made available can also help.

When rigging up and rigging down, the use of an extra or temporary crane driver should be considered which will cut down on delays.

iv. Back Loading

In offshore environments delays are frequently encountered because the rig has to back load equipment before loading the well test equipment.

The operating company and the rig operators need to work together to coordinate the removal of equipment taking up space on the offshore platform but no longer needed for current operations, such as excess tubing. Often, the ship that transports the well test equipment is also used to transport the excess equipment back to the base.

The arriving ship will be loaded with the well test equipment; the rig will clear space by shifting the backload onto the occupied ship. A bit of a juggling act takes place where the well test equipment is loaded out of sequence and placed in any free space, leading to some of the well test equipment being placed in unsuitable and unsafe spots. For instance, smaller equipment like the choke manifold is stacked on top of tubing or drill collars.

With this shuffling, the well test equipment is finally loaded on the rig/platform but it is scattered around the facility and needs to be re-positioned according to the proper layout.

If the crane is needed for routine operations while occupied with loading process, it causes further delays with positioning the well testing equipment.

The loading of well testing equipment is generally considered the commencement of rig up, by company personnel, but the well testers are not in charge of their equipment rig up until they are in their final position. This emphasises the importance for company personnel on getting the equipment positioned according to the agreed layout. The timing should be recorded in the well tester's paperwork.

It makes it much easier to reduce these delays if enough space is made for the well test equipment.

v. Pressure Testing

After the equipment is rigged up the pressure testing sequence can take several hours, and time for this should be allowed for.

The time for pressure testing is often compounded by the possibility of leaks from rig pumps and associated equipment supplying pressure instead of the well test equipment.

Prior to carrying out a pressure test, a pressure/function test of the pumps should be performed to verify that they are fit for purpose. The well test supervisor and company representative should witness and sign off on the performance test. The connected flowlines to only the pumps should be tested as well. This ought to be carried out before the well testers are ready for pressure testing their equipment.

NOTE: *Chicksans are often used by the rigs, but these are banned in well testing operations due to safety reasons.*

Having a tested rig pump can save wasted hours of pressure testing phantom leaks.

vi. Time Factoring

The time needed to intricately rig up the well testing and other service companies' equipment should always be calculated from the moment **all** the equipment and crew, whichever is last, arrives on site.

Planning should account for loading and transit time, which should be identified in TWOP, because there are frequent problems with short rigging up periods. This is a safety concern that needs to be rectified to avoid well testing crews working extended periods and potentially resorting to cutting corners.

2.4 WELL TEST AND OTHER PERSONNEL

There are several proposed requirements for the level of experience of the well testing staff, which vary from one service company to another depending on crew availability.

What is missed is the term *crew*, and it is unreasonable to expect any one person to be an expert in all related fields. All that really matters is sufficient experience within the crew.

There are exceptions to this rule when it concerns specialities; job functions that are outside the mainstream well test function. The following personnel need to have all the relevant experience.

- Job Supervisor
- Data Acquisition Engineer
- Sub Sea Engineer
- Sampling Specialist
- Slickline and E-Line Operators

On a well test, various service divisions of a company are found, e.g., Slickline, and their personnel are often uninvolved in the test when they rig up the well test crew assist them as the current site host. To avoid complications in the chain of the command and ensure safer operations it should be written into the well test contracts that all same company service personnel report to the well test supervisor as a single focal point for the company. In case of other service companies' involvement, the well test remains the present host to operations under the leadership and coordination of the attending company representative.

All the additional visiting operational crews would be expected to meet the same HSE criteria and training standards as the well test crew hosting the site. With certain qualifications or required training for scopes of work on site, the qualified well test crew may assume this role on behalf of and for the benefit of visiting service providers, if absent among them, further cementing the site host role adopted by the well test personnel.

Most service companies operate a "Safety Passport" system which should have up-to-date records of ALL the training an individual has received. This ought to be available at any time for inspection.

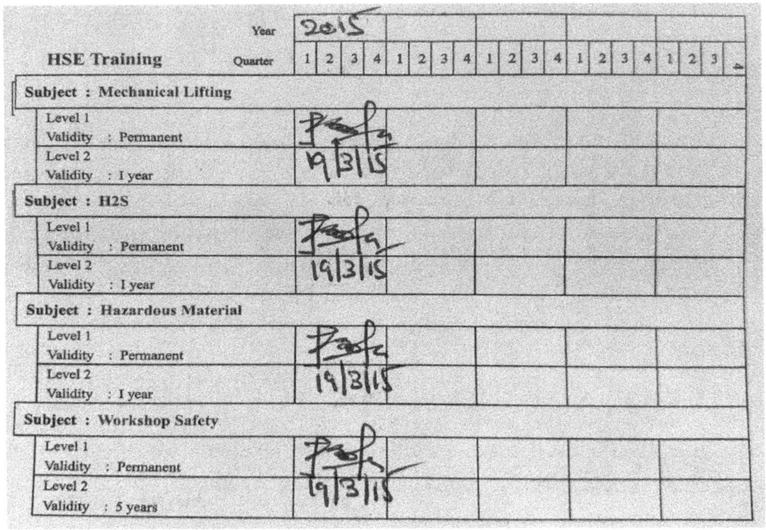

HSE Training	Year	2015																		
	Quarter	1	2	3	4	1	2	3	4	1	2	3	4	1	2	3	4	1	2	3
Subject : Mechanical Lifting																				
Level 1 Validity : Permanent																				
Level 2 Validity : 1 year																				
Subject : H2S																				
Level 1 Validity : Permanent																				
Level 2 Validity : 1 year																				
Subject : Hazardous Material																				
Level 1 Validity : Permanent																				
Level 2 Validity : 1 year																				
Subject : Workshop Safety																				
Level 1 Validity : Permanent																				
Level 2 Validity : 5 years																				

FIGURE 2.1   Example of a Page in the Safety Passport.

i. Senior Supervisor/Job Coordinator

A Senior Supervisor should have at least ten years relevant experience in all the areas and services where they are responsible. If there are any relevant experience gaps, they may be tolerated if a second or shift supervisor compensates for it with the necessary expertise.

The Senior Supervisor will have

a. Participated in every stage of the test from planning to execution. Involvement in all safety matters and liaising with clients, especially with TWOP, procedural, and safety meetings. Accountability on the safe operation of equipment and the provision of all relevant certifications
b. Good organisation and communication skills are essential, as well as the ability to deliver formal presentations to the client and other relevant personnel
c. Led and organised crews in all elements of the job while liaising with all other companies and service providers
d. Been well-versed in all HSE functions and the ability to perform a job safely, ensuring all operational aspects of safety are carried out, demonstrating the ability to deliver a concise safety briefing to the staff under his supervision at the start of every shift

Ultimately, they are responsible for all aspects of the job with focus on safety, performance, and reporting.

ii. Supervisor

A Well Test Supervisor with at least five years' experience can be in charge of the less complex jobs. A supervisor's experience and knowledge may not be as extensive, but they can take on the responsibilities from a Senior Supervisor in their temporary absence. The supervisor is often chosen to compensate the senior supervisor in their weaker or less experienced areas.

They should be experienced in all HSE functions, able to conduct a job safely ensuring all operational aspects of safety are carried out. They should be able to deliver a concise safety briefing at the start of every shift.

iii. Crew Chief/Senior Technician

Working under the instruction of a supervisor, a Crew Chief must be able to work independently and be completely familiar with all the equipment they are expected to utilise during the well test. A crew chief should be an effective communicator and possess the necessary experience to advise, assist, and mentor junior members of the crew. A crew chief's expectations are:

a. Should prepare and perform all the standard operations of a well test, BS&W, etc., without instruction
b. Capable of operating and maintaining all well testing equipment with no or minimal guidance or supervision, while adhering to the company's HSE and maintenance policies and procedures
c. Can assist with the training and instruction of junior personnel in line with company policies and procedures
d. Well-versed in all HSE functions and the ability to perform a job safely, ensuring all operational aspects of safety are carried out. Can be asked to deliver a concise safety briefing at the start of the shift or at least contribute towards it

iv. Operator/Technician

A Well Test Operator should be able to operate and maintain majority of the well test equipment with minimal guidance or supervision. The operator must be familiar with their responsibilities on the test and comply with the company's safety and HSE guidelines, having passed certification and in the possession of all relevant certification.

a. Capable of using the equipment alone and unsupervised with the equipment for reasonable periods of time
b. Capable of taking samples, monitoring instruments, and recording data
c. Able to verify that the equipment is in good working order and functioning properly

v. Data Acquisition Engineer

The position of Data Acquisition Engineer in well testing is the most underestimated job, requiring in-depth knowledge and broad expertise in:

In:	With:
Well testing	Crew Chief (minimum three years)
Computers and instrumentation	Engineering background
Reporting and data	Experienced with reporting
Telecommunications	Networking and data transmission
Language	Fluent in English
Reservoir and well engineering	Downhole gauges used or merged, well completion, and reservoir knowledge

Often the service providers will substitute an experienced computer operator for a data acquisition engineer. But without the essential knowledge in all these subjects any individual would face difficulties as a focal point for the company representatives.

A data acquisition engineer will monitor the well performance, and with a well testing background they will be able to comprehend what the well is doing or what is happening with the equipment.

Example: WHP increasing, flow rates decreasing – can indicate the choke is plugging or hydrating.

The well is not being monitored if the data acquisition engineer is just a "number cruncher." With the right experience the well test makes technical sense and proceeds smoothly.

Since the data acquisition engineer will be the one that will produce the final report, if he lacks experience in well testing or reservoir engineering, significant data and events will be missing from the report.

vi. Off Duty Personnel

When a well test is due to commence on a rig or site, it is the company representative's duty to ensure that there is clean and well-maintained accommodation available for unit's crew.

The rig or other "permanent" service company's crew tends to "spread out" to take up all the available accommodation or simply "acquire" all the facilities. As a result, well test crews are forced to share accommodation alongside other crews and service hands.

Although this may not appear to be a problem, it actually represents a HSE violation and requires attention. Personnel from different working shifts put together become disruptive for one or the other, and no one in the room will get the full undisturbed rest that HSE dictates. This can result in overexertion which will lead to accidents and mistakes.

To avoid these inconveniences, the operating company should make sure that service personnel are accommodated separately with all the facilities intact . . .

a. The rooms need to be locked and the key with the responsible person
b. Efficient blackout curtains around each bunk. (very important offshore)
c. Fresh bed linen
d. In-room changing area and lockers for storage. (adequately organised for crews on shift or on leave)

The rooms should be assigned according to shift patterns, even though this may not always be practical, so people can get a good rest without being disturbed by neighbouring shift changes.

The act of flowing a well is one of the riskiest operations that may be performed, hence the personnel and crew involved with this activity should be accorded the respect that is due. This involves having adequate accommodation; without proper rest it can be a potential accident waiting to happen. This becomes even more sensitive during spotting of equipment and rig up, with the demanding manual labour, without proper accommodation available for them.

Having a test crew shuttling from one location to another is not always practical. The operating companies should consider personnel needed on-board in case of an emergency.

vii. Department Head/Co-Coordinator

In charge of overseeing well testing procedures and making sure that all equipment are up to standard, adequately maintained and fit for purpose. Additionally, the holder of this position must be aware of and comply with the statutory requirements imposed by the country, region, and operating company.

While leading by example, one of their main roles is to assign well test personnel to specific roles, while:

a. Ensuring all field and operational staff are aware of and implementing all company programs, processes, and procedures from both operational and HSE aspects. All staff must have completed the appropriate training before arriving to the location
b. Ensuring there are sufficient resources to carry out the day-to-day operations of the department, fully trained personnel, facilities, transport, PPE, etc.
c. Liaise with the human resources department (HR) regarding employment, discipline, and termination of personnel
d. Liaise with the HSE department on all safety issues and training. All staff must be up to date on operational methods and HSE procedures; this includes PPE
e. Ensuring that incident and reporting protocols are carried out and implementation of appropriate corrective measures
f. Participation in HSE policy and safety meetings and joint liaison with HSE to the operating company representatives

g. Holding all staff accountable for safety actions; make sure that these duties and responsibilities are included in their job descriptions
h. Ensuring that all well test staff receive the necessary technical and associated training and carry out random inspections of maintenance
i. Conduct routine and unannounced site visits to ensure all company policies, practices, and procedures are in place and followed

viii. HSE Personnel

Whether HSE Personnel are site or office based they would have their duties entail:

a. Assist the well testing management in all matters relating to HSE
b. Liaise with operating company to ensure compliance of all HSE requirements, issues, and training
c. Responsible for relaying regular updates on all HSE related issues to all staff and advancements in training requirements
d. Keep track of all job HSE records. Preparation of reports to any incidents or accidents and making sure that all involved personnel have an incident-related refresher course to prevent it from happening again
e. Attend well test training courses to understand the involved processes and assist in TWOP, HAZOP and LOPA meetings
f. Enrol in any operating company required safety courses to facilitate their purpose on site
g. Conduct routine site inspections – onshore and offshore – and check for HSE and well testing procedure compliance

2.5 EXPERIENCE

There should not be any "green hands" as part of the main well test crew, unless it has been expressly agreed with the operating company. This is crucial and unnegotiable in HPHT, H_2S, and other high flow rate tests.

Green hands undoubtedly need to gain experience; their attendance on a job should be in the junior position, shadowing a professional, and under continuous supervision. Furthermore, it is the duty of management to ensure they are allocated to jobs of lowest risk until they are qualified to progress. When attending their first job in a responsible capacity then an experienced operator (old hand) should accompany them to both support and assess the new operator's performance.

If a crew chief is "breaking out" or is considered weak, the support crew around him should compensate by having a full crew chief on their shift.

The scheduling of the test affects the operational experience of the assigned well test crew. If the operating company calls out the service on short notice, the service company might not have enough experienced staff ready for mobilisation. This depends on the organisation of the operating company's representative or engineer. The company representative or engineer typically has a bias towards the drilling activities, allowing other operations to slip under the radar if not addressed.

While this bias is understandable and necessary during the drilling phase, the company representatives should also have a degree of training in the limits, procedures, and scheduling of a well test, in order for the operation to run smoothly in terms of time and manpower. A few oil companies (e.g., Chevron) once had a specialised department supplying both company representatives and reservoir engineers for well test environments, but this practice has again "gone by the wayside."

It would be beneficial if the Company Representative and Senior Rig Personnel attend a familiarisation course with the well test service provider to understand the equipment and the processes involved.

In order to guarantee that the proper level of operations can be carried out, the option of a standby well test crew would prove useful. To avoid the standby crew being un-gainfully employed, the service provider would like to charge for this option, and operating companies are rarely prepared to pay.

2.6   LANGUAGE

Every member of a well test crew should be able to communicate in a common language, verbally and in writing, with adequate fluency. This is a safety concern as much as a professional necessity.

Among the junior positions absolute fluency is not critical but the senior staff must have a thorough understanding of the language involved. With the competency system in existence today, language should be included as a component. The junior operators must be able to write and describe the associated well testing activities that will be used in preparing the detailed reports.

The language barrier also draws attention to the possible drawbacks of solely relying on inadequate or impractical competency courses.

3 Time

The use of accurate time is so obvious that it is often overlooked.

The importance of time is in the correlation of data with sequence of operations. The transfer of data from different sources emphasises the significance of time as a common factor, whether between distinct production services such as fracturing and surface well testing or downhole gauges and surface well testing or the same production service but using different systems, i.e., manual and electronic readings.

3.1 STANDARD TIME

With modern computer technology comes the internet and a time function known as Coordinated Universal Time (UTC) that can be readily accessed providing an internet connection is available. The term UTC refers to a time zone where all countries' timing centre and standards are in sync with one another. Ideally, the local UTC time should be used as an independent time standard for all well tests when possible. Alternatively, all watches and time instrument on site must be synchronised.

Whichever method is employed, there can only be a single time reference, otherwise this leads to confusions, inaccurate readings, inaccurate results, and calculations.

When a new well is perforated, a well test can be thought of us as begun. However, this is only the start; other processes have already begun prior to the well perforation or flowing the well.

Perforating is a reference to the installed gauges, either memory or permanent, if used, often pre-programmed and started days before the perforation. The gauges used are frequently provided by multiple manufacturers or service providers, which increase the likelihood of time inaccuracies and further reading errors or mismatches.

SO, AT WHAT TIME DO THE INSTALLED GAUGES OPERATE?

A time that is specified by the operating company, and ideally this should be from a separate, precise, digital clock. In most cases today the time can be taken from a wristwatch, and the wristwatch's owner may not always be there for the entire period of the test.

It isn't the best idea to use a computer in the company office; depending on the settings of the computer it can drift over time. Even worse anyone can change a computer clock so it cannot be regarded as a reliable source of time.

The recommended practice is to set a location time and insist that all time measurements are synchronised for individuals on site. In an offshore environment, the radio room regularly has a precise time standard.

  DOI: 10.1201/9781032623689-3

As a result, all activities and measurements are based on a common time source, making it simpler to cross-reference and correlate data between different services and instruments. Any localised time changes, such as British Summer Time, must not apply – which is a risk with internet connected time; once the start time has been set it must remain fixed and adhered to until the well test has finished.

When interference tests are conducted in different wells in the same reservoir, timing becomes much more crucial. With these tests the time of the pressure disturbance in one well and the resultant effect in the observation well use synchronised time, allowing reservoir engineers to accurately determine reservoir parameters.

Time is another reason for manual reading sheets while well testing despite having a computerised data acquisition system. Having a physical record with written time is less likely to be lost or corrupted than anything electronic.

Moreover, manually preparing a separate data sheet and record allows a cross check of the electronically gathered data and flow rates. The data reported in the manual reading sheets data must be from independent sensors than the ones used with the data acquisition system to confirm readings and identify any transducer/ sensor errors.

Manual readings will vary if there is a difference in the referenced time between individual members of a well testing crew and changes from shift to shift. Table 3.1 shows the errors in manual readings if there is a difference with referenced time. Flow rate units have been ignored as a common factor between the numerical values. Its apparent that if the times are not properly synchronised, there can be significant flow rate measurement errors. As the flow rates increase so do the magnitude of inaccuracies and errors as illustrated in Figure 3.1.

TABLE 3.1
Flow Rate Error Table

		Flow Rate				
		1,000	**2,000**	**3,000**	**4,000**	**5,000**
Time Error	5 Seconds	0.058	0.116	0.174	0.231	0.289
	10 Seconds	0.116	0.231	0.347	0.463	0.579
	15 Seconds	0.174	0.347	0.521	0.694	0.868
	30 Seconds	0.347	0.694	1.042	1.389	1.736
	1 Minute	0.694	1.389	2.083	2.778	3.472
	2 Minutes	1.389	2.778	4.167	5.556	6.944
	3 Minutes	2.083	4.167	6.250	8.333	10.417
	4 Minutes	2.778	5.556	8.333	11.111	13.889
	5 Minutes	3.472	6.944	10.417	13.889	17.361

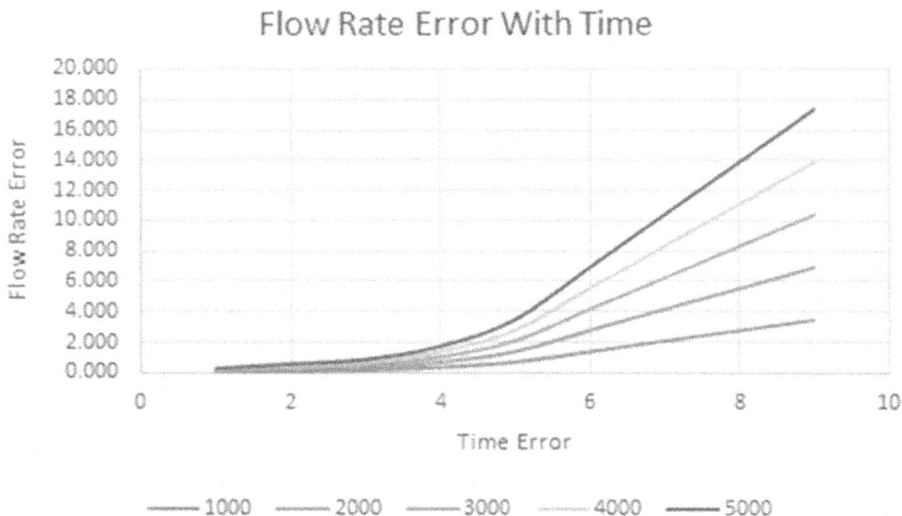

Flow Rate Error With Time

FIGURE 3.1   Graphical Flow Rate Error.

Normally with current data acquisition systems the time is constant, but there have been instances where operators reset the time at the beginning of their shift. The fact that electronic time can be changed forwards or backwards creates gaps in the data. On a computer it allows for the overwriting of old data, which makes this a more serious problem than just a simple time inaccuracy. The inaccuracies found in Table 3.2 reflect how this results in errors.

These are classic examples of how practices and procedures not followed cause problems, yet they are easily avoided with a standard test time set and strict instructions; then neither operator should adjust the data acquisition time.

The easiest way to determine time errors is to look at the *stable* fluid flow rates during crew shift changes as in Table 3.2. It is assumed that the shift changes at midnight and midday, but the same detecting principle applies at other time shift changes.

TABLE 3.2
Time Calculation Error

	Oil Rate	Water Rate		Oil Rate	Water Rate
23:00:00	3210.55	200.22	11:00:00	3210.55	200.22
23:30:00	3210.49	200.19	11:30:00	3210.49	200.19
0:00:00	**3204.40**	**198.01**	**12:00:00**	**3216.40**	**202.01**
0:30:00	3210.52	200.20	12:30:00	3210.52	200.20
1:00:00	3210.55	200.22	13:00:00	3210.55	200.22
	Night Shift Data			Day Shift Data	

From the night shift data, a flow rate change is obvious at midnight with a significant decrease compared to the average flow rate readings. Conversely the rate increases by a similar magnitude when the day shift take over at midday.

TABLE 3.3

Difference between Flow Rate Readings

Period	Flow Rates	Change
Midnight	3204.40–3210.49	– 6.09
Midday	3216.40–3210.49	+ 5.91

Table 3.3 displays the flow rate difference between the last reading of one shift and the start of another. The fact that they are close values and observed only then is a good indication that it is a time mismatch between the two shifts; from that error onwards the new shift continues with 30-minute intervals and returns to the more accurate flow rates. These discrepancies must be corrected as soon as they are noticed and rectified among all personnel.

3.2 WATCHES

With the younger generation of well testers and engineers the simple wristwatch is fast becoming unpopular in preference to the power of the PDAs, phones, and other portable smart devices.

While it is not the intention to discredit these devices it must be pointed out that they cannot be used in any of the hazardous zones, unless they are intrinsically safe or certified, otherwise they pose as a fire and explosion risk. The smart watches with advanced functions are an even worse unnoticed hazard, by incorporating transmitters, Bluetooth, etc.

Why?

i. They do not have a certificate of compliance that any electrical or electronic equipment is compulsory to have for being utilised in a hazardous zone

ii. The emitted electromagnetic radiation from built in wifi communications systems is potentially powerful enough to ignite any gas or gas leaks

This applies to all devices including remote displays that are transmitted from any data acquisition systems.

While the detail, application, and content can change from country to country, a simple truism is the best rule to follow:

If the electronic device, cable, or battery powered system is in or passes through a hazardous area it must have the relevant safety certificate attached to it (normally a metal plate) attesting its suitability.

If the certification is not on the device, it cannot be used, it is as simple as that.

There are certified laptops and tablets that allow them to be used in hazardous areas, but these are very expensive and rarely used. If in doubt – check.

The use of uncertified devices in hazardous areas should be considered a violation of HSE regulations and disciplinary action should be taken.

A basic waterproof digital watch (at the least) should be a part of the well tester's personal equipment. These should be handed out by the service provider specified in the individual job descriptions and in the TWOP; otherwise, manual readings and observations cannot be taken reliably.

The use of a watch fastened to a clipboard – or a clipboard with an attached plastic watch – helps to standardise time.

3.3   EVENTS

A well test report has to include information regarding any event, activity, or job that affects any of the well characteristics (pressure, flow, etc.) regardless of which service provider performed them. This makes it obvious why a standard time must be used by all personnel involved in the test. (This will be dealt with in further detail in the section under reports.)

3.3.1   PRODUCTION MONITORING TESTS AND TIME

The purpose of production monitoring, also referred to as gas oil rate (GOR)/pressure gas oil rate (PGOR) test is to assess the performance of a well that is connected to a gathering station or production facility. These tests are typically carried out using well testing packages mounted on a trailer.

During these tests the expected flow rates to the gathering station – or equivalent facility – will change. It is critical that both the well testing unit and the production company correlate time between them.

In order for all the data to be synchronised and validated by the gathering station, the reference time with the well testers must be set to the facility's equipment, often electronics (PLCs etc.).

The operating company will often designate a specific "test loop" within the gathering station and it is common practice to simultaneously flow the intended well through the facility's test loop and the well testing equipment, i.e., in tandem. In these cases, the operating company can monitor the performance of the well and record the results.

When the well test reports are submitted, both sets of data are compared. For the sake of these comparisons, a standardised time's importance can be acknowledged.

Operating companies have refused to pay for tests and imposed penalties on service providers based on comparisons made leading to perceived errors.

3.3.2   MEMORY GAUGES AND TIME

The vast majority of memory gauges do not use real time but calculate time using an elapsed or delta time principle. The actual time is then generated against data points by inputting the starting time corresponding to when the battery was connected to the memory gauge.

This is a major source of errors.

If the memory gauge operator does not have synchronised time, then all the readings are shifted relative to the other operational readings. When the batteries are connected, the presence of a company representative/engineer is strongly recommended, to confirm the time and ensure adherence to the scheduled test time.

When memory gauges were first developed there was a limited memory storage capacity, but technology has advanced. With these advancements the method of using a dead time for running the gauges in hole should be considered obsolete, especially with advances in battery and power consumption, making memory gauges a better tool than in the past.

Most gauges use an electromagnetic type pulse at start up, to indicate being powered on, and when readings are taken, they are typically recognised by a visual/audible module/amplifier/verifier that comes with the gauges. If a sample rate of 60 seconds is set, the gauge will emit a pulse every 60 seconds that is converted into its visual/audible form by its amplifier.

The gauge would now be emitting data pulse for every reading it takes, and a complete verification can be carried out by verifying the timing of multiple pulses, from each gauge as necessary.

This does not mean that multiple amplifiers have to be supplied, although a spare is desirable, as the pulses can be compared to the standard time.

If the times do not match then these gauges should be re-programmed.

NOTE: *Some gauge software allows for time shifting and in turn the flexibility of aligning the timing, for tandem gauges, with each other and the gauges with another sequence of events.*

3.3.3 Slickline, Electroline (E-Line), and Coiled Tubing Operations

This time synchronisation applies to all services companies; Slickline, Electroline, Coiled Tubing, etc.

Most of these services rarely keep accurate records of their events, as their operations often rely on different parameters other than time, i.e., sequence of events, depths, and physical objectives. Yet any tasks they perform affect well testing objectives, and all their activities and precise times should be precisely recorded. Events should be recorded from initial rigging up to final rig downs, with particular attention being paid to any activity that affects the well parameters, e.g., opening the swab valve. After completing the job, a full copy of the events performed should be given to the well test supervisor for inclusion in the well test report. These records are correlated with acquired data in well test reports along with their effects on results, or they are simply informative.

The Electroline (E-Line) Lubricator is a prime example of external activity affecting the well pressure. The gap around the Electroline cable and the sealing tubes should be sealed using an injection head, relying on an Injection Head to seal the space around the Electroline cable and the sealing tubes. If the tube sizing is wrong, too big, or the grease injection pressure is too high, it will effectively pump up the well. The downhole gauges detect this and may be erroneously diagnosed if the event is not reported in detail.

If other service providers are collecting and recording their own data from their transducers, these have to be reported. The possibility of discrepancies between the sensors is a variable to consider. It is advised that the sensors used for reported readings have an accuracy check, preferably utilising the well testing's deadweight tester, which identifies any possible discrepancies that need rectifying.

3.3.4 TIME/EVENT CONCLUSION

The quality of data and its interpretation would drastically improve if all companies and their respective engineers, supervisors, testers, and technicians referenced the same time and kept accurate records of all activities that affect the well parameters. Correlating between data and events would be made easier, leading to a better understanding of the well's performance.

3.3.5 PRIOR TO STARTING THE JOB

All personnel on site must have left their mobile phones and other electronic devices in the safe area, are using the necessary PPE, and are using simple waterproof watches for timekeeping.

Before beginning any operations, even pressure testing of equipment, the data acquisition system must be initiated and recording data. It must be synchronised to local standard time, with personnel synchronising their watches to the data acquisition system.

3.4 MANUAL READINGS SHEETS

Throughout this book the importance of manual readings is frequently emphasised, so they are never underestimated or ignored.

The majority of tests in well testing use a data acquisition system, hence the argument "There is data acquisition so we do not need to carry out manual readings" – **Wrong**!

How can we be certain that the data acquisition system is accurate or even correct? Blindly accepting its results without question or examination, has, does, and will lead to major problems.

The well test crew's major events and time-consuming tasks are divided among:

  i. Rigging up and pressure testing
 ii. Opening up and establishing stable flow
iii. Rigging down and leaving

These are the most human-power intensive times of the crew, and during stable flow through a separator very little is required, besides monitoring the process and taking readings with enough time to carry out manual readings and record them on manual readings sheets.

Going back to first days of well testing there was generally a crew of 5 to cover a 24-hour period:

1 Well Test Supervisor (Floating)
2 x Crew Chiefs (1 per shift)
2 x Operators (1 per shift)

There were no computers or data acquisition systems, and all the readings were gathered and recorded manually. As a standard, during stable flow periods, readings were taken every 30 minutes and used for manual calculations using tables and a calculator to determine the flow rates. There was enough time to get it all done.

After major events such as well opening, shut in, or major choke change, the wellhead pressure (WHP) rapidly changes; to track these changes with enough frequency to plot a sensible curve, Table 3.4 suggests the common practice in data recording rates from time of event.

TABLE 3.4
WHP Recording Rate

Delta Time Minutes	Data Recording Rate Minutes
0–15	1
15–30	5
30–60	15
Onwards	30

Note: Delta time is time since the event.

Table 3.4 is a general guideline; the interval times may vary if the pressure or flow rates are rapidly changing.

When opening or shutting the well, readings every 1 minute are taken for 15 minutes, using a deadweight tester, and then the frequency of taking readings gradually decrease until they are every 30 minutes after the first hour.

This is not only possible, but it has been done. There is no excuse for the well test crews to avoid manual readings on a whim. Non-compliance should be grounds for disciplinary action.

It should be re-iterated here that the manually acquired data are used as back up records on gravities, BS&W, choke and orifice sizes, pressures, temperatures, etc., and to verify the electronically measured data on the data acquisition system.

The data acquisition system may produce inaccurate results due to the following:

   i. Incorrect sensor position
  ii. Poor calibration
 iii. Damaged sensor
  iv. Fault in the data acquisition system
   v. Wiring error
  vi. Configuration fault
 vii. Complete system or power failure

These faults can develop prior to or during the test leading to errors, unless they are picked up on by the crew simply walking round the unit and writing the manual data on the manual reading sheets at least every hour.

Manual reading sheet readings MUST come from passive sensors, completely independent and *unconnected* to the data acquisition system, like the deadweight tester and the Barton recorder. As distinct readings they will serve to corroborate the data acquisition system data. This also includes the much-needed monitoring of burners.

Often personnel would prefer to use a pressure dial gauge to read the wellhead pressure, which has always been considered an unacceptable practice by capable well testers and engineers. The deadweight tester is considered the only true standard pressure reference available on site, and its use is essential and failure to do so is a disciplinary offence.

All other information MUST also be written on the manual readings sheet including:

- Choke changes
- Chemical injection
- Switching flow
- Orifice plate changes
- Burner or export status
- Major events from other service companies

The general rule is any event that changes the magnitude or direction of pressure or flow should be recorded.

The manual reading sheets should be handed into the data acquisition lab after each set of readings to verify the data. To have an operator use a radio and to dictate readings to the data acquisition engineer, expecting it be written down in the office, is a poor excuse and impractical. The well test operators are in charge of the manual readings sheet.

3.4.1   FLUID DATA

The well test operators are expected to collect fluid samples for properties and contaminants. Unless after changing a choke, the frequency of these measurements is *every hour* throughout the flow period.

Liquid samples – on the hour

- BS&W
- Gravity
- Salinity
- Shrinkage

Gas samples – on the half hour.

- Gas gravity
- Contaminants

Meter factors

Combined and/or normal meter factors should be taken after a suitable period from a choke change when there is a stable flow and fluids surface under the new flow conditions.

3.4.2 OPERATOR COMPANY PERSONNEL

Company representatives should randomly witness if the readings are collected accurately and HSE procedures are followed. Revising the manual reading sheets would ensure

   i. That they are being filled in
  ii. Recording the well performance

Failure for the crew to carry out manual readings should result in disciplinary action. There are NO excuses.

3.4.3 DATA ACQUISITION

After receiving the manual readings sheet, a data acquisition engineer is responsible for ensuring that:

    i. Any changes to the well or equipment parameters are entered into the system immediately and any retrospective corrections are made
   ii. Any difference between the manual readings sheet and the electronic data acquisition system are detected and any discrepancies addressed
  iii. The manual readings sheets are clearly timed and dated, legibly written without scribbling, kept clean and tidy
  iv. The manual readings sheet is kept in the file dedicated to the current well and job, in its original or scanned form, in case any queries arise after job completion and report submission. Barton and other recorded charts should be stored as well in the same file and format

4 Safety

Safety is paramount in all well testing activities, and all the well test crew must receive enough training in all aspects of safety and be fluent in procedures before getting into operations in the field. There are no short cuts; all procedures must be followed correctly.

Operator training will include – but is not limited to – all aspects of:

- Basic well testing
- HSE introduction
- Permit to work
- Manual handling
- Banning the use or possession of mobile phones etc.
- PPE
- Self-contained breathing apparatus
- First aid
- Buddy system
- Sour (H_2S) gas
- Job safety
- Environmental
- Housekeeping
- Tool and machine
- Others included in the company safety program

A "Safety Passport" style document, which should be kept current and in the operator's possession at all times while on location, will identify all the safety courses taken and passed. The well test supervisor is in charge of making sure all crew documents are up to date. The crew's documents would be randomly checked while on site by safety officers.

Some operating and service companies follow the policy that inexperienced well test personnel, generally with less than six months' experience, should not be employed on sour tests where exposure to gas concentrations could be lethal.

4.1 BASIC ELECTRICAL SAFETY

All instruments and electrical/electronic devices used in a well testing unit must have their safety certificate specifying if they are intrinsically safe or explosion proof. This implies that the equipment will have a certifying plate or sign on the device. Without this the device cannot be used. The certificate or sign determine the level of safety and the relevant hazardous area/zone it can be used in.

Essentially, with an intrinsically safe device, the power is limited and electricity can be conducted by standard flexible cables, whereas explosion proof (Exd) is

     DOI: 10.1201/9781032623689-4

normally heavy-duty power requiring armoured junction boxes, glanding, and cabling. Both options cannot be mixed on or for the same device.

When Exd electrical devices are used, their certification is invalidated if any of the internal wiring protrudes from the armour, conduit, gland, or junction box. Any holes, cracks, or damage to a conduit should also be corrected. Under no circumstances can unprotected cabling be allowed in a hazardous area. These faults should be rectified only by a certified electrician as soon as possible; otherwise, each scenario presents an explosion/fire hazard from flammable gas exposed to the Exd boxes.

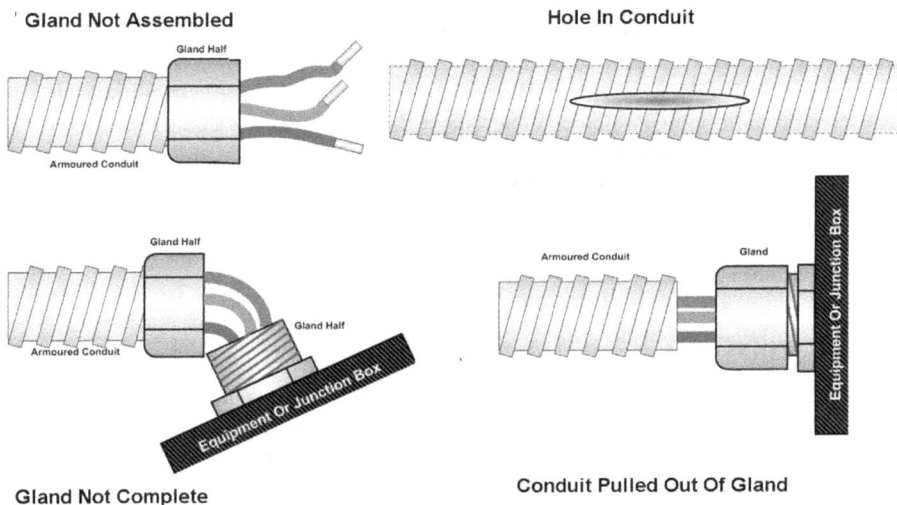

FIGURE 4.1   Illustration of Unprotected Cabling.

The use of plastic conduit instead of a metal conduit is not permitted. The conduit's main function is to shield the inner cable from physical damage and crushing. The certified flexible metal conduit, such as the approved Kopex Ex range, should be used; plastic conduits do not accomplish either purpose.

The use of electrical tape is *not certified* for cabling or fittings in any Exd applications, and the fittings have to be individually certified. Any extension lead type cables must also use a certified armoured cable; regular cables are not certified for use in any of the electrical areas on site.

In all cases a suitable electrical earth must be used.

Each major piece of well testing equipment must have an earth point, typically a threaded section and nut that allows the earth wire to be permanently connected (which should be kept clean and free of rust and paint), so that a heavy-duty earth wire (6 mm^2) can be connected to it and linked to a suitable ground point. This is done to prevent the buildup of static electricity that can result from the well's discharge. This includes vessels and laboratories mounted on trailers.

4.1.1   MOBILE PHONES

Mobile phones are a source of ignition . . . it's as simple as that.

Most of them lack a safety certificate, ATEX, CSA, etc., and solely by this fact alone their use is prohibited in a hazardous area. The electrical safety rules fundamentally state: unless the device is classed as simple apparatus, which a phone is not, it has to have passed tests and regulations to ensure it is safe for use in hazardous areas.

Certified devices will have a safety plate installed on them detailing their level of safety and the certifying standard.

Bluetooth-enabled phones or devices are more dangerous in hazardous areas. Bluetooth is a wireless protocol that uses an electromagnetic wave, with frequencies from 2.402 GHz to 2.480 GHz and capable of causing electrical sparks, for transferring data between devices over short distances. To put this in perspective a microwave oven operates with a 2.45GHz wave frequency, right in the middle of the Bluetooth operating band, and it often gives off sparks.

With gas nearby, an explosion is possible with mobile phones and smart devices, if they cause a spark like a microwave does. It becomes even riskier and more probable where operators are periodically sampling gas.

4.2   ZONING

All well/production sites will be classified into areas that are based on the potential for significant concentrations of combustible gases or vapours to amass around each piece of equipment on the location.

The hazardous areas have been classified by the American Petroleum Institute (API) as zones under three headings Zone 0, Zone 1, and Zone 2. The lower the zone the more hazardous the potential danger.

The majority of well testing companies adopt the approach that the zones do not overlap within the confines of the area defined by the well testing layout drawings. The service company should keep a record of the zone restrictions for the country or location of installed equipment and the well test supervisor and seniors should be familiar with these restrictions and their implementation.

If in doubt, then any area or zone should be immediately treated as Zone 0. Any classified zones should be confirmed with the operating company on account of zone restrictions and their location.

4.2.1   ZONE 0

A location or area where a hazardous atmosphere may be present constantly. The most important fact about a Zone 0 area is the continual presence of flammable atmosphere in a concentration that's within the flammable confines for the hazardous substance (gas, vapour, liquid, etc.).

In land operations the pit below the wellhead is categorised as a Zone 0 area.

4.2.2 ZONE 1

An explosive atmosphere is likely to occur under unusual operating conditions or circumstances. The Zone 1 classification is given to any zone or area where any hazardous, flammable, explosive, or volatile substances (gases, vapours, etc.) are processed or stored. The classification is an ignitable concentration present in sufficient volume to pose a risk during routine activities or operations.

i. Christmas Trees

The area surrounding the Christmas tree is classified as Zone 1 for the risk of volatile fluids escaping while regularly operating its valves.

The Christmas tree may be classed as Zone 2, if it is not in use and the valves cycled–closed, until valves are cycled open, and it's reverted to Zone 1.

ii. Choke Manifold

Valves are regularly opened on the choke manifold and associated data header; during the well test for sampling (gravity, BS&W etc.), volatile fluids may escape, classifying the choke manifold and immediate area as Zone 1.

iii. Tanks

Any tanks used to store volatile fluids are classed as a Zone 1 due to the potential emission of flammable gasses near them and their vents.

iv. Electric Pumps and Associated Equipment

Electric-driven equipment such as transfer pumps must be certified for use in Zone 1 if operated in the area, otherwise certified for Zone 2.

4.2.3 ZONE 2

In these areas any flammable, volatile, or explosive materials are handled, processed, and stored under documented controlled conditions. This is an area in which an explosive atmosphere will not occur in routine operations and, if one does, it will not last long before it is rectified.

i. Separator

Well test separators in their steady state fall under the classification of Zone 2 since normally they will only release flammable gases in abnormal conditions, such as a leak, whilst any type of gas or fluid sampling upgrades the separator to a Zone 1.

ii. Diesel-Driven Pumps

Diesel-driven transfer pumps are suited for Zone 2 only if they are equipped with all the relevant safety devices: automatic shutdown devices, spark arrestors, inertia starters, or special electrical starters.

iii. ESD System

The emergency shutdown device system is usually classed as a Zone 2 device, but this can be upgraded to Zone 1 depending on its operational condition and specifications.

iv. Manifolds

The oil and gas manifolds, also known as diverter manifolds, are usually classed for Zone 2, but they can be upgraded to Zone 1 dependent on their operational conditions and specifications.

v. Piping or Flowlines

All piping and flowlines are classed as Zone 2 and any area around them are equally classified.

The indirect heater must be located in Zone 2 because it uses a naked flame to heat well effluent. Equally the steam exchanger is restricted to Zone 2 since its surfaces can reach high temperatures.

4.2.4   SAFE AREA OR CLEAN ZONE

A safe area is where no flammable, volatile, or explosive substances are handled or processed.

Often the well testing office, commonly known as the lab, is considered a safe area. This is only true if it is outside the zone areas of the well test equipment. If the lab is within any of the zones, it should be fully certified and pressurised to suit.

All living quarters and break areas are considered safe areas and should be situated accordingly. A workshop would also be found in the safe area, but with any welding activities, appropriate gas detectors should be operational.

4.3   INSTALLATION

The actual layout and installation of a well test package is totally dependent on the available space for rig up. The space will be identified during the pre-job inspections and planning phases.

If the correct planning is undertaken the service company will provide a thorough layout drawing showing the location of each package component, the zoning, and their tie up.

Figure 4.2 lays out the major equipment of a well test package, with the recommended distances between them and zoning boundaries. This should only be used as a general guideline rather than definite and certified design.

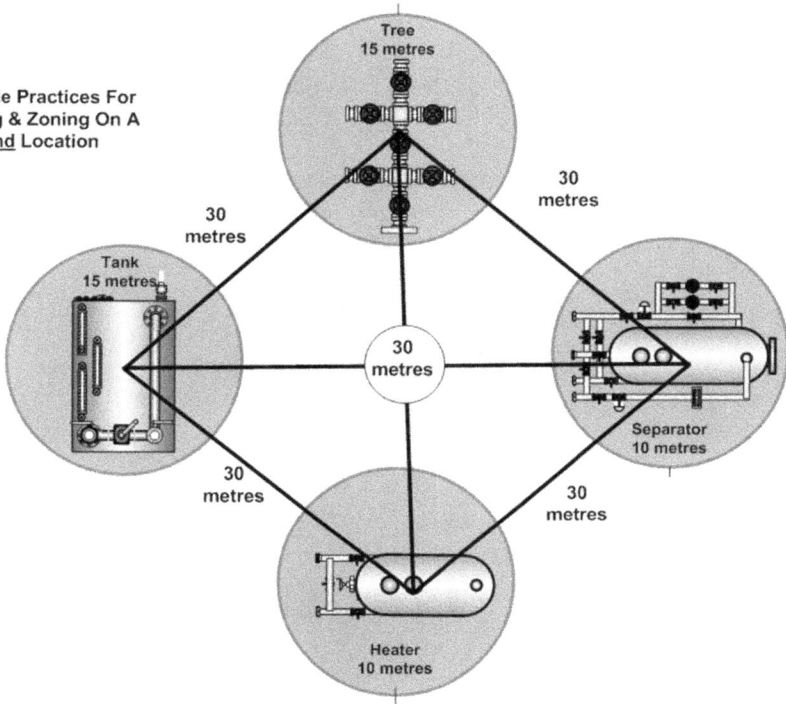

FIGURE 4.2 Land Distances and Zoning Example.

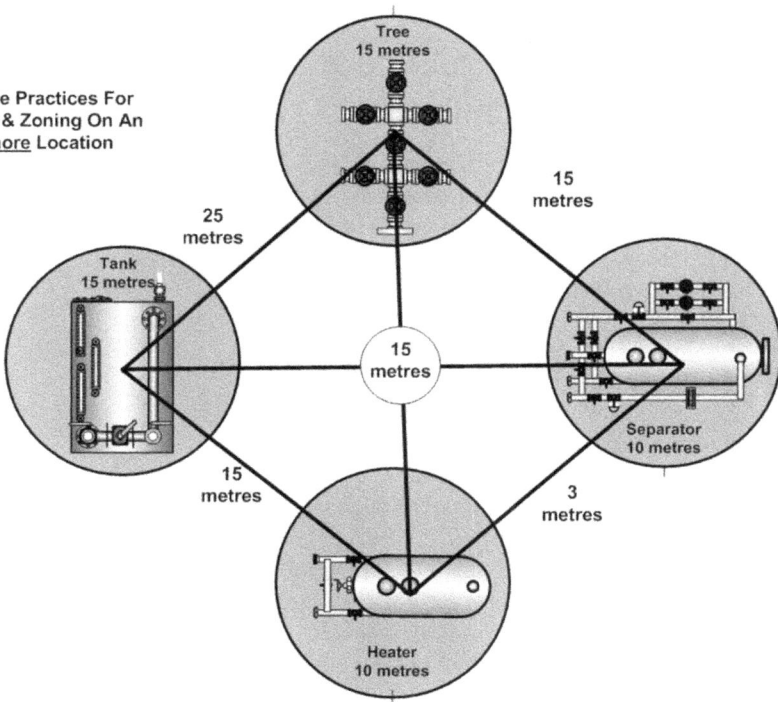

FIGURE 4.3 Offshore Distances and Zoning Example.

If these suggested distances between equipment cannot be achieved due to a tight space, then all equipment are considered as Zone 0 and all safety precautions should be taken.

Each whole piece of equipment needs to have a distinct serial number in order to track it. This applies to the major well testing unit components, such as separators, etc. and connecting items such as flowlines, elbows, and crossovers.

4.3.1   PIPEWORK IDENTIFICATION

Major well test equipment can have the identification either on a metal plate (preferred) or stencilled on the body, whereas all the pipework should have a permanent metal band with the identification data stamped on it.

Details stamped on the pipework band as a minimum should include:

- The maximum working pressure (MWP)
- The date the test was performed
- The actual pressure applied
- The pipework service standard/sour gasses
- Certifying body
- Test certificate number

NOTE:

   i. Any pipework/crossover etc. with a missing certifying band cannot be used
   ii. All flowlines should have a tie down ability

4.4   PRE-JOB ARRANGEMENTS

A pre-job or TWOP type meeting between the operating company and the well testing company should take place to confirm the final configuration.

The well testing company will carry out a site visit to determine where and how the equipment will be spotted/positioned and the facilities required.

This will establish:

- Equipment positioning
- The hazardous zone areas
- Safe access/exits to and from the equipment
- Any detection equipment required
- Flowline routing and tie downs
- Mounting and positioning of burners and booms (if required)

Once completed the well testing company should draw a layout diagram, detailing all the necessary information regarding the planned well test. This should be passed to the operating company for initial approvals.

Upon initial agreements of the test aspects a full TWOP should be exercised with all the involved companies. Following the layout, the process will be discussed and amended according to the operational requirements and in consideration of any other services and companies.

During the TWOP all safety, HAZID, and HAZOP studies will be discussed and approved. Also:

   i. If on land all access routes should be specified and discussed. Offshore emergency procedures proceed as specified by the rig operator
  ii. All documentation referring to meetings should be included in the service company/operating company/rig file
 iii. The primary focus of every meeting must be safety

4.4.1 ROUTINE MAINTENANCE

All well test equipment is subject to maintenance and inspection before, after, and during well test operations.

The maintenance is carried out to keep groups of "job ready" equipment that can be mobilised to potential and existing jobs with minimum delay. This also ensures that equipment can be switched from one set to another as needed.

The equipment will be maintained according to the company maintenance procedures and certified by an appropriate independent certifying authority. In some cases, the equipment is certified for periods up to five years until a major inspection, but where H_2S is involved, the certification can only be for a single job.

In the event that a piece of equipment doesn't meet the design requirements, the certifying authority can effectively de-rate it; e.g., if 10 kpsi (10,000 psi) line pipe does not meet the specification, it can then be de-rated to 5 kpsi (5,000 psi) or a lower rating if it complies with the lower rating specifications.

A folder with all the documentation relating to inspection, pressure tests, and thickness tests should accompany the well testing package throughout its operational period. The documentation should be available at all times to all personnel, which should also be examined by the operating company representative as soon as it arrives on location.

The overall subjects of maintenance are – but are not limited to – the following:

- A full inspection of the equipment and a list of obvious defects produced
- Function testing of all moving parts and regulatory devices
- All repairs, maintenance, and replacements as required
- Internal inspection of relevant equipment
- With a major service involved a full thickness test is required as well as recertification of all pressure bearing equipment
- Thickness tests may be required if the equipment had abrasive fluid flowed through it for an extended period of time

- All pressure bearing equipment should be pressure tested and documentation (recorder charts) kept in the equipment folder
- Any calibrations, such as flow-meters, back pressure valves, and recording instruments, should be performed and documented
- All slings and associated lifting equipment should be inspected and replaced as necessary
- Verification of all accessories, chokes, orifice plates, surface sampling, etc., is present and in the proper quantities
- After completing all necessary tests, the equipment should be corrosion protected and painted in accordance with company standards
- The equipment need to be organised in groups and transported with the necessary documents

4.4.2 CREW SAFETY

Upon arriving at the location, the entire crew should be given a full orientation of the site by the designated safety personnel.

If the site is unattended by any operating company personnel in a production site or an offshore satellite, then the well test supervisor assumes the responsibility for the safety induction. The well test supervisor in buddy system configuration and using appropriate PPE, gas detectors, etc. should assess the location and decide if it is safe to proceed. If there is any uncertainty, they should cease all operations and contact the authority.

Example

- A mobile well test crew were tasked to perform a production monitoring (GOR/PGOR) test on a remote well site containing H_2S
- After arriving on location, the supervisor inspected the location and his personal H_2S alarm sounded as he approached the wellhead. Bubbles could be seen coming out of one of the flanges on the wellhead
- The supervisor moved the crew to a safe location and reported to the gathering station the leaking wellhead. The crew were told to stand down and were allocated a different well
- The well was then scheduled for maintenance

All PPE and safety equipment must be checked and approved before any operations commence. A safety meeting should be held after any orientations to establish any issues or questions.

4.4.3 CREW SAFETY MEETINGS

Before any operations begin, the job supervisor or his designated representative should conduct the safety meeting referred to as a "Toolbox Talk."

Whoever conducts the meeting should identify and address the three primary causes of accidents:

**I Didn't Think
I Didn't See
I Didn't Know**

It should be made clear that safety is everybody's responsibility and that any crew member must raise any concerns or lack of knowledge at the safety meetings, and record it in the safety meeting signed report. In addition to any shift, hand over details and issues shared by the previous shift supervisor.

Standardisation of time among all personnel and consistent with the data acquisition system is established and reviewed in these meetings.

It is normal practice that only the well test crew attend the safety meeting, but if the well is to be flowed for an extended time, then the attendees should also include:

- Company personnel
- Rig personnel (driller and crane driver as a minimum)
- Any other service providing personnel associated with the test
- The company representative can request other members to attend, as seen fit

The safety meeting will cover all aspects of the well test and upcoming operations, detailing any specific actions and tasks. Before any rig up or pressure testing tasks the safety talk will be broadened and cover details. If any chemicals or hazardous substances are used, they are to be discussed including reference to their safety data sheets.

All attendees will sign a register specifying that they have attended, understood, and agree to comply with all the topics discussed during the safety meeting. These signed registers are saved in the job folder and will be returned to the service provider's base or head office for future reference under contractual obligations.

With mobile well testing units on a reservoir or concession testing campaign, safety variables are relatively constant. It is good practice and training exercise to allow the crew members to take turns in conducting the safety meeting among their peers, under the supervisor's guidance. Every member of the crew should participate as part of their training and to ensure the individual's understanding of safety.

A concern with routine or recurrent well tests, with relatively constant risk variables, is that the safety talk minutes are repeated among the crew. The base coordinator or HSE officer should routinely review the safety minutes to ensure true and efficient safety meetings are held. If necessary, they should advise the supervisor on ways to include new risk and safety material in the meetings and avoid repetition.

To add further value to the safety meeting, in attendance of the crew, the well test supervisor may pick a crew member and ask them to explain a certain task or its methodology to make sure of their working and safety knowledge, to have the knowledge shared among the crew and use it as a team-building exercise.

4.4.4 PPE

A supervisor's major responsibility is to make sure that everyone on site has and is using the correct safety apparel and equipment. All personnel are responsible for their own PPE and any defects with their co-workers' PPE.

This will involve inspection of:

- Safety footwear
- Eye and face protection
- Specific safety clothing

Torn, worn, and defective PPE should be avoided and personnel will wear proper and appropriate PPE to suit operations.

4.4.5 STOP WORK AUTHORITY

Arriving on site the well test crew and personnel should be familiar with the location's stop work authority and comply with it. In case no such policy is set, well test staff would adhere to their own company's stop work program.

The well test supervisor is in charge of making sure that during each shift, at least one crew member fills out an observation ticket. During the safety briefing held at the start of every shift, the observation tickets can be revised and alternative crew members assigned the responsibility.

4.4.6 PRESSURE TESTING

All well test pressure equipment must have valid annual survey certificates that are available on site for inspection and include a full hydrostatic test to the equipment's specified test pressure (TP). These documents and certifications should be inspected before equipment mobilisation to site.

On site and after equipment rig up it is mandatory to pressure test the connected equipment to detect any possible leaks. Pressure testing operations **must** be preceded by a safety talk by the well test supervisor or their designated representative, outlining the procedures involved and placing a strong focus on safety and adherence to rig or location safety procedures. Attendance is required from all relevant personnel from the rig and operating company.

Warning signs should be in place and the pressure test area isolated. All unauthorised personnel must not enter the area unless permission is granted by the well test supervisor. This is a matter of safety; without these restrictions in place and granted access, any of the site personnel may wander into the area assuming it is safe for them to do so. Respecting and adhering to isolated areas is critical.

During the pressure testing sequences each member of the well testing crew will be assigned to a monitoring station with clear responsibilities, while personnel not involved must leave the isolated area.

The pump and all of its flowlines, which constitute the pressurising source, should have been tested to highest expected pressure intended for the well testing equipment prior to commencing the pressure test sequence. This is to ensure that any drop of pressure during the pressure tests are not from the pump setup. A well test chart recorder and/or data acquisition system should record the integrity of the pump setup, as well as the equipment pressure testing sequence.

Any temporary equipment connected to a well or related equipment that is subjected to well pressure, even briefly, must undergo pressure testing using a non-volatile liquid (glycol can be used in cold conditions). All pressure tests should be witnessed and signed off by the operating company representative, attesting to the satisfactory completion of the equipment pressure test when satisfied.

4.4.6.1 Pressure Test Objectives

The objective of pressure testing is to ensure that surface equipment will not leak or fail under pressure while in use. A leak during operations may lead to an escape of hydrocarbons at the least and may result in loss of life in a worst-case scenario.

A written plan or sequence of events for the pressure test should be revised and signed off by the operating company representative before pressurising.

This will detail:

- Sequence number
- Equipment/section of setup
- Maximum pressure
- Pressure test duration
- Any exceptions

4.4.6.2 Pressure Testing General Points

- Prior to pressure testing, all personnel must be made aware that a potentially hazardous operation is about to start and restrictions are in place
- The separator, although a lower pressure, poses a greater hazard due to its volume capacity
- A work permit system must be issued
- A safety talk must be held for all personnel involved; this includes pump operators and other on-site personnel
- The well testing equipment, including flowlines from the pump, which would be pressurised, must be cordoned off and warning signs posted around the vicinity
- It is mandatory that all non-essential personnel depart the area
- Only personnel essential to monitor the operation can be present within the cordoned off areas and their number kept to a minimum
- In order to prevent any accident, pressure should be instantly bled off if any unauthorised personnel intrude into the area
- A crew member (commonly the senior or supervisor) should be outside the barrier with open communication and in line of sight with the pump operator to relay instructions. No other personnel should issue instructions to the pump operator, unless there is an emergency requiring bleed off of pressure

- As well as communicating with the pump operator, open communication should exist between the pressure testing areas and the focal point crew member. Therefore, if any issues arise, the pressure can be bled off to lower the risk of accident and injury
- The pressure test sequence should be clearly defined and adhered to
- The well test supervisor will make decisions on how the pressure is applied, bled off or suspended
- For each series of tests, the defined pressure-testing program should be adhered to
- While pressure testing is ongoing, no non-essential welding operations should be carried out
- There should be no unnecessary crane lifts across the well test areas. The senior well tester on shift should approve any necessary crane lifts

While the pressure test area is cordoned off, senior personnel should lead by example and adhere to the rules and respect the isolated areas, otherwise junior staff will likely underestimate the seriousness of the situation. During TWOP the pressure testing process and sequence would be discussed with emphasis on safety and the well test supervisor's final authority.

4.4.6.3 Pre-Test Safety Meeting

Prior to any pressure being applied or flushing lines in preparation for pressure testing, a safety meeting must be held for all personnel involved within the vicinity – or on deck – to visually identify the setup and procedure, following which well test personnel would be involved in a pre-test briefing. If a shift-change is due during the pressure testing, the handover must be done when pressure is bled off and pressure testing halted, until a shift handover is conducted and a pre-test briefing with the arriving crew members organised.

A qualified and well-experienced supervisor or senior should be present throughout the pressure testing stages and assume command of the operation and vicinity, while liaising with the operating company representative and other relevant personnel.

Personnel are required to sign an attendance register to account for their presence and acknowledge their understanding of the rules, procedures, and conditions of the process.

Within the safety meeting the following should be emphasised:

- The dangers of pressure and explosive decompression
- Unless absolutely essential, no personnel should approach the flowlines or equipment. The buddy system is in effect and co-workers must keep an eye on any personnel inside the pressure test barriers
- The role of nominated well test personnel stationed around the area to monitor for potential leaks and stop personnel from trespassing
- The prohibition of tightening or approaching unions or flanges in any way while the setup is under pressure
- The barrier is at an adequately safe distance for equipment isolation

- Any access to the testing area is blocked off and restricted, including stairs, walkways, or open space, with adequate warning signs in place
- These barriers should stay in place while the well test is in progress

4.4.6.4   Duties of Nominated Senior or Supervisor

- Responsible for all pressure testing related activities
- Will ensure that the testing equipment is fit for purpose
- Will certify that the well test equipment is suitably configured for the proposed well testing job and no post-pressure testing changes to equipment will be made. Identifying the pressure testing sequence and confirming all valves are in the correct position for each pressure test section, etc.
- Ensure all pressure chart recorders have fresh test charts inserted and ready for use
- Any data acquisition equipment is fully functional and connected, all connecting needle valves for sensors are open and **data is logged** before and during all phases of the pressure test
- The pressure lines are tied down to avoid any whip injury during testing
- The equipment must not be left unattended at any time
- The area around the well test equipment must be clear of all and any obstacles to facilitate inspection or emergency evacuation
- All pre-test checks and tests have been carried out
- That the safety procedures are continually followed
- All necessary permits have been signed and issued, with the necessary rig personnel informed
- Immediately stop pressure testing and bleed off pressure to safe levels at any sign of a leak or suspected leak
- Immediately stop test and bleed off pressure to safe levels if communicated to by staff
- Nominate a well test crew member to record the complete sequence of events on the manual readings sheets for the pressure test

4.4.6.5   Pressure Testing Using Third-Party Equipment

When on a rig, either onshore or offshore, it is recommended that the pressure test pump and associated lines are pressure tested prior to proceeding with pressure testing the well test equipment.

This is due to the fact that the pressure test pumps are employed for other tasks and might not be able to maintain pressures for prolonged periods of time as necessary for pressure testing.

Additionally, pumps used in pressure testing of equipment require regular testing and certification on their suitability for the purpose. The pump and associated equipment documentation package should entail these references.

4.4.6.6   Tied Down Lines

All lines must be tied down and secured using a recognised safety mechanism, to prevent any whipping of the equipment.

Equipment must also be completely earthed to a set point using the appropriate cable size.

4.4.6.7 Monitoring the Pressure Test

It is imperative to have recorded proof that a successful pressure test has been carried out; simply declaring "it's okay" is no longer sufficient, regardless of the applied pressure's magnitude being perceived as significant or otherwise.

A standard synchronised time reference is critical, to make sense of the recorded pressure readings and the sequence of events for the stages.

i. Pressure at Pump

There should be a calibrated chart recorder connected to the pump and in use, preferably in parallel with a suitable electronic data acquisition system, to monitor the applied pressure directly from the pump. The pressure chart recorder should be suitable for its purpose in terms of pressure range, operable, well maintained, and with a valid calibration certificate.

The mechanical chart recorder and data acquisition system should be set up and recording data before the pressure testing starts. This establishes an atmospheric pressure baseline and a non-zero reading at this stage is an indication of some sort of error, such as calibration.

On completing the pressure test, pressure within the system must be bled down to zero while the data is recorded. This confirms that the pressure has been released but also confirms the functionality of the recorder and electronic sensors against zero readings.

ii. Pressure at Well Test Equipment

The well test company should also have a similar recorder – recognised practice is electronic and mechanical recorders in tandem – mounted at the choke manifold.

The separator would already be equipped with a multi parameter chart recorder, normally a Barton, on its gas line. Also present, as standard secondary measurement, are the electronic transducers connected with the Data Acquisition System. All of these adhere to the standards, have their performance verified, and comply with the required certifications.

The two tapping lines for the differential pressure, used on the Barton and electronic pressure sensor, should be isolated from any pressure to avoid any damage. Only the direct pressure tapping should be open and recorded by both devices.

The data acquisition system should be set to produce plots of pressure vs time during the entire test. The scaling of these plots should be such that the upper scale setting cannot mask any potential leaks, and the lower scale setting cannot artificially magnify temperature effects to a degree of misinterpretation.

At the end of pressure testing, these recorded charts and plots should be dated, annotated with sequence of events, and signed by the relevant personnel. The plots generated from the data acquisition system should also be printed and included in the dedicated folder with the recorder charts.

In the unlikely event of an accident, these charts and plots will demonstrate that all necessary care and compliance were taken before performing well testing operations.

4.4.6.8 Pressure Safety Valves

Pressure safety valves (PSVs) and relief devices should be replaced by blanking plugs before applying testing pressure (body test). If they are not removed then the maximum applicable pressure is 90% of the equipment's rated maximum allowable working pressure (MAWP)

4.4.6.9 Pressure Testing Procedures

The well testing company should submit company specific safety procedures and guidelines for hydrostatic pressure tests before any pressure testing exercise. Generally, the procedure and guidelines entail the following:

i. Prior to filling up the system with the pressure testing liquid, it is essential to flush and clean the equipment and flowlines with fresh water from any residual hydrocarbon. This task, in particular, applies to any rig pumps used before connecting them to the well test equipment

ii. All air must be vented out before applying pressure in the system. Valves at the highest points in the system are opened to allow air to vent and check that there are no dead-ends in the system that can trap any "air pockets." The presence of air, which is compressible, will give false readings during the pressure test, and this is avoided by filling the system with flowing water or other allowed liquid, a relatively incompressible substance

iii. All spills and overflows must be thoroughly cleaned and dried so that leaks and drips caused during the pressure test are easily detected without confusion

iv. Before ANY pressure is applied, all personnel must be informed to mandatorily leave the area, with no exceptions. This is followed by a rig/location addressed announcement over external loudspeakers, as available

v. The test pressure (TP) must be applied as gradually and as slow as possible. The initial pressure step should be attained and held for a period of usually five minutes, without constant pumping to preserve pressure or compensate for pressure drops

vi. Pressure must be released if there is a visible leak, indicated by a drop in pressure on the pressure pump chart recorder or any of the other monitoring instruments

vii. Pressure should be slowly bled off to zero or maintain an agreed on safe pressure, while checking the returns for indications of trapped air

viii. **ONLY** when the pressure is completely bled off and verified can any remedial work be carried out

Once the corrective work is completed the pressure testing process is repeated from the first step

ix. When the initial test, normally lowest applied pressure, has been successfully passed, the hydro test pump operator will alert all personnel that pressure is about to be applied

x. Pump pressure is increased for the second stage equipment pressure test. Once the desired pressure is attained it should be held for the secondary holding period, normally 5 to 15 minutes. This is usually enough time to verify that no leaks are present. If there are any visible leaks, the process is repeated from step vi

xi. Continue applying pressure in accordance with the predetermined stages, with each pressure step held for an adequate period of time to ensure there are no visible leaks, until the final stage. At this point, if there are visible leaks, the process is repeated from step vi

xii. To ensure the integrity of the rig up and well testing system, when applied pressure has stabilised at test pressure value, the test system (equipment under test) and pressure gauges should be isolated from the pressure pump source. By eliminating the pressure pump at the targeted test pressure, we ensure that the pump does not unintentionally compensate for any pressure drops due to leaks

xiii. Upon completion of the pressure test stages, commence bleed down sequence

4.4.6.10   Completion of Pressure Test

Upon the successful completion of the pressure test.

i. Bleed off all pressure in a controlled and safe manner
ii. Ensure no pressure is trapped in any lines before attempting any rig down operations
iii. Open any valves at the high points and drain the fluid from the equipment that has been pressure tested from the lowest points to ensure maximum drainage
iv. Disconnect and remove the pressure test pump lines and store safely
v. Reconnect any equipment removed for pressure testing purposes, separator relief valves, etc.
vi. Close the work permit
vii. Annotate all charts and DAS plots and store them in the dedicated folder
viii. All documentation to be signed off and filed

4.4.6.11   Pressure Testing with Gas

Pressure testing with gas on well testing equipment is extremely dangerous due to the potential of explosive decompression involved with a compressed gas. This makes it an unpopular method for pressure testing and left for dedicated purposes such as against the front valves of the choke manifold.

If performed, a pre-test safety talk should precede it, usually by the supervisor from the dedicated gas pumping company, outlining all the processes and procedures that will take place. The dangers of gas testing must be stressed.

The gas used must be non-flammable and inert. Nitrogen is the preferred option here as it is one of the inert gasses.

The equipment used must first pass a hydro pressure test described in the previous section, before any pressure testing using gas can be attempted. However, because gas will locate a leak far more easily than a liquid, the tolerance for hydro pressure test leaks can be lowered.

Explosive decompression has a considerable risk of harm; hence it is crucial that the lines are securely fastened, tied down, and verified to that affect.

Here, safety is critical. No personnel should be allowed near the equipment when using gas for pressure testing; all monitoring should be via dedicated instruments. Temperature dips frequently indicate leaks. All non-essential personnel should evacuate the area and muster in the designated safe area.

As with all operations, safety procedures must be followed. As a matter of importance all and any potential pressure has to be completely bled off before any remedial work is performed.

5 Well Test Equipment

A well testing package typically uses standardised equipment; the one-size-fits-all approach which applies to over 90% of well test packages in the service sector. However, the following conditions call for specific equipment or specifications:

i. H_2S or other hazardous gasses
ii. Corrosive liquids
iii. Hazardous liquids
iv. High pressure high temperature (HPHT)

This chapter is intended to give an **overview** of the basic well testing equipment and their operability; it is not meant to serve as an all-inclusive manual.

Equipment drawing and illustrations are an example of generic design. Types, sizes, and designs vary according to the manufacturer, pressure rating, and flow. Manufacturers' manuals should be consulted for specific equipment.

The well testing, pressure bearing equipment, and their components are built to individual and collective standards. The documents listed in Table 5.1 include some of these fundamental standards. This cannot be considered a complete list of standards as they vary from country to country.

TABLE 5.1
Equipment Standards

Equipment Code	Equipment Type
API 6A	Flow heads, surface safety valves, choke manifolds, upstream heater chokes, flowlines
API RP 14E or ASME B31.3	Low-pressure flowlines downstream of heat exchangers or choke manifolds
API 12K	Heaters and steam exchangers
API RP 14C	Surface safety systems
API 14A & 14D	Surface safety systems, shutdown valves, and ESD systems
API 16A	Specification for drill through equipment includes API hubs
ASME Section VIII Div. 1	Boiler and pressure vessel code for pressure vessels.
NACE MR-0175	For all H_2S service equipment

All well test equipment MUST have all the correct valve handles and associated control mechanisms correctly installed on the equipment at all times.

DOI: 10.1201/9781032623689-5

5.1 OPERATING THE EQUIPMENT

It should be emphasised that only the well test crew, working under the supervision of a senior well tester, are qualified to operate any of the well testing equipment. **The operating company representatives or personnel cannot instruct** a junior well tester to operate or make changes to the equipment. All requests for changes to the equipment, excluding emergency events, must go through the senior well tester working the shift.

In the event of any kind of emergency, anyone may and should operate the emergency shut-down system, which should be a simple pull-button mechanism strategically placed around the well test equipment.

Separator Example

There have been numerous instances of inappropriate sequence of operations, when the choke was increased while the well was flowing through the separator. The surge in pressure caused the orifice plate to be blown down the gas line and into the back pressure valve leading to the pressure relief valves to lift and vent the separator pressure.

It is a major repair after an orifice plate has been blown down the line. The gas line, including the orifice box and back pressure valve, has to be completely disassembled in order to remove the plate and rectify the damage. The back pressure valve may have to be replaced as well as any damaged components in the orifice box. After completing repair, the system will have to be pressure tested. All this results in considerable delays and costs.

The separator has the risk of rupturing or exploding, if the vent from the separator is improperly piped and does not permit full discharge.

Choke Manifold Example

The most frequent occurrence is when the company/rig personnel insist on bleeding pressure off or regulating pressure using the choke manifold valves. These are gate valves and have two positions – **either open or closed – and NO position in between.**

The seal faces are damaged by erosion as a result of this unusual choking technique, and the valve is rendered unusable because it can no longer maintain pressure.

The front valve of a choke manifold is seen in Figure 5.1, and the damage was brought on by the well pressure being bled down while employing a partially open gate valve.

The erosion is clearly visible on the matching parts. These are the seal faces of a 15,000 psi (15 kpsi) valve that can no longer hold or maintain pressure. Due to the probable location's remoteness and unlikely replacement on site, there will be delay and lost time.

In this type of down time the company will look to charge the well testing company for equipment failure. However, if such damage was caused by personnel, other than the well testing crew, demanding to operate the equipment – who should be charged for the lost time and the replacement of the seats? One part of the valve costs over $2,500 and this does not take into account the time and labour in repairing the valve.

It is critical that everyone on site and at any TWOP meetings understands that the physical operation of the well testing equipment lies in the hands of the appropriate service provider.

Damaged Valve Gate   **Damaged Body Seal Ring**

FIGURE 5.1   Damaged Valve Gate.

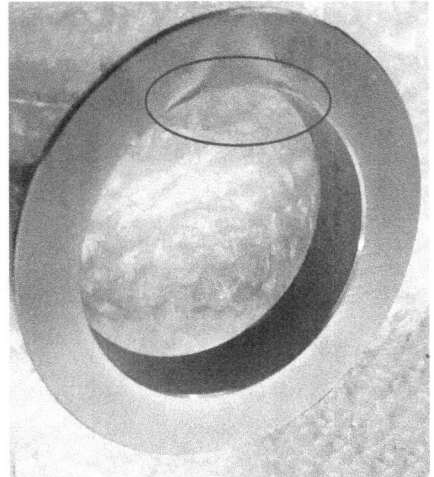

5.2   EARTHING

The entire well testing package must be earthed on each individual major component basis using the proper single core specific earth cable with a green sheath to indicate earth. To adhere to IEC requirements the earthing cable size should not be less than 6mm². In essence, an earth cable must be linked to a proper earth point from each major piece of equipment in the well testing package.

On land locations, rather than using an earth stake it is more effective to earth the system to the well's casing using a G cramp or to a metal fence post that has been driven into the ground. This a more fundamental application for mobile testing units because their tyres insulate them from the ground.

Proper connectors and clamps should be used that are unpainted and free of rust.

5.3   TIE DOWN

All pipework should be tied down using a recognised tie down system. This does not apply to pipework on board a mobile testing unit trailer, but any pipework out of the trailer unit should be tied down. This is to prevent pipe whipping.

Less damage occurs from high-pressure leaks in high-pressure armoured hoses, Coflexip type, than in high-pressure leak in fixed flowlines. Therefore, it is safe practice to use a high-pressure armoured hose to lower the risk of whipping hazard under failure.

5.4   EQUIPMENT LAYOUT

Figure 5.2 depicts a general and basic well testing equipment layout. In the later sections of this book, more details will be given on the setup and individual equipment.

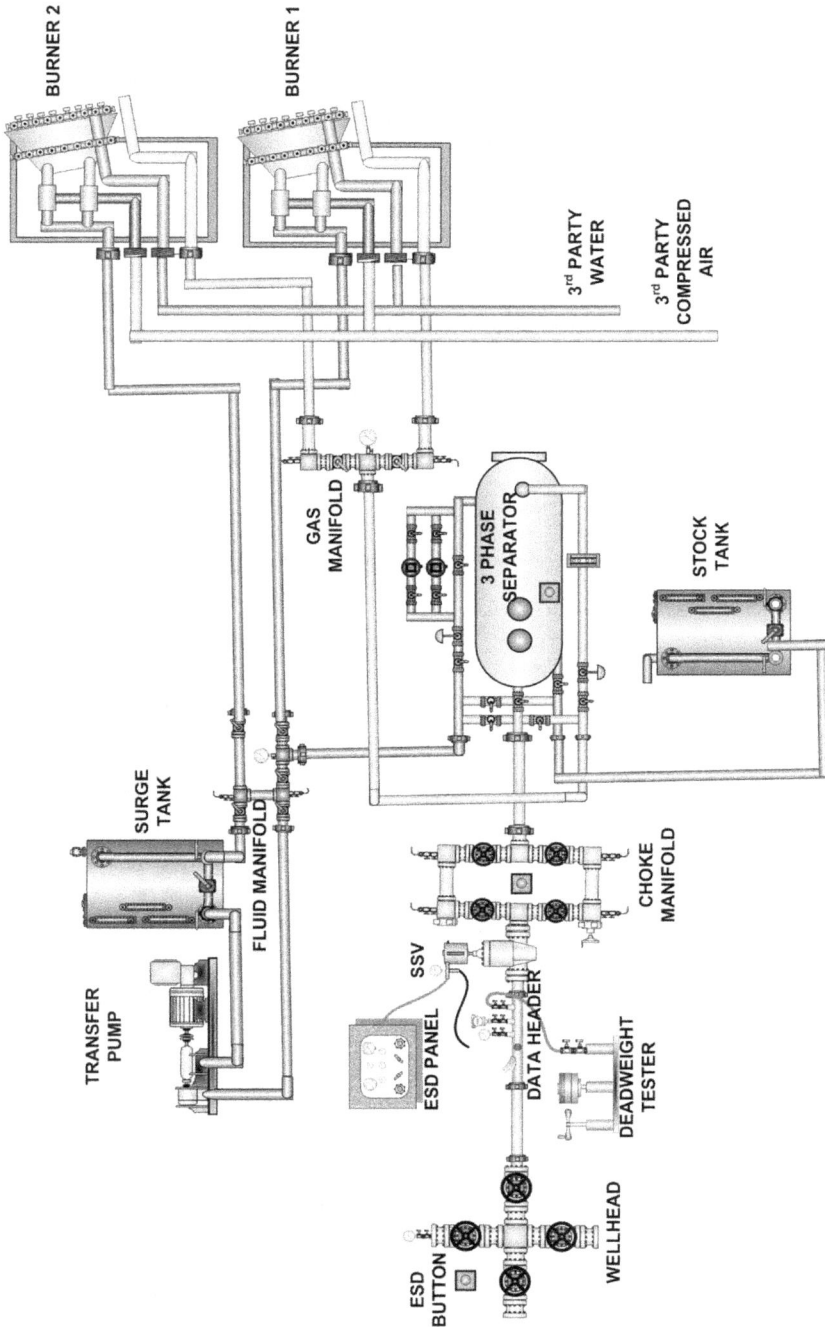

FIGURE 5.2 Basic Well Test Layout.

5.5   FLOW LINE RIG UP

Every flowline on a well test rig up should be straight and at the same level. The bends, elbows, and flow diversion cause back pressure, restricting potential flow rates. Besides flow diversion pipework, the reduction in flowline area, crossovers etc. will also result in flow reduction and erosion.

When gas flows are involved, any change in flow direction can also cause hydrate formation which can also cause flow restriction and eventually flow blockage.

When using offshore burners, all the flowlines usually have to be elevated to connect with the corresponding flowlines on the burner boom. The elevation is usually accommodated by use of small pup joints and elbows that divert flow up to 90 degrees (shown in Figure 5.3), which is the largest flow direction change. Such a directional flow change would dramatically support hydrate formation, especially on the gas line.

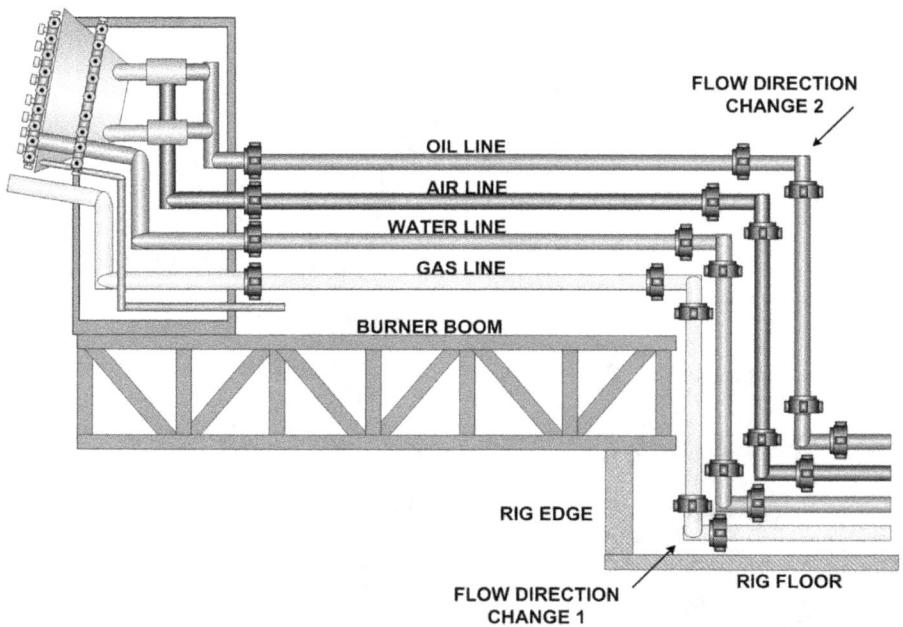

FIGURE 5.3   Rig Connection to Burner.

Hydrates can accumulate at these elbows, depending on the flow rates and water content, effectively reducing, even completely blocking, the available flow area. A blockage can see the separator pressure communicated to downstream flowlines, of a lesser pressure rating, possibly causing flowline rupturing.

Flow direction changes cause higher erosion rates, especially with high-pressure gas, that carries any denser particles directly into the inner walls of the pipework.

Straight flowlines limit the risks.

5.6 HYDRATES

In well testing a hydrate is conceptualised as an ice-like structure filled with gas molecules, equal to hydrocarbon gas, extremely flammable.

Hydrates can – and do – form spontaneously if the requisite pressure and temperature conditions are met; a drop in pressure is often not necessary.

The refrigeration effect, Boyle's Law, is the most common type of hydrates formation referred to in the industry. Based on a pressure drop, across a choke or other flow restricting equipment, it can create the optimum pressure and temperature conditions for a hydrate to form.

Hydrates can occur in both oil and gas flows when light hydrocarbon gases, such as methane, ethane, CO_2, and H_2S, merge and combine with free water, under specific pressures and temperatures. The presence of the heavier end hydrocarbons gasses, butane and heavier, tend to prevent or inhibit the hydrate formation.

The process of monitoring and recording the produced gas gravity, any impurities, and BS&W is given additional importance because hydrates can only form in the presence of specific gases. Later portions of this book go into detail on how these processes are carried out.

In rare but plausible scenarios, hydrates have the ability to drastically restrict the flow area across any choke, further constricting the flow, lowering flow rates, and building up pressure.

Ultimately, if not addressed, this can cause the well to be shut in due to a total blockage, or an orifice plate can be blown out of the Daniel box from a pressure surge. Wellhead pressure, flow, or routine monitoring with manual readings can all easily detect or identify this.

5.6.1 HYDRATE PREVENTION BY CHEMICAL INJECTION

Hydrates can be prevented or inhibited by injecting a liquid inhibitor into the well flow. However, caution must be exercised because some of these fluids are regarded as hazardous and volatile, necessitating adherence to safety data sheets and procedures.

Injection of these fluids will help prevent and/or inhibit the formation of hydrates by effectively lowering the freezing point, an "anti-freeze" approach where Mono-Ethylene, Tri-Ethylene Glycol, or similar agents are used to lower the freezing point of the free water in the flow, thus preventing hydrate formation.

Methanol was used but has been banned due to safety issues.

NOTE: *Refer to the Katz tables for hydrate formation data.*

Since the highest-pressure drop occurs at the choke manifold, often indicated by icing or freezing on the body of the choke, the chemical is normally injected upstream of the choke manifold through the data header.

The standard pump type used for chemical injection is the Texsteam (refer to Figure 5.4), usually the Model 500. The Model 500 Texsteam can pump up to pressures of 12,000 psi, depending on the plunger fitted in the body. The plungers range from 0.25 to 1.25 inches and the continuous fluid pump rate varies from 16 to 500 gallons per day.

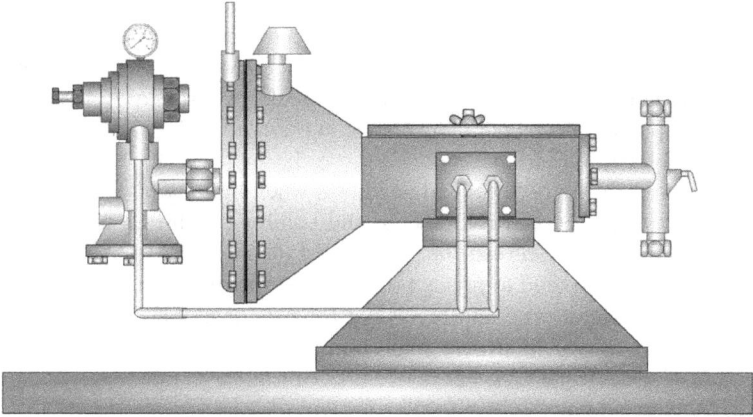

FIGURE 5.4 Typical Texsteam Pump.

For all configurations and performances, refer to the Texsteam operating manual because the injection pressure can decrease as the plunger size increases. Plunger assembly (Figures 5.5 and 5.6) can be modified to alter the Texsteam configuration.

If chemical injection is being used, we then also need to know the injection rate and volume being injected into the flow. The Texsteam manual gives graphs that translate the counted strokes from the Texsteam into gallons/day (GPD), so an estimation of rate and volume can be calculated.

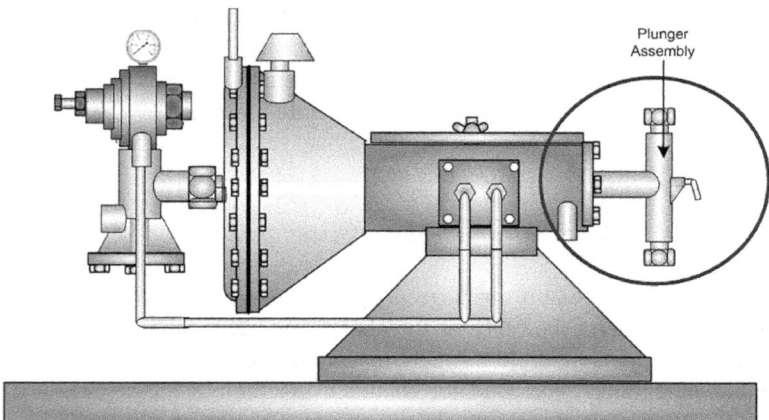

FIGURE 5.5 Location of Texsteam Plunger Assembly.

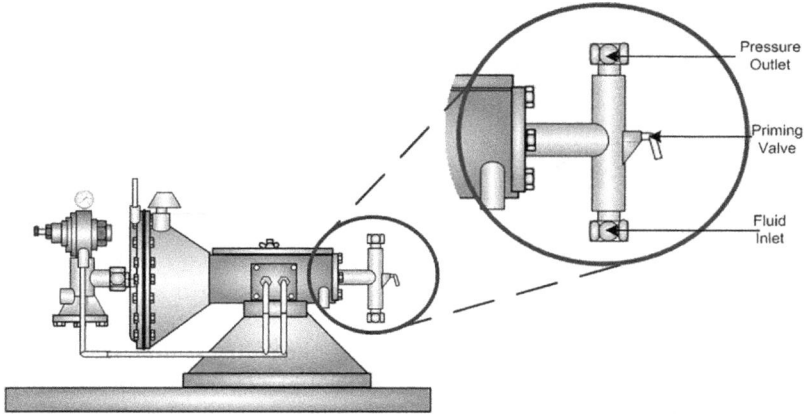

FIGURE 5.6   Texsteam Plunger Assembly Details.

If accuracy is needed, as in the case of subsea injection, there are high-pressure miniature turbine meters that can be used to measure the injected chemicals, which can be monitored by the data acquisition systems.

An alternative method is to monitor the amount of liquid left in the chemical tank/ reservoir. In some configurations a rectangular reservoir is built into a skid with the Texsteam (as shown in Figure 5.7), which makes measurement relatively easy. Other systems use the original barrel or drum the inhibitor was supplied in (Figure 5.8).

The two methods of housing the fluids require precise height measurements taken from the sight glass at regular intervals, normally every 30 minutes to begin with, recorded on the manual readings sheets. In the case of the reservoir, it is a fairly simple calculation to convert level height into volume, but in the case of a barrel the calculation is slightly less straightforward. The sight glass approach has the added advantage of allowing the monitoring of remaining fluid, so the reservoir can be either filled or replaced before it is depleted.

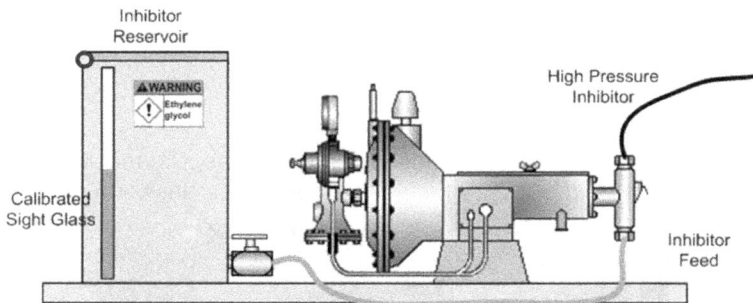

FIGURE 5.7   Injection Skid with Reservoir.

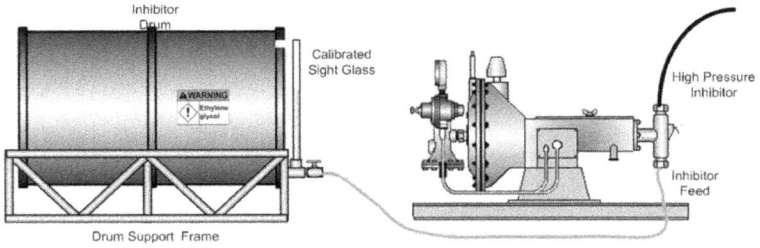

FIGURE 5.8   Chemical Injection from Barrel.

The chemical injection tapping is normally the last one on the data header configuration to minimise any interference to the instruments or sampling. This minimises
the pressures introduced by the chemical injection being picked up by sensors, and
samples are void of the injected chemical. (Illustrated in Figure 5.9 and expanded in
Figure 5.10.)

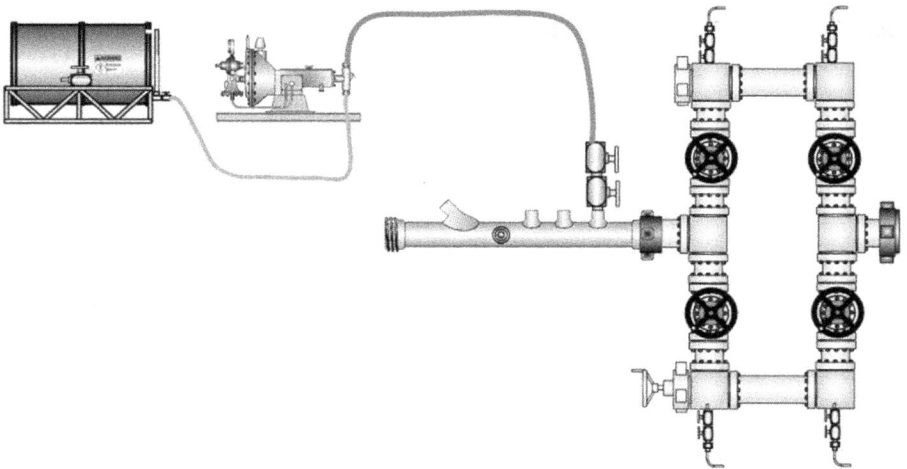

FIGURE 5.9   Chemical Injection before Choke Manifold.

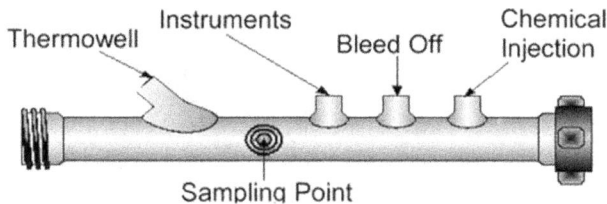

FIGURE 5.10   Data Header Tappings.

5.6.1.1 Volume Calculation of a Fluid in a Barrel

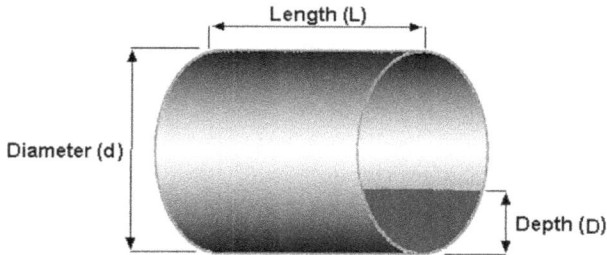

FIGURE 5.11 Barrel Volume Schematic.

Calculation of the liquid in the barrel

$$R = \frac{d}{2}$$

$$Volume = L\left(R^2\left(\frac{R-D}{R}\right) - (R-D)\sqrt{2RD - D^2} \right)$$

Where
    R = the radius of the cylinder
    D = the depth of the liquid
    L = the length of the cylinder

    The same units must be used for all measurements for this calculation, giving the volume results in those cubic units. If the measurements are in inches, then the calculated volume will be in cubic inches.

    The calculated volume remaining in the barrel should then be converted into either gallons or barrels; the difference between records on regular intervals is the injection rate per interval.

5.7 DATA HEADER

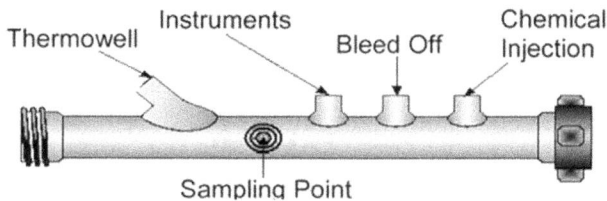

FIGURE 5.12 Data Header Layout.

The data header is typically used for instruments, monitoring, sampling, and chemical injection. It can be employed wherever access to the flow is necessary throughout the well test package, not just at the choke manifold.

In case of upstream of the choke manifold, the data header is generally positioned before the SSV so if the valve closes the upstream pressure can still be monitored.

Hammer unions have a pressure limit of up to 10,000 psi. Therefore, where 15,000 psi (high-pressure) rated data headers are used, the standard hammer unions are either replaced with flanges or are fitted with Grayloc type clamp connections. A high-pressure data header must be equipped with autoclave-type fittings/valves capable of withstanding equal pressures.

Either version of the data header pressure rating may have up to eight ports. When a high-pressure data header is used each port should only be used by a single instrument installed; instrument manifolds are prohibited to avoid further complexities and increased risk. However double needle valve configuration should be used at all times for pressure tapings.

The thermowells supplied with the electronic temperature transducers are typically used in data headers, presenting a potential hazard. These thermowells often do not have a relevant safety certificate for pressure ratings and H_2S compatibility.

When sand or proppant are anticipated the thermowell port can be blanked off to prevent the erosion of the installed thermowell and the release of pressure and flow to the atmosphere.

5.8   THE CHOKE MANIFOLD

The choke manifold is the single most important piece of equipment on a well test package as it can both isolate and control the well flow. It is critical that it is in good operating condition, maintained regularly, has adequate valve seals, and is collectively able to hold the maximum pressure expected from a well.

A standard choke manifold consists of four valves as shown in Figure 5.13. The Diamond Choke Manifold is often favoured as it is cheaper to manufacture with only four solid blocks while the standard manifold has six. From a well testing perspective the standard option is the better version as it gives more options on fluid sampling and bleeding off (illustrated in Figure 5.14).

The valves on the choke manifold and safety valve can be bi-directional, depending on the manufacturer, which is an advantage frequently overlooked or unknown. As a result, if a valve fails a pressure test, the valve may be unbolted and spun 180 degrees for a new seal face against the applied pressure. This needs to be verified with the type of valves used.

All bleed offs MUST be carried out using the adjustable choke and NOT by the gate valves. This malpractice of using the choke manifold's valves would render them unusable for their intended pressure isolation and control function.

The vast majority of the high-pressure valves are gate valves and as such have a set number of turns to fully open and close. Personnel who operate these valves must be aware of the number of turns required to fully open and close each valve.

Knowing the number of turns required to open or close a valve makes it possible to account for an obstruction across the gate's path. This is indicated by the valve not operating over the full number of turns.

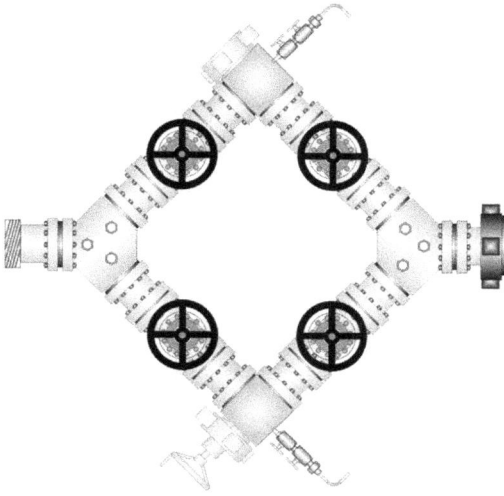

Diamond Choke Manifold

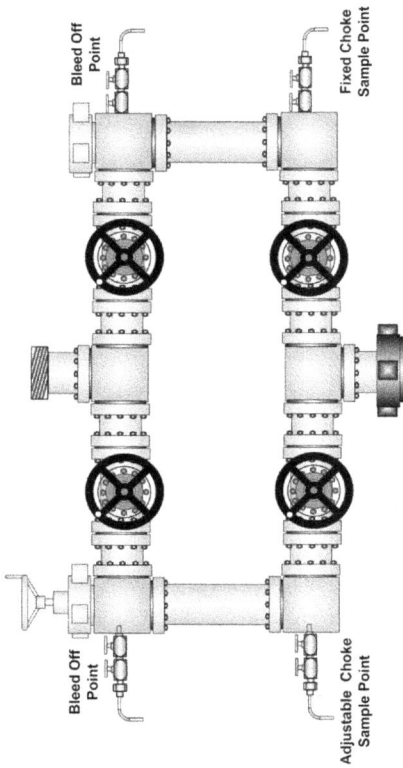

Standard Choke Manifold

FIGURE 5.13   Choke Manifold Schematics.

FIGURE 5.14   Standard Choke Manifold Sampling and Bleed Off Points.

Under no circumstances should the valves be forced open or closed, as this can cause the stem to shear and render the valve inoperable.

Like with all valves, turning clockwise closes the valve and anti-clockwise opens the valve. Unless shutting the well in at the choke manifold, the basic rule of operation is ensuring a corresponding valve is open (opening) before closing (beginning to close) the corresponding valve. This also applies to fluid manifolds, gas manifolds, and separator manifolds.

If the downstream choke pressure is continuously monitored, during stable flow this can indicate that the choke cutting is beginning to erode or cut out or plug.

Choke cutting out

Wellhead pressure decreasing and flow increasing

Choke plugging or hydrating

Wellhead pressure increasing and flow decreasing

5.8.1 Basic Gate Valves

Well testing gate valves normally have the following maximum operating pressures.

5,000 psi
10,000 psi
15,000 psi

Higher pressure range valves are available but not usually as standard.

NACE MR-01–75 H_2S service is often the certification standard for the presence of H_2S.

With the valve open the gate is in the upper chamber, allowing the fluid to flow. With the valve closed the gate is across the flow path so no flow is possible and a seal is achieved. (illustrated in Figure 5.15)

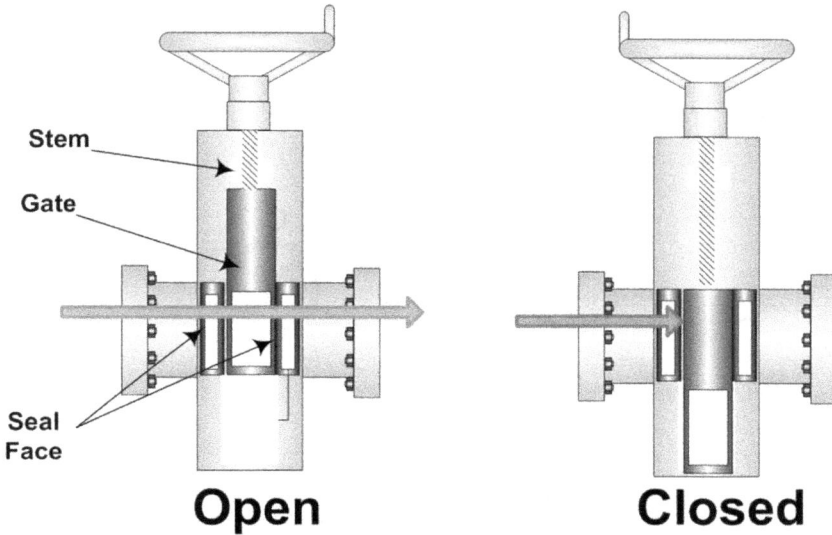

FIGURE 5.15   Basic Gate Valve Operation.

The gate is a polished solid block of steel that seals on additional seal faces inside the valve body, illustrated in Figure 5.16.

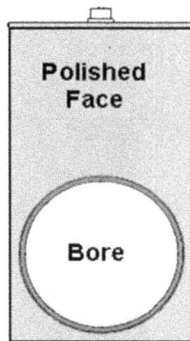

FIGURE 5.16   Basic Gate Face.

It needs to be re-iterated that gate valves only have two positions – fully open and fully closed – with no positions in between and are not designed for bleeding off or choking pressure. The adjustable choke is used for this purpose. The adjustable choke, on the other hand, is not designed to hold pressure and no amount of tightening will result in a pressure seal; it will only damage the choke cone and seat.

The valves used on the choke manifold are usually the same as the ones used on the emergency shut-down (ESD) valves.

The inlet of the choke manifold is referred to as the front of the choke manifold. Therefore, valves A and C, depicted in Figure 5.17, are the front valves and are the main isolation valves for the entire well testing package. The front valves should always be sequenced to close before the rear valves, except in emergency conditions. Following this sequence avoids pressure trapping between the valves.

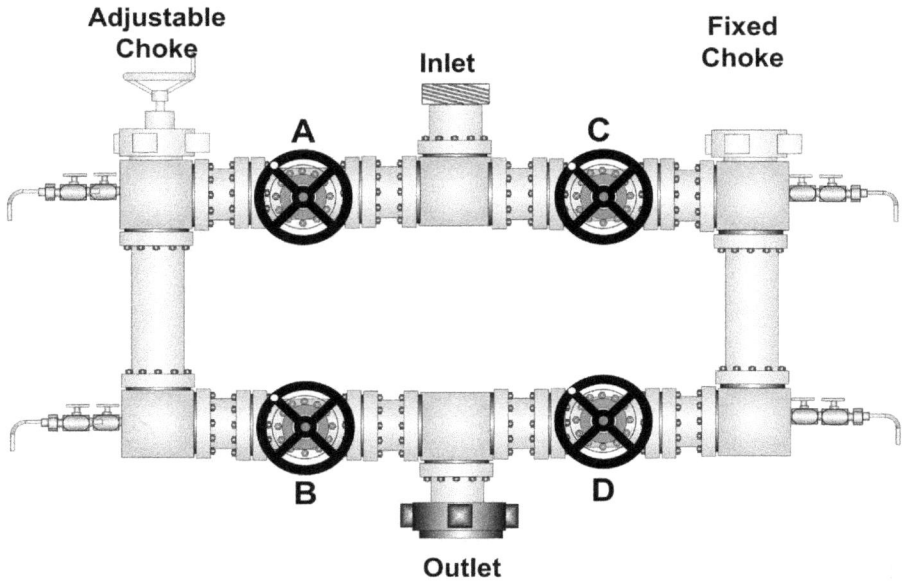

FIGURE 5.17   Choke Manifold Valves.

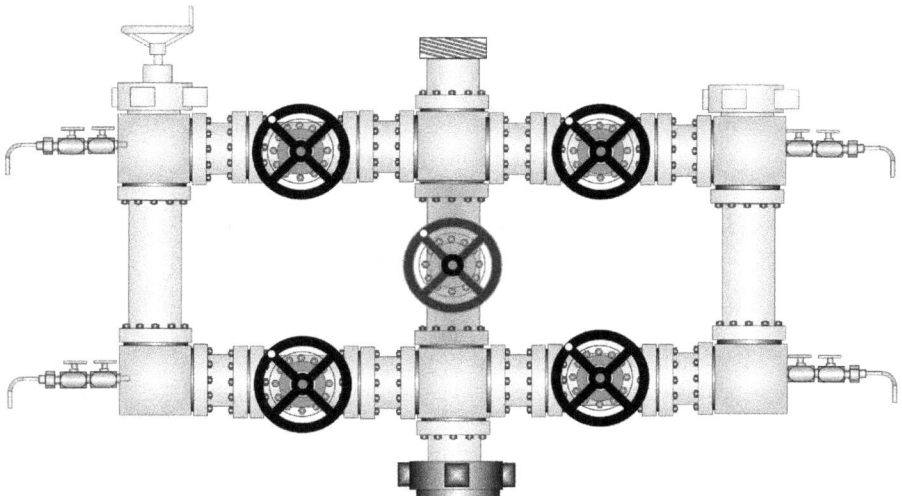

FIGURE 5.18   Choke Manifold with Built-In Bypass.

The rear valves, B and D, are a backup to the integrity of the primary (face) valves and should not be used as the primary isolation, other than in an emergency.

The standard rule now is that two pressure isolation valves must be in place for safety reasons. Therefore, the previous five valve choke manifolds (Figure 5.18), with a built in bypass, are no longer permitted for use with pressure isolation, because the central valve only has one valve for isolation, circled, this particular choke manifold design is prohibited. The central valve should be removed from the configuration and replaced with appropriate blind flanges.

5.8.2 Trapped Pressure

It is possible to trap pressure between the valves if the valve sequences are not followed. Pressure can kill and maim.

Well test operators must constantly be alert for the presence of trapped pressure within any of the equipment or pipework, by checking dial gauges or other connected pressure instruments. When changing chokes, for example, pressure may be trapped between the two valves on the choke manifold.

In order to facilitate safe operation, the correct rating valves, plugs, gauges, and pressure pilots must be installed on the choke manifold or ESD system. All the bleed off points should be fully open and all pressure bled off before attempting to change chokes.

When bleeding off pressure, the operator should be wearing full PPE and stand to one side upwind of the bleeding off point, instead of directly in front of it, to avoid any splashback or gas effects.

Before attempting to change a fixed choke or the adjustable choke bean, it is imperative that all related bleed off points are fully open and that there is no trapped pressure within any area of the choke manifold.

To prevent fluid spills, bleed offs and sampling should be carried out with a "drip tray" or suitable fluid catcher below the bleed off area. If the fluid is volatile the drip tray should be emptied immediately after completing the task.

5.8.3 Chokes

The choke manifold normally has two chokes that are usually expressed in 64ths of an inch, unless in a metric system, then millimetres (mm).

    i. Adjustable choke
      a. Main use for cleaning up and bleeding off
      b. Normally used for short periods
      c. Accuracy depends upon zeroing before use
      d. Not as accurate as fixed chokes
    ii. Fixed choke
      a. Main flow period choke
      b. Used after required choke size has been determined by the adjustable choke

Choke manifolds can have two fixed chokes, instead of one of each, as is the case in long term production, but this is rare in well testing. Even more rare and unlikely are dual adjustable chokes.

It must be emphasised that the choke size shown on the micro sleeve of the adjustable choke is not precise indication. The actual adjustable choke size depends on two major factors:

i. Any damage or erosion to the adjustable choke bean or the adjustable choke stem
ii. The zeroing of the adjustable choke stem against the adjustable choke bean

The adjustable choke is imprecise at best. This is why when switching between adjustable and fixed chokes or trying to match chokes on production with an adjustable choke, results are different. This difference is ignored when it is insignificant, and if there are major differences, then both chokes and their respective installations should be checked.

An adjustable choke usually range from 0 to 96/64" or 0 to 80/64" dependent on its brand, model, and manifold pressure rating.

Without the movable components found in an adjustable choke that are buffered with flow and induced vibrations, fixed chokes **must** be used for all flow measurements. Table 5.2 lists the standard set of fixed chokes and their sizes, supplied with a choke manifold configuration.

TABLE 5.2
List of Fixed Chokes Normally Available

64ths	Fraction	Inches	mm	Qty
Inch	Inch			No.
8	1/8	0.125	3.175	1
12	3/16	0.188	4.763	1
16	1/4	0.250	6.350	2
20	5/16	0.313	7.938	1
24	3/8	0.375	9.525	2
28	7/16	0.438	11.113	2
32	1/2	0.500	12.700	2
36	9/16	0.563	14.288	2
40	5/8	0.625	15.875	2
42	21/32	0.656	16.669	2
44	11/16	0.688	17.463	2
48	3/4	0.750	19.050	2
52	13/16	0.813	20.638	2
56	7/8	0.875	22.225	2
64	1.00	1.000	25.400	2
72	1 1/8	1.125	28.575	2
80	1 1/4	1.250	31.750	1
88	1 3/8	1.375	34.925	1
96	1 1/2	1.500	38.100	1
128	2.00	2.000	50.800	2

An entire set of undamaged or eroded chokes should be available for use on site. They should be verified in quantity and condition before mobilising the unit. Any missing or damaged chokes should be reported and replaced. Table 5.2 may not always apply to 10,000 and 15,000 psi choke manifolds, since the maximum fixed choke size can be limited depending on make, model, and maximum pressure rating.

Fixed chokes must include a specialised spanner for inserting and removing the choke bean in and out of the choke manifold body.

5.8.3.1 Fixed Choke Construction

A fixed choke is basically a plug with a known fixed bore through the centre as shown in Figure 5.19; it is often referred to as a choke bean. It has a steel body with a ceramic or tungsten carbide inlay. The ceramic is a harder substance and more resistant than standard metals to erosion effects of sand and debris produced from a flowing well. Chokes with solid steel as a fixed bore exist, but these chokes wear out very quickly and should be regularly checked.

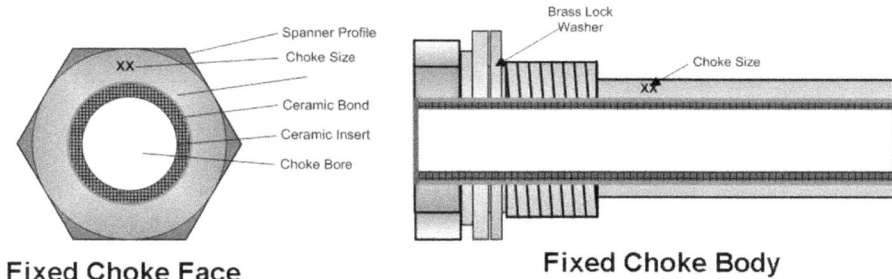

Fixed Choke Face

Fixed Choke Body

FIGURE 5.19 Fixed-Choke Schematics.

Fixed chokes should also have a brass/copper washer after the threads as depicted in Figure 5.19. The inclusion of the brass washer helps prevent the choke vibrating and coming loose from its mount in the choke manifold.

An adapter can be fit into the choke manifold to interchange the bodies of fixed and adjustable chokes.

Another type of fixed choke has a half or partial ceramic lining as shown in Figure 5.20. The rear end of this choke design is more prone to erosion, as is the unlined or fully metal choke.

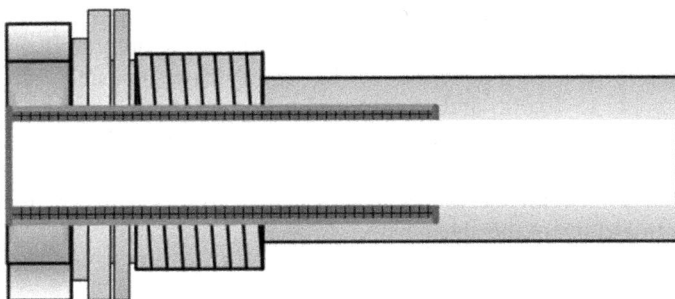

FIGURE 5.20   Half-Lined Fixed Choke.

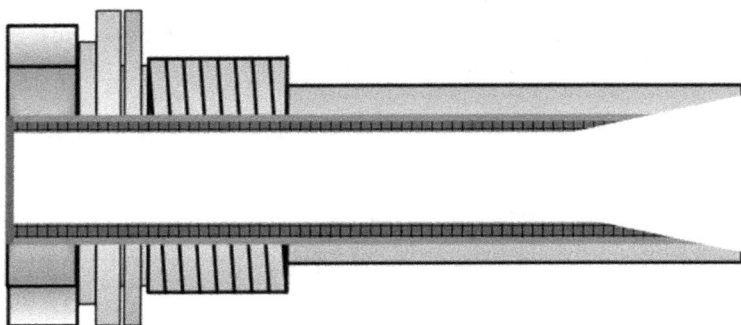

FIGURE 5.21   Fixed Choke with Erosion at Rear.

Erosion is not limited to the rear of the choke body but can also occur at the face of the choke, as illustrated in Figure 5.22.

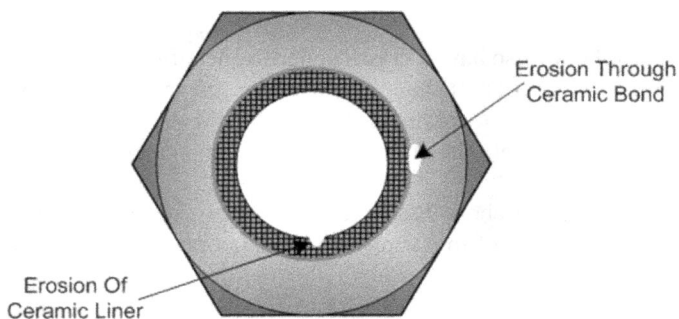

FIGURE 5.22   Erosion at the Face of the Choke.

Erosion typically occurs at the bond or on the ceramic liner. As a result, the chokes ought to be inspected when the equipment arrives on location and when a choke is changed or replaced. Damaged chokes should not be reused.

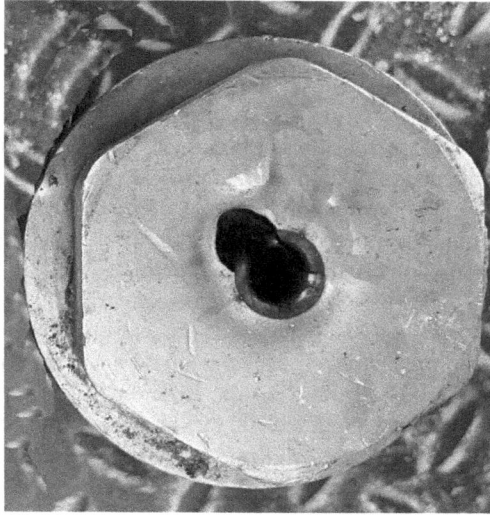

FIGURE 5.23   Erosion Caused by Flow across a Ceramic Bond.

5.8.3.2   Insertion and Removal of Fixed Chokes

Under no circumstances should a choke manifold be left without a choke in place. This should be the relevant choke bean pertaining to the type, adjustable or fixed. Their installation aids in protecting the internal threads from damage and rust.

The internal thread in the choke body needs to be cleaned and lubricated with proper grease before the job commences. This also applies to the chokes themselves, cleaned, thread-checked, and greased prior to insertion.

Under **NO** circumstances should a well be flowed through a choke manifold without a choke in place, even at its biggest fixed choke size. Without a choke installed while flowing the well the internal threads will erode and no chokes can be installed, an irreparable damage.

Whenever a fixed choke is installed, its size and condition should be verified by two crew members and its size recorded on the manual readings sheet, as reference, prior to flowing the well through it.

Before opening or installing a choke, there must be no **pressure within the body of the choke**, so the relevant bleed off needle valves should be open to ensure any pressure is bled off. The choke assembly has a securing bolt holding the union in place; no pressure should be in the manifold either before the securing bolt is removed.

It is crucial to use a non-sparking hammer when loosening the fixed choke bonnet, especially if hydrocarbons have been bled off and consequently a potential explosive atmosphere exists.

The correct PPE is essential in all choke operations.

Before attempting to knock the choke bonnet off, it is necessary to ensure there is no trapped pressure and any locking screws are released and removed. After freeing the choke bonnet with a non-sparking hammer, the bonnet should not be removed abruptly but slowly opened and gently shaken so any trapped fluids may gradually escape. Care should be taken when the bonnet is taken off the threaded section, and nobody can get their hands trapped under the union; when it is completely unscrewed, it could fall and hurt someone with its sudden weight.

It is common for 15,000 psi (15 kpsi) choke manifolds to have a fitted handling sub, shown in Figure 5.24, to assist in handling the extra weight of the hammer union and drop risk. The handling sub is a simple device with a tube mounted on top of the choke manifold and a solid rod fitted inside. In order to prevent the choke bonnet from falling when it is disconnected, a chain or linkage is fastened from the centre of the bonnet to the rod.

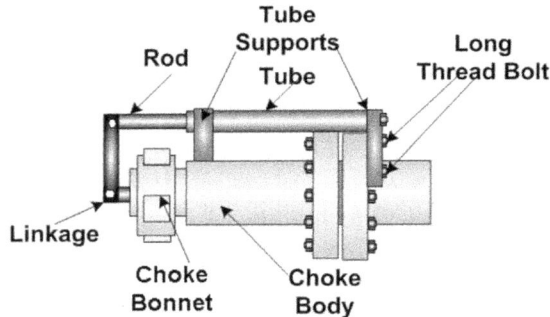

FIGURE 5.24 Fixed-Choke Support.

The choke bonnet is loosened with a non-sparking hammer and fully undone by hand, physically vibrating it as it is being unscrewed to gradually release any trapped pressure. Once the assembly is undone the rod slides along the tube supporting the weight.

5.8.3.3 Choke Installation

The fixed choke thread usually matches that of the adjustable choke, allowing for both sides of the choke manifold to be used as a fixed choke if required or with the adjustable and fixed choke bean interchanged. Figure 5.26 illustrates cross sections

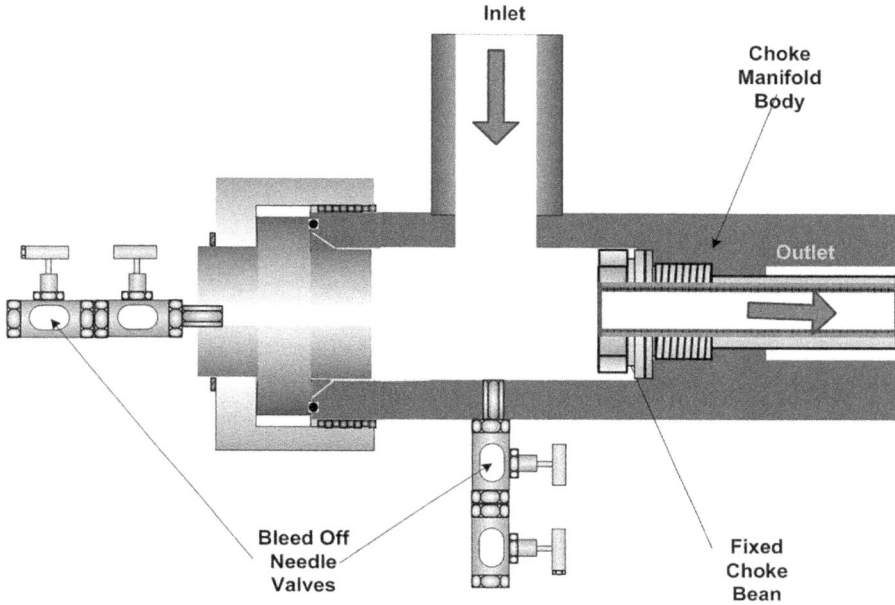

FIGURE 5.25   Fixed-Choke Assembly.

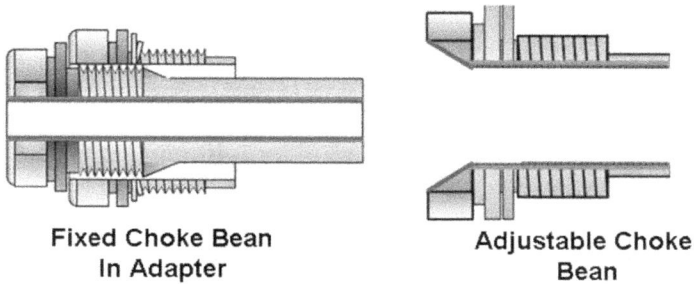

**Fixed Choke Bean
In Adapter**

**Adjustable Choke
Bean**

FIGURE 5.26   Choke Beans.

of a ceramic-lined fixed choke mounted in a fixed-choke adapter and an adjustable-choke adapter.

When installing an adjustable choke, its choke stem must be backed off to a fully open position to avoid the possibility of the choke tip contacting the seat or choke bean and resulting in self-damage. The ceramic used in adjustable chokes is usually Tungsten Carbide, which gives them very good anti-erosion characteristics, but it is also a very brittle crystalline material.

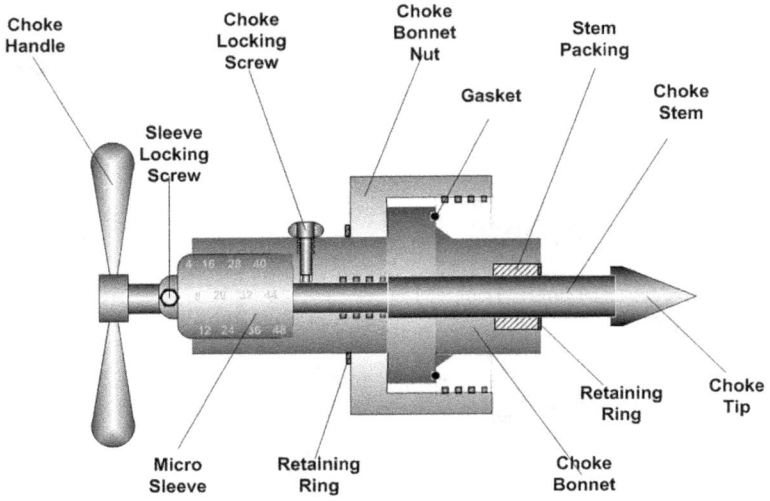

FIGURE 5.27   Adjustable Choke Removable Assembly.

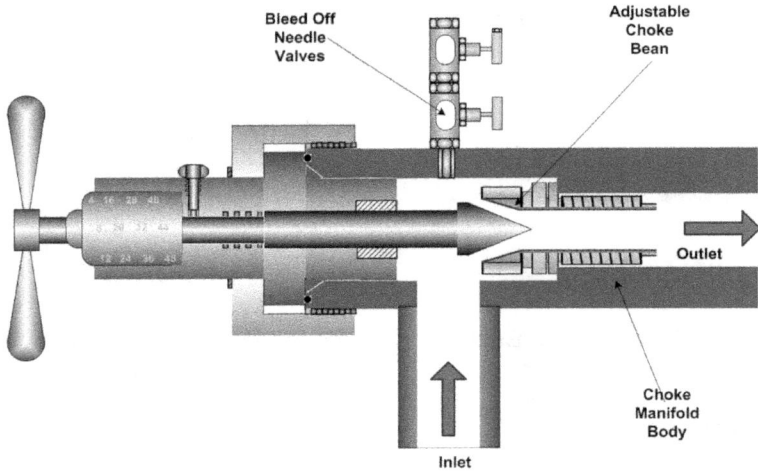

FIGURE 5.28   Adjustable Choke Assembly.

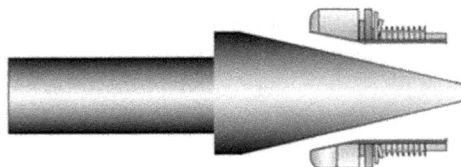

FIGURE 5.29   Open Adjustable Choke.

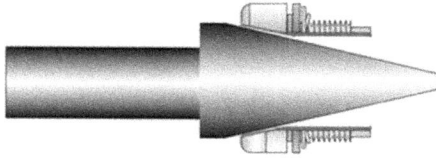

FIGURE 5.30   Fully Closed Adjustable Choke.

With this material's brittleness, if the adjustable choke stem is over tightened, forced into the choke bean, it can cause the Tungsten Carbide liner to shatter and break. As shown in Figure 5.31, the Tungsten Carbide liner is shattered and broken as a result of over-tightening the adjustable choke stem.

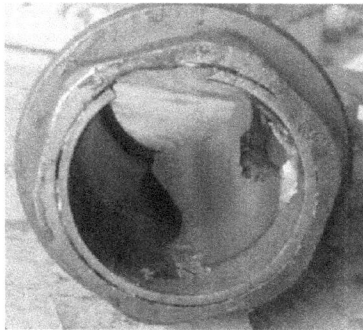

FIGURE 5.31   Damaged Adjustable Choke Bean.

It cannot be repeated enough:

An adjustable choke will not hold pressure and should not be used as a closing mechanism under any circumstances.

5.8.3.4   Choke Inspections

As the fixed chokes are changed more frequently than an adjustable choke they are inspected more often. However, the adjustable choke needs to be inspected and zeroed before the start and during the test, if removed or replaced, especially with high flow rates or high BS&W readings.

High gas flow rates will cause more damage to all chokes due to the flow velocity, significantly contributing to erosion. Therefore, during test periods, it is critical that all chokes be regularly inspected.

It is also good practice to lock the adjustable choke when it's in use during flow periods; because it is adjustable, the vibrations from high flow rates can cause it to move. Even when locked, the slight vibrations could still have the adjustable choke open/close even slightly.

Being affected by vibrations and due the adjustable choke's imprecision, it is recommended to divert the flow through a fixed choke as soon as possible. There will often be a difference in pressure and flow rate when the flow diverts, due to imprecise adjustable choke size. Failing to zero an adjustable choke will make the differences greater.

5.8.3.5   Opening a Well on Choke

When opening a well:

  i. The choke manifold valve at the back **must** be fully open before opening
 ii. There must be a fixed choke in place
iii. The adjustable choke must be open – it can be just slightly open, cracked open but not closed

If the adjustable choke is not slightly open then the pressure can lock the adjustable choke in place, either damaging it or locking it in place.

To overcome the pressure lock the well must be shut in and the back and front valves on the choke manifold closed. All pressure before and after the adjustable choke must be bled off before attempting to release the choke from its position.

The adjustable choke must not be forced open or closed under any circumstances.

5.8.3.6   Switching Chokes

This is a minimum two-man operation and cannot be carried out by a single person. Both personnel **must** be members of the well testing crew and familiar with the equipment and sequence for a safe operation.

If in any doubt one valve must be open before closing the other, otherwise the well is shut in.

Before switching chokes, the separator must be bypassed, otherwise the orifice plate can be blown down the gas line into the back pressure valve. This can result in the separator venting or even rupturing.

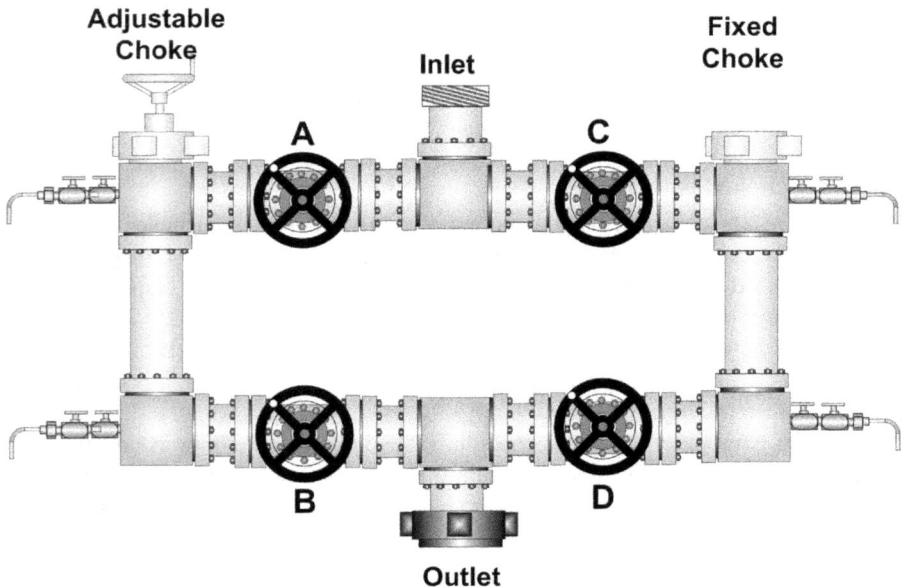

FIGURE 5.32   Choke Manifold with Identified Valves.

Steps to switch chokes with a flowing well referring to Figure 5.32:

Flow should not be diverted between different sized fixed chokes unless under instructions and with experienced personnel. The flow should be diverted to an adjustable choke for the required choke size, by increasing or decreasing the variable choke, before switching to the fixed choke size. All these operational steps should be recorded on the manual readings sheet.

   i. Ensure that the correct chokes are in the appropriate side or the adjustable choke open to the same choke size as on the fixed side. The choke size should be witnessed and recorded on the manual readings sheet prior to installation
  ii. Open both back valves on the choke manifold, valves B and D
 iii. Allow pressure to equalise and stabilise. Shut any open needle valves
  iv. Position personnel at the front valves A and C
   v. Slowly crack open the closed valve until fluid movement is heard or felt
  vi. Only when step iv. is completed, start opening the closed valve and closing the open valve counting the turns. (This should be done quickly with no pauses till complete)
 vii. Verify the turns match the model of the valve
viii. With the opened valve fully open, back off the valve a quarter turn
  ix. With the closed valve fully closed, open the closed valve a quarter turn
   x. Using the quarter turn step, close the relevant back valve (B or D) to match the closed front valve

Only after the switch can the adjustable choke be increased or decreased as required.

It is not possible to determine the position (open/closed) of a gate valve by sight, so by turning a valve a quarter turn either clockwise or anti-clockwise, the position of a valve can be determined. This is a safety precaution and employed:

* After a valve is closed turn the handle a quarter turn anti-clockwise
* After a valve is opened turn the handle a quarter a turn clockwise

When checking whether a valve is open or closed by this method, we simply have to turn the valve handle clockwise and observe:

* If a valve stops on a quarter of a turn the valve is closed
* If a valve continues to open after a quarter of a turn, then the valve is open

   To clarify the status with using a quarter of a turn: When the valves are closed a quarter of a turn does not crack them open; it is only to aid in the operation of the valves.

Besides assisting in identifying whether the valve is closed or open, leaving the quarter of a turn option will also help prevent the valve locking. In no circumstances should a valve be fully open or forced completely shut when testing.

Always restore the valves to the original quarter a turn setting.

5.8.3.7 Dual Chokes

With high flow wells it is often necessary to use two fixed chokes in parallel, not a fixed and an adjustable, installed on both sides of the choke manifold as depicted in Figure 5.33.

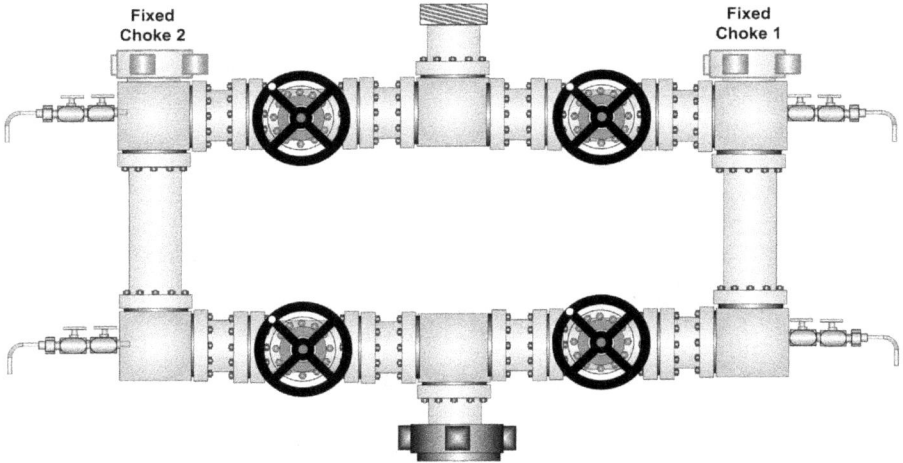

FIGURE 5.33 Dual Fixed Chokes.

Using two chokes in parallel does not mean that the size of each choke in 64ths is simply added directly, but the area represented by each choke is added and then the equivalent **Total** choke diameter is calculated from the total area. It is not recommended to use two fixed chokes of widely different sizes.

Area of Single Fixed Choke Equation

The area of a single choke is given by the equation.

$$A = \pi r^2$$

Where:
A = area (64ths of an inch)
r = radius of choke (64ths of an inch)
π = pi
(An accurate estimation of π = 355/113 = 3.141592920353982)

However, the choke size is given as a diameter, and the equation becomes:

$$A = \pi \left(\frac{d^2}{4} \right)$$

Where:
A = area (64ths of an inch)
d = diameter of choke (64ths of an inch)
π = pi

Equation for Area of Dual Chokes

With the area of two chokes our formulae become:

Total area = area choke$_1$ + area choke$_2$

$$\left[\pi \left(\frac{d_t^2}{4} \right) \right] = \left[\pi \left(\frac{d_1^2}{4} \right) \right] + \left[\pi \left(\frac{d_2^2}{4} \right) \right]$$

Where:
A = area (64ths of an inch)
dt = total diameter (64ths of an inch)
d1 = diameter of choke1 (64ths of an inch)
d2= diameter of choke2 (64ths of an inch)
π = pi

This equation can be simplified through common factors on both sides of the equation so it becomes:

$$\left(\frac{\pi d_t^2}{4} \right) = \pi \left(\frac{d_1^2 + d_2^2}{4} \right)$$

Further simplified by removing the 4 and π to become:

$$d_t^2 = (d_1^2 + d_2^2).$$

Taking the square root of both side the final equation becomes:

$$d_t = \sqrt{(d_1^2 + d_2^2)}.$$

Where
dt = total diameter (64ths of an inch)
d1 = diameter of choke1 (64ths of an inch)
d2 = diameter of choke2 (64ths of an inch)
$\sqrt{}$ = square root

If we take a choke manifold with two 48/64″ fixed chokes the true area of the choke from the above equation

$$d_t = \sqrt{(48^2 + 48^2)}.$$

$$d_t = \sqrt{(2304 + 2304)}. = \sqrt{(4608)}. = 68/64″$$

5.8.3.8 Estimating Flow Rates across a Choke

This practice has been in use for many years and as the title implies it is an estimation. If these algorithms were accurate there would be no need for separators or more recent MultiPhase Flow Meter (MPFM) systems, so effectively it is a best guess. With established fields an algorithm may have been developed that is specific to that field but not be applicable to others.

i. Reporting Estimations

Where the choke estimations are cited in a final report it must be made clear that it is an estimation stating the equation used and that the results should not be relied upon. Results will differ significantly when the well is diverted through the separator.

ii. Oil Flow across a Choke Estimation

To estimate oil flow (Q_o) the following equations can be used

Gilbert's Equation

$$\frac{P \times \left(C^{1.89} \right)}{435 \times ((G/1000)^{0.546})}$$

Ros's Equation

$$\frac{\left(P + 14.73 \right) \times \left(C^2 \right)}{17.4 \times G^{0.5}}$$

Where
   P = upstream pressure (psig)
   C = choke size (64ths)
   G = GOR (scf/bbl)

iii. Estimation of GOR Based on Depth

The common factor between the equations is the gas oil ration (GOR) which is generally unknown. It can be generally estimated based on reservoir depth/100, but this introduces further uncertainty.

iv. Estimation of Gas Flow across a Choke

To estimate gas flow across a choke (Q_g) there are certain pre-requisites.

   a. The flow must be single phase gas
   b. Under critical flow conditions

There are two versions of the same equation.
The basic equation is:

$$Q = \frac{C \times P}{18}$$

This equation makes the assumption that the flowing gas has a specific gravity of 0.6 and a temperature of 80 degrees Fahrenheit.

The equation is expanded if the gravity and temperature are known:

$$Q = \frac{C \times P}{\sqrt{(G \times T)}}$$

Where

Q = estimated gas flow in Mscf/day (Thousands standard cubic feet per day)
P = upstream pressure in psia
G = gas gravity
T = gas temperature in degrees Rankine
C = choke coefficient (unitless)

Degrees Rankine = degrees Fahrenheit + 459.67

v. Choke Coefficient Calculation

$$C = 0.0819 \times Choke\,Size\,(64ths)^{2.0854}$$

Some data acquisition systems can facilitate these equations. Given that people tend to assume everything on a computer is accurate, it should be emphasised that it is only an estimate.

5.8.3.9   Critical Flow Definition

Critical flow is a relationship between the pressure upstream of the choke and downstream of choke. The relationship between both pressures induces critical flow when the flow velocity across the choke exceeds the velocity of sound (sonic speed); at this flow velocity any disturbances downstream of the choke are stalled at the choke, unable to overcome the flow velocity (sonic). This configuration prevents any downstream of choke activities being transposed upstream of the choke or into the well and/or reservoir.

Critical flow is a pre-requisite to well testing operations; otherwise, the operations are not properly representative of the well conditions. Out of critical flow, any disturbances downstream of the choke would affect well conditions during the test.

Critical flow calculations rely on a number of variables such as the fluid density, which becomes even more complicated in a multiphase flow regime. For the sake of simplicity in field operations the equations have been simplified to:

$$\frac{P_U}{P_d} \leq 0.55$$

Where

P_u is the upstream pressure of the choke
P_d is the downstream pressure of the choke

5.9   EMERGENCY SHUT DOWN (SURFACE SHUT-IN VALVE – SSV)

An emergency shut down system is often referred to as an ESD and the associated safety valve is often referred to as a surface shut-in valve (SSV).

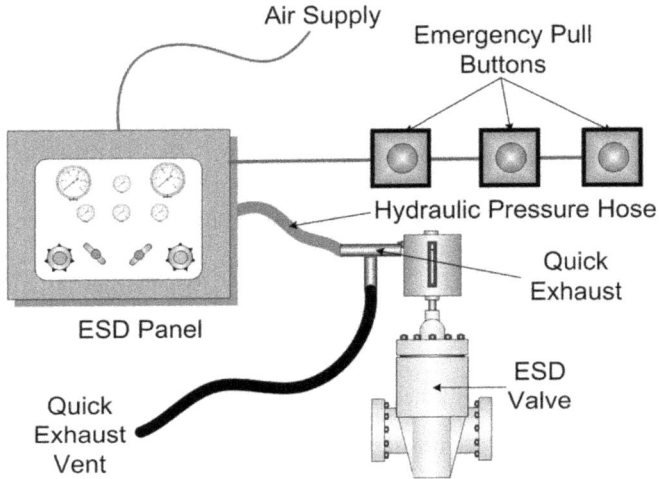

FIGURE 5.34   Emergency Shutdown Device Schematic (ESD).

The purpose of the ESD system and SSV is to isolate the well in case of emergency. The pull buttons associated with the ESD, fixed on panels, are strategically positioned around the well site, normally at the wellhead, choke manifold, and separator.

In case of emergency any personnel on site can access the closest emergency pull button panels, and pulling the buttons initiates a sequence of pneumatic and hydraulic pressure reliefs that, in a matter of seconds, return the SSV into its naturally closed state. This shuts in the well almost immediately.

The emergency shut down buttons have changed from a push button option to a pull button option to meet current safety regulations, avoiding unintended push of the buttons.

Under **NO** circumstances can there be any valve between the ESD and the SSV, as it circumvents the safe action of the system. Use of valves in this line should be considered a misconduct with disciplinary action.

5.9.1   Surface Safety Valve (SSV)

An ESD valve or SSV usually consists of two parts:

5.9.1.1   The Valve

This is usually identical to the gate valves installed on the choke manifold but with an actuator instead of the normal manual handle.

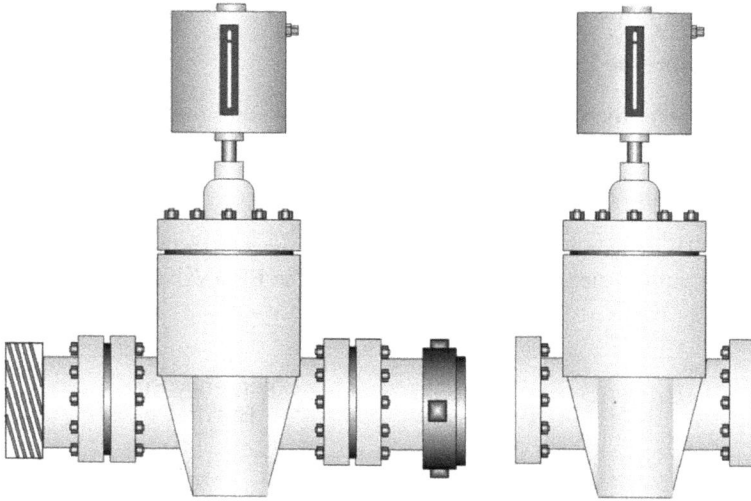

FIGURE 5.35   ESD Valves and Actuators.

5.9.1.2   Actuator

This is either a hydraulic or pneumatically operated plunger type assembly. When pressure is applied to the top of the actuator it forces a plunger down, opening the valve.

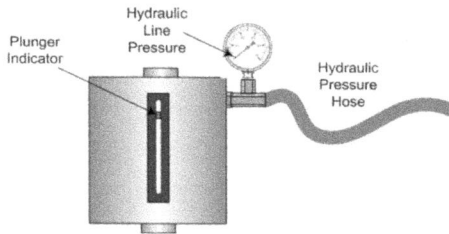

FIGURE 5.36   Actuator with Dial Gauge.

Continuously applied pressure in the actuator holds the valve open; when the hydraulic pressure is bled off an internal heavy-duty spring raises the plunger and closes the valve.

The hydraulic actuator has a reverse acting assembly that relies on the applied pressure to keep the valve open so with any failure the valve will automatically close.

A "glass" panel is put on the actuator body to provide operators with a visual cue to assess the position of a plunger and, consequently, whether the valve is open or closed. An operator can check whether the actuator is open or closed from an indicator included into the glass panel that is attached to the plunger.

A T piece installed on the hydraulic pressure inlet/hose to the actuator would allow use of a dial gauge or DAS sensor to monitor the hydraulic pressure on the actuator. The pressure (or lack of) on the gauge or sensor would indicate whether the valve is open or closed, respectively.

Most of the actuator systems are hydraulic with a standard valve.

The ESD system is controlled and monitored by a pump panel that:

i. Maintains the Pressure

Typically, a high-pressure pump linked to a sensor and regulator that maintain a constant pressure.

ii. Releases the Pressure

This is usually a hydraulic pneumatic switch connected to the shut-down buttons, which when pulled releases the pressure holding the ESD valve open.

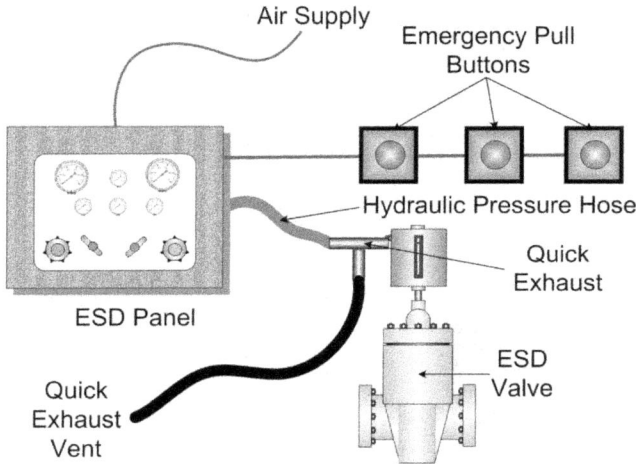

FIGURE 5.37   ESD Connections Schematic.

An ESD panel should have the following within its design framework:

iii. ESD Hand Pump

A high-pressure hand pump can be used to pump open the ESD valve, in the event of any failure.

iv. ESD Air Reservoir

In the event of an air failure, there is sufficient air to keep the system operational until the air supply can be restored.

v. ESD Hydraulic Oil Reservoir

As a buffer and backup in case of leakage, etc.

It is important that there are **no valves** on the hydraulic pressure hose. If any valve is present and closed, then the system will not function as intended. Either by stopping hydraulic pressure from opening the valve or preventing the pressure from being bled off, the valve remains open. Similarly, non-return valves would obstruct hydraulic pressure in either direction.

Utilising a quick exhaust system will enable the hydraulic oil to be bled off quicker and reduce the valve's shutting time. Even if a quick exhaust system is not used, the exhaust vent should return the fluid to a non-pressurised container. Any back pressure will affect the timing or delay the shutting of the valve.

In the hydraulic lines, especially with a quick release system, there must not be any valves, fittings, joints, or reducers. Any of these fittings could introduce a reduction in the cross-sectional area of the hose and effect closing and opening times. The larger the vent line bore, the faster fluid volume is exhausted – proportionally reducing hydraulic pressure – and the quicker the valve shuts. Other factors such as viscosity of the hydraulic oil will affect the closing times.

The SSV valve is held open by a pneumatically operated pressure pump in the ESD panel. Pulling any of the ESD buttons bleeds off the air to the ESD panel's pump, deactivating it, which causes all the applied hydraulic pressure to the SSV to bleed to zero. This action closes the SSV.

ESD systems often employ a quick exhaust system on the SSV to vent the hydraulic pressure quickly. Without a quick exhaust fitted the pressure is bled down via the hydraulic pressure hose and back into the reservoir, which is often slower, prompting a larger bore hydraulic hose with no restrictions. After installing the ESD valve, a test should be carried out to determine how long it takes the ESD valve to close. This should be in seconds rather than minutes.

The system configuration can be expanded to integrate other triggers for shutdown such as hydraulically actuated by pressure operated pilots (Hi-Lo pilots), manual intervention via a control panel, and even integration with the data acquisition system based on monitored pressure or even calculated flow rates.

In any case, before the ESD system can be used, it should be certified for performance and pressure testing and then independently function tested and verified by one of the operating company representatives.

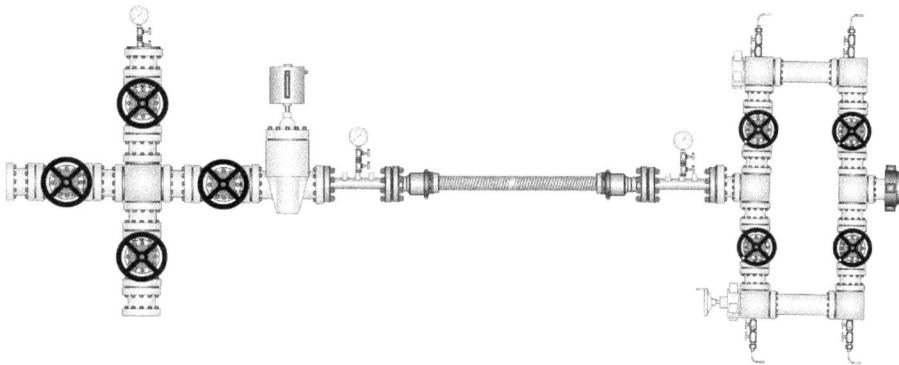

FIGURE 5.38   SSV Installation.

Figure 5.38 shows a high-pressure system using an armoured hose, which can be replaced with standard metal pipework as specified by the test design. The SSV is installed on the flow wing valve of the wellhead to completely isolate the well testing package from pressure in an emergency.

NOTE: *The inclusion of two data headers, before and after the armoured hose, are to monitor the pressure across the hose and to facilitate pressure bleed off in the case of a blockage or hydrate. The same should apply for standard pipework.*

The data headers can also be used for chemical injection, sampling, or any data acquisition sensors.

5.10 DEADWEIGHT TESTER (DWT)

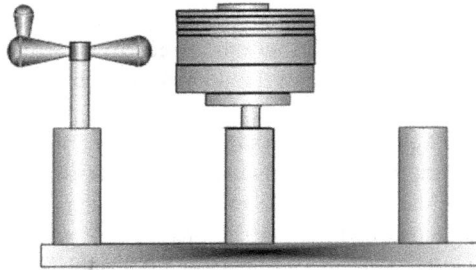

FIGURE 5.39   Portable Deadweight Tester.

The deadweight tester (DWT), Figure 5.39, is the primary source of pressure measurement on the location, for its reliability and minimal calibration or correction requirements. When used as it stands the deadweight can measure to a resolution of 1 psi.

The normal pressure ranges for a DWT is 0–5,000 psig, or 0–10,000 psig and should be used regularly for measuring wellhead pressure and validating the data acquisition system readings.

15,000 psig deadweight testers are available but rarely used out of safety concerns, being physically close to a 15,000 psi pressure system. The 15,000 psig deadweight testers are more often used in calibration.

Many well testers will say a DWT isn't needed since other transducers and measuring devices replace it; they fail to realise that a deadweight tester is used to function test or calibrate the dial gauges, chart recorders, and electronic sensors and as such the truest pressure reference. It should always be connected to read wellhead pressure and functional.

Company representatives and supervisors should witness readings taken with a DWT during the job, particularly before opening the well and after shutting in. Further standards and details are given on the use of DWT and its readings in the instrumentation chapter.

5.11 THREE-PHASE SEPARATOR

The three-phase separator serves as the primary component of a well test package since it separates and measures liquids and gas. There are different designs for the vessel and associated plumbing, but they serve the same purpose.

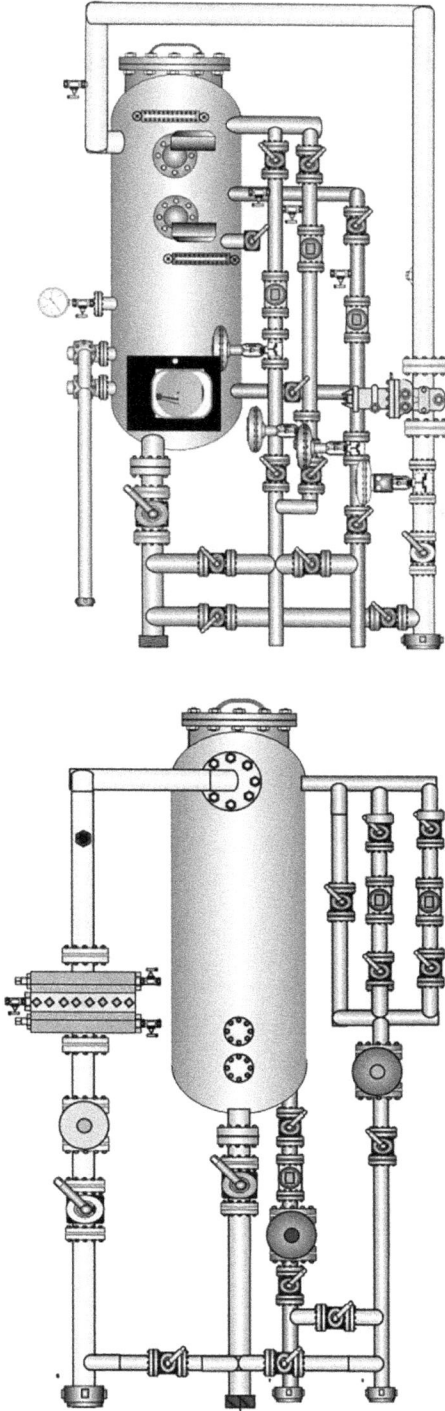

Three Phase Separator Top View

Three Phase Separator Side View

FIGURE 5.40 Three-Phase Separator Schematics.

5.11.1 WELL TESTING FLUID MANIFOLD

The valves on both the gas and fluid manifolds are normally a ball valve construction and as such cannot hold the same pressure as the gate valves on the choke manifold and SSV. The pressure rating of these valves should be at least the same as the maximum working pressure of the separator.

Similar to the choke manifold, one valve needs to be open before the other can be closed.

5.11.1.1 Gas Manifold

FIGURE 5.41 Gas Manifold.

The gas manifold, Figure 5.41, is located at the gas outlet of a separator and its function is to divert gas flow from the separator from one burner/flare stack to the other in the event of wind direction change. It can also be used to split the gas flow to two burners/flare stacks simultaneously.

5.11.1.2 Oil Manifold

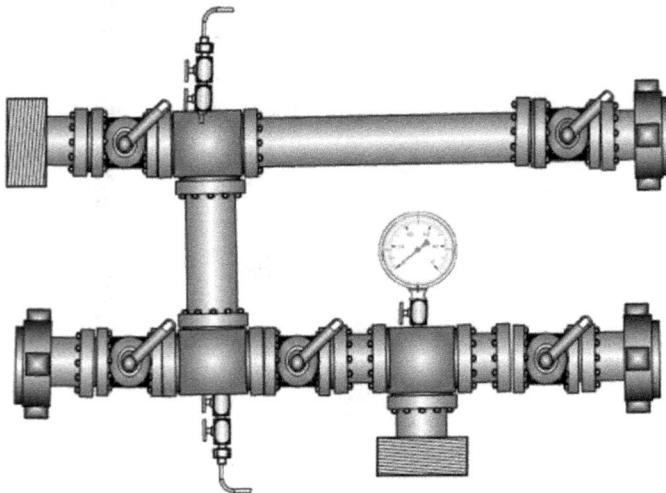

FIGURE 5.42 Oil Manifold.

An oil manifold can be used with either the oil or water phases separately. Connected to the oil outlet of the separator it enables the switching of oil flow between the burners and/or tanks, as well as switching the flow of oil from one burner to another and/or splitting the oil flow to two burners simultaneously.

When oil flow is switched to a tank for storage purposes, the tank is considered the last destination. However, if the tank is being used temporarily, a transfer pump would be required to flow the oil from the tank to one or both burners.

Emptying the tank of its fluid content, to different destinations, is usually carried out when there is no flow and the well shut in. With an integrated fluid manifold on the tank, it makes the process simpler and independent of other equipment.

5.11.2 TRANSFER PUMP

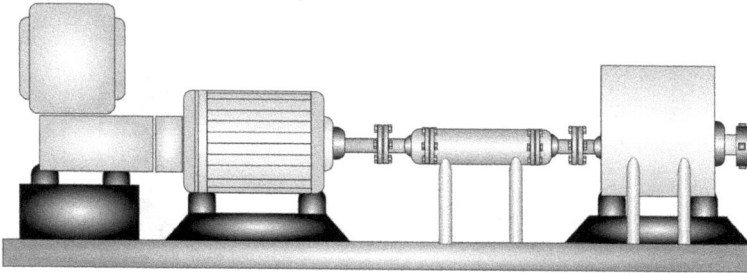

FIGURE 5.43 Transfer Pump.

Transfer pumps are used, as the name suggests, to transfer non-gaseous fluids (dead oil or water) from one place to another. On offshore locations this is usually electrically powered and onshore a diesel-powered version is used. The diesel-powered pumps must be placed in a safe zone, with fitted spark arresters and safety devices.

The electrically powered transfer pump is avoided onshore, because of its motor start-up configuration which can trip all the fuses, which is risky with remote setups. There are adaptors to reduce this initial current spike, but a qualified electrician should check before connecting the transfer pump for the first time.

5.12 TANKS

As illustrated in Figure 5.44, there are two basic types of tanks used in a well testing package. Both tank types should have separate vent lines to a safe area and under no circumstances these vent lines be incorporated with the separator's venting systems.

i. Surge Tank

This calibrated pressure-bearing tank can hold both pressurised gas and liquid phases. Before its use, an installed pressure relief valve would be set at the tanks maximum working pressure or lower.

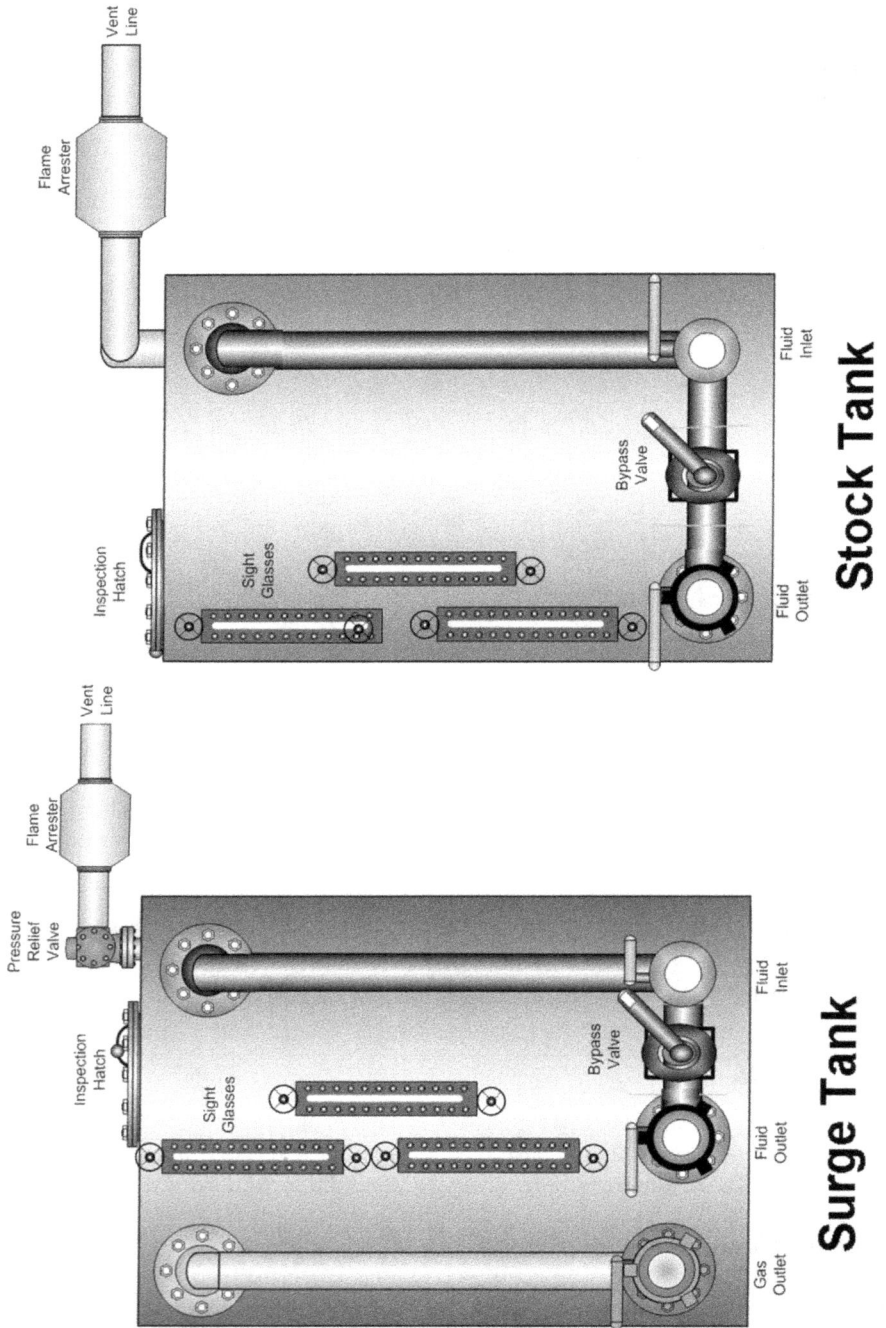

FIGURE 5.44   Tank Schematics.

A surge tank can be used as a second stage separator or knockout pot to lessen the liquid content of the gas component or phase of the flow. These options usually have a Daniel box or pitot tube system fitted to measure gas flow.

ii. Stock Tank

A calibrated and non-pressurised tank is for the storage of non-pressurised liquids and serves as a reference for checking and calibrating both the oil and water fluid meters.

5.13   WELL TEST BURNERS

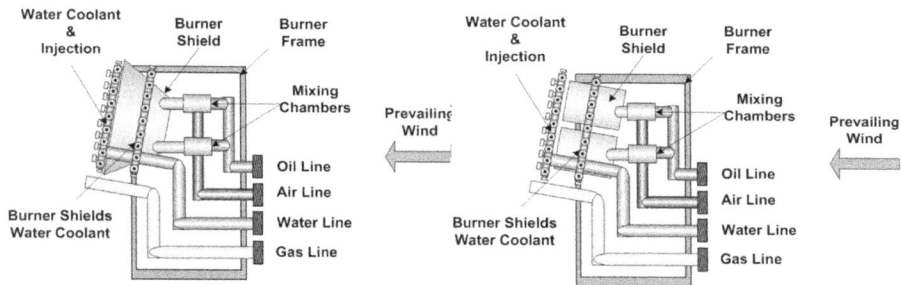

FIGURE 5.45   Well Test Burner Schematics.

The well test burner is specifically designed for burning well fluids while producing the least amount of smoke and fallout. Different types and brands are available to choose from.

The most common fault with burners involves the lack of compressed air for atomisation. If no atomising air specification is defined, or if in doubt, employ multiple certified compressors with associated check valves and safety equipment. The burner manual, its requirements, and compressor ratings should be revised in sync before mobilisation.

5.14   AIR COMPRESSORS

Standard air compressors are utilised on land operations, while in offshore environments specialised air compressors are required.

General air compressors used for instrumentation purposes are often unsuitable for burner operations, due to air volume restrictions. In any case they must have all the safety options and have non-return valves installed and required manifolds. For the purposes of maintenance and to allow refuelling a minimum of two air compressors should be on site. Check for the required air capacity with the burner operations manual.

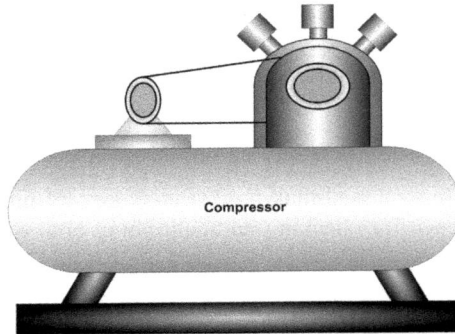

FIGURE 5.46   Air Compressors.

5.15   HIGH PRESSURE AND 10 KPSI

High pressure, commonly referred to as HPHT (high pressure high temperature), although this is not always the case, refers to pressures above 10,000 psi. (10 kpsi)

When considering a potentially high-pressure well, equipment and pipework from the wellhead to the choke manifold must be adequate, and high-pressure rated. Downstream of the choke manifold pressure is reduced to normal operating pressures, as illustrated in Figure 5.47, and the standard equipment may be used.

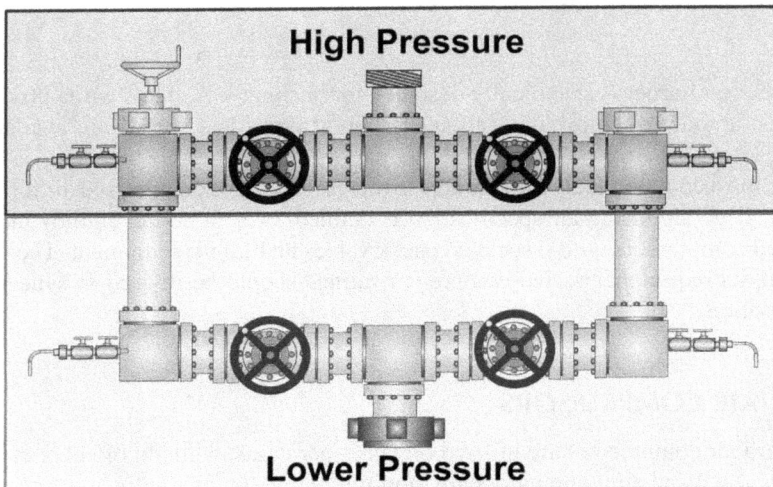

FIGURE 5.47   Pressure Regimes on the Choke Manifold.

The generally accepted standard *prohibits* the use of hammer unions with pressures greater than 10 kpsi and a different kind of connecting mechanism has to be employed, such as:

   i. Flanges
  ii. Grayloc
 iii. Techlok
 iv. Destec

All these use a bolting mechanism with either a clamp or a flange, to connect the pipes.

The high temperature of well flow equally influences the type of equipment used, primarily with the temperature ratings of flanges and seals. It is imperative that the ratings of all junctions, joints, and seal mechanisms are verified before use with higher than normal temperatures.

5.15.1 TECHLOK CLAMP

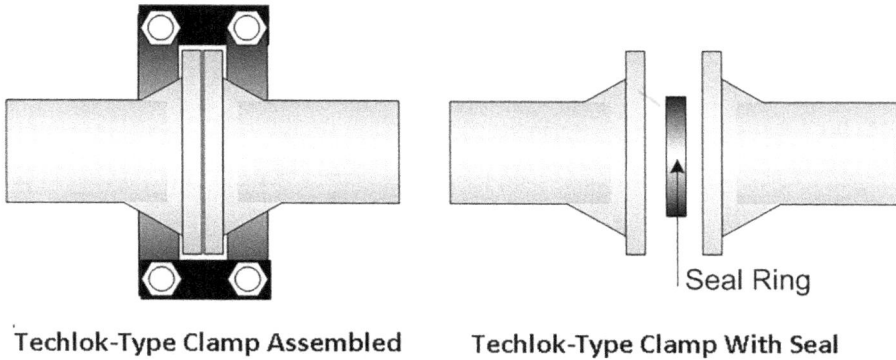

Techlok-Type Clamp Assembled   Techlok-Type Clamp With Seal

FIGURE 5.48   Techlok-Type Clamps.

Similar to other piping clamps, Techlok uses a bolting mechanism in order to connect the pipes. The clamps use a seal ring in a machined groove. Once the clamp or flange is tightened, the seal rings cannot be reused, so the service provider should have spares available on location.

FIGURE 5.49   Flange Details Stamped on Edge.

Flanges differ and only matching flanges must be used, their details stamped on their edge as shown in Figure 5.49.

5.15.2 FLANGES AND GASKETS

With flanges, the gaskets or seal rings should be replaced every time they are undone. The gasket or seal rings are vulnerable to compression damage if the flange bolts are tightened in the incorrect sequence.

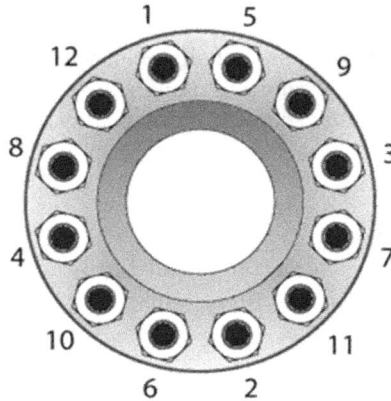

FIGURE 5.50   Flange Bolting Sequence.

Figure 5.50 shows the sequence for tightening bolts on a 12-hole flange. The number of bolts vary depending on flange size and pressure ratings; nevertheless the bolt tightening diagonal sequence principal applies. The flanges are initially loosely fastened by hand before being gradually tightened, in sequence, until they are entirely tight.

The final tightness is commonly achieved by using a "flogging spanner" and a hammer. Figure 5.51 shows the flange with a connected "flogging spanner." This is the *only* spanner type that should be used with a hammer since it is reinforced throughout its construction and has a designated lug for the hammer to strike. Much like with hammer union lugs, only non-sparking hammers should be used. The tightening of the nuts and bolts follows the same sequence.

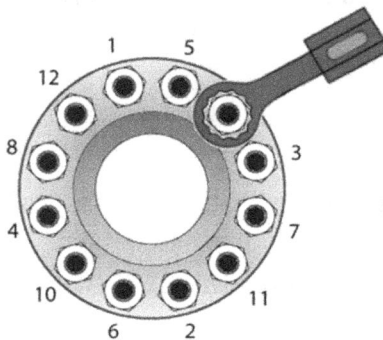

FIGURE 5.51   Flange and Flogging Spanner.

For safety, a lanyard or flexible line, as shown in Figure 5.52, should be attached to the flogging spanner to prevent impact accidents, if the spanner slips off of a nut and shoots into the air. Offshore, if the spanner flying off doesn't cause damage it would most likely be lost to the sea.

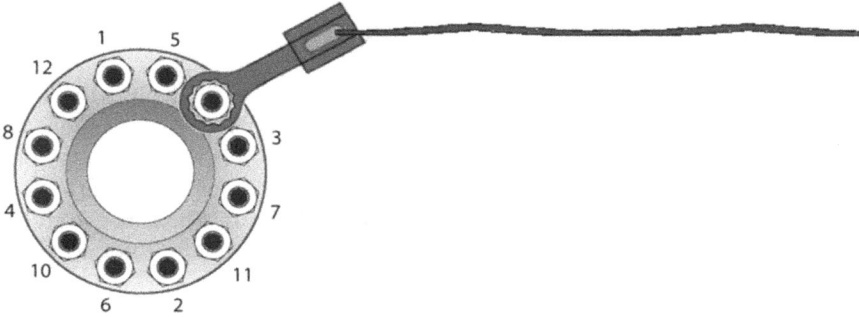

FIGURE 5.52   Ring Flogging Spanner with Safety Lanyard.

The "ring type" flogging spanner has been depicted in the figures, but there are other options such as an open-ended flogging spanner, shown in Figure 5.53. The open-ended spanner, while a valid tool, is more likely to "fly" off the flange when hammering, making it less popular. Regardless, use of a safety lanyard is recommended in all cases.

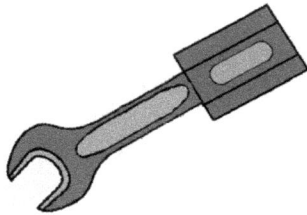

FIGURE 5.53   Open-Ended Flogging Spanner.

The safest choice is a pneumatic spanner or torque wrench, but these are not always made available.

5.15.3 ARMOURED HOSES

With pressures exceeding 10,000 psi (10 kpsi) a Coflexip type armoured hose is often used in place of a metal pipeline. This is preferred for safety; if a hard iron pipe fails in any way, the result is an explosive decompression, but with an armoured hose it is a safer gradual type of pressure release.

Coflexip is a main supplier of armoured hoses and has been the common term to refer to armoured hoses. The armoured hose shown in Figure 5.54 is with flanges; Grayloc connections can also be used.

FIGURE 5.54   Typical Armoured Hose.

The drawback with armoured hoses is their resistance to chemical reactions with the well fluids' contaminants (H_2S, CO_2, etc). Their specifications and compatibility must be checked and verified against the job requirements before mobilisation to site. Armoured hoses rely on the integrity of their internal liner withstanding potential chemical attacks from the well fluids. The concentration of these hazardous contaminants could cause failure fairly quickly if not accounted for properly.

It is common practice to insert a data header on the end of the armoured hose that connects to the wellhead. Additionally, a data header is often placed on the other end of the hose, connecting to the choke manifold, allowing any trapped pressure within the armoured hose to be easily bled off.

FIGURE 5.55   Armoured Hose and Data Headers.

Connecting data headers on both ends of the armoured hose, as in Figure 5.55, makes it easier to connect the hose to fixed flanges such as on a production tree.

A flange configuration armoured hose is the most commonly used with well testing, primarily due to the availability and cost of replacement seals. Pressure and temperature ratings of the gaskets should equally comply with the expected flow.

Armoured hoses should be treated and prepared in the same manner as standard pipework, including all high-pressure inspection and testing procedures. The issued test certificate should state the liner involved. The armoured hose specifications should be submitted and revised during TWOP, and once accepted as fit for use, the specification and testing documentations accompany the unit on site.

5.15.4 Hammer Unions

The seals in hammer unions, which are like "O" rings, are a form of elastomer; these must be replaced to suit the components in the fluid flow as well as fluid temperature. To account for these variables, it is necessary for the well test company to be informed of expected temperature and any corrosive fluids or gasses present at any of the flow stages.

H_2S and CO_2 are the primary gasses that require specific seals. Seals are most vulnerable to CO_2, while other acids can also cause harm. The hammer union construction should conform to the requirements of ASTM and AISI standards and meet or exceed NACE MR-01–75 and API RP14E for H_2S applications.

A hammer union, sometimes known as "Weco Union," are of three main components:

    i. A male part
   ii. A female part
  iii. A nut part

Hammer unions are manufactured in two design options:

    i. Threaded
   ii. Weld neck

Due to potential leaks with the threaded option from repetitive hammering, the weld neck unions are the preferred option. In well testing, welded hammer unions are mandatory in the presence of H_2S or other hazardous fluids.

With the well testing unit moving from one well to another, correspondingly produced fluids change and may include toxic gasses; then all hammer union fittings must be weld neck type and welded in place. Welded thread connection unions are not permitted.

Threaded unions can be used in non-hydrocarbon applications, water, air, etc.

The female half – or thread half – is the external threaded portion that's welded to the end of the pipe. The hammer union nut is loose on the pipe over an arrow-shaped profile, commonly referred to as the male half.

In well testing setups the flow direction is towards the wing half as shown in Figure 5.56, while pumping operations have the flow direction reversed, towards the thread half.

The male section inserts into the female, and when tightened, the male profile squeezes onto the elastomer ring in the female half, as shown in Figure 5.57, completing the assembly and effectively creating a seal. To seal and lock the pipe into place, the nut is hammered and tightened onto the female part, Figure 5.58.

The size of a pipe restricts the volume of flow; pipework should be adequate in size for expected flow rates.

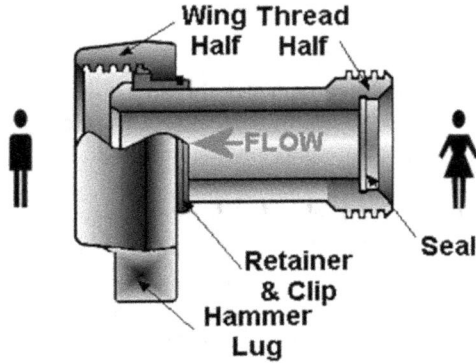

FIGURE 5.56   Hammer Union Components.

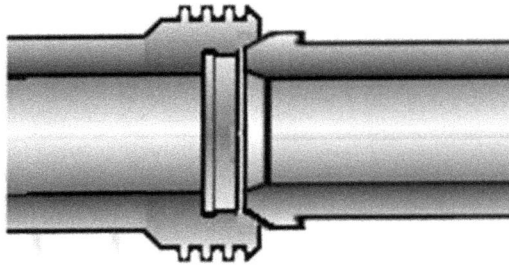

FIGURE 5.57   Hammer Union Junction.

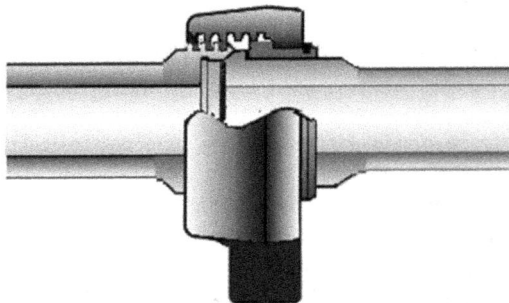

FIGURE 5.58   Hammer Union Joined Together.

Table 5.3 lists size and pressure specifications of hammer union pipes. Pipes of similar nominal diameters may be connected with each other, but their pressure ratings mismatch. For example (as highlighted in Table 5.3) a 2-inch 1502 wing half will make up with a 2-inch 602 or 1002 thread half hammer union identification, but they obviously have different pressure ratings. A connection is feasible between them but the lower pressure rating pipe will fail under pressure, explosively.

Source of hammer union data from multiple manufacturers of hammer unions.

TABLE 5.3

Example Chart of Hammer Unions Size and Pressure Ratings

Figure	SWP	Nominal Pipe Diameters (Inches)									
100	1,000	2	2 ½	3	4	6	8	1			
200	2,000	1	1 ¼	1 ½	2	2 ½	3	4			
206	2,000	1	1 ¼	1 ½	2	2 ½	3	4	6	8	10
207	2,000	3	4	6	8	10					
211	2,000	1	1 ¼	1 ½	2	2 ½	3	4			
400	2,500	5	6	8	10	12					
400	4,000	2	2 ½	3	4						
600	6,000	1		1 ½	2	2 ½	3	4			
602	6,000	1	1 ¼	1 ½	**2**	2 ½	3	4			
1002	10,000	1	1 ¼	1 ½	**2**	2 ½	3	4	5	6	
1003	10,000	2	3	4	5						
1502	15,000	1		1 ½	**2**	2 ½	3	4			
2002	20,000	2	3								
2202	15,000	2	2 ½	3							

The wing nut of the hammer union will have the type of union cast into the back of the wing as shown in Figure 5.59.

FIGURE 5.59 Hammer Union Wing Identification.

Service companies use a colour code to identify the working pressure of the pipe. Therefore, the colour of the wing will depend originally on the manufacturer and then service companies' coding.

If there is any uncertainty regarding the compatibility between unions, the information on the metal ring installed on the pipework should have the necessary details.

5.15.5 HAMMER UNION ELASTOMER SEALS

Hammer unions heavily rely on the rubber seal mechanism, partly an elastomer seal and partly the machined housing for the elastomer. It is essential that both these components are in good condition.

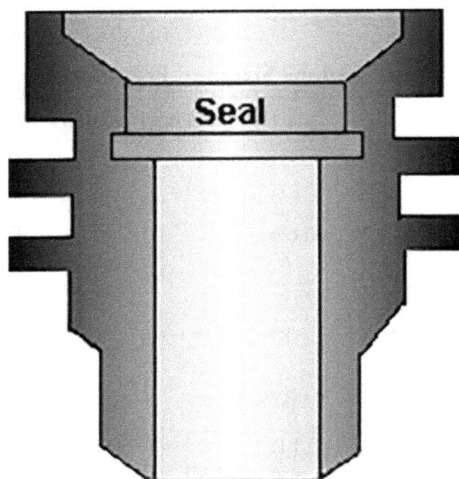

FIGURE 5.60 Hammer Union Seal Ring.

Figure 5.60 illustrates the seal ring with a "Top Hat" type profile. It is crucial that the profile in the union is both clean and free from rust and pits on the metal faces. Prior to use, both the elastomer and the profiles should be thoroughly cleaned.

Example of temperature ratings of common elastomers used in hammer unions listed in Table 5.4:

TABLE 5.4
Example Table of Typically Used Elastomers

Material	Temperature Rating	Hardness
Buna(N) (nitrile)	-35 °F to + 450 °F (-37.2 °C to + 232.2 °C)	70–90
Viton	-15 °F to + 400 °F (-26.1 °C to + 204.4 °C)	55–90
EPDM	-60 °F to + 400 °F (-51.1 °C to + 204.4 °C)	30–90
Kalrez	-10 °F to + 615 °F (-23.3 °C to + 323.9 °C)	75–95
Teflon	-250 °F to + 450 °F (-156.7 °C to + 232.2 °C)	50–65

As a general statement, the harder the elastomer seal the greater the resistance to gas permeation; on the other hand, the harder the elastomer the less effective it is at sealing.

5.15.6 Hammer Union Lugs

Hammer Union connections are often made up in a hazardous area; so hammers used must be of the non-sparking material. These would be nonferrous alloys – bronze brass, copper-nickel alloy, copper-aluminium alloys, or copper-beryllium alloys.

FIGURE 5.61   Standard Profile of a Wing Nut.

Prolonged hammering of these union lugs, especially with a steel or iron hammer, will distort the lugs' profile to a shape difficult to strike with the hammer and likely to cause accidents from misplaced hammer blows.

The use of non-sparking hammers prolongs the life of the hammer union. The hammer head material would be softer than the wing nut lug and absorb the brunt of the damage instead of the lug.

FIGURE 5.62   Impact Damaged Hammer Unions.

Figure 5.61 is an example of HSE safety directives being ignored, by both well test and other crews, where a steel hammer capable of producing sparks is used instead of a non-sparking hammer. This type of damage should be picked up by a HSE inspection and is reason enough to fail equipment on its yearly inspection. Hammer unions with this type of lug damage should be replaced as soon as reasonably possible and should never be sent to the field.

Well test crews resist the use of non-sparking hammers, because they are lighter than the steel hammers which makes them take longer to hammer a union up. Unfortunately, weak training is to blame; personnel should understand that vigorously hammering a union does not always guarantee a seal.

With the several duties of well test supervisor's position, thoroughly inspecting equipment before it is shipped to the location may not always be possible. The responsibility falls on all well test crew to bring any equipment damage and defects to the attention of the safety officer or in turn the company representative.

5.15.7 Non-Sparking Hammers

Unacknowledged or unrecognised manufacturers of non-sparking hammers make them a risk. Figure 5.63 shows "supposed" brass non-sparking hammers supplied locally; it is apparent the hammer has endured major damage. The cracks and crystalline structure define the composite as an amalgam of materials rather than genuine metal. This is a dangerous state of a brand-new hammer that could not complete a job and a serious risk with chunks of it flying off.

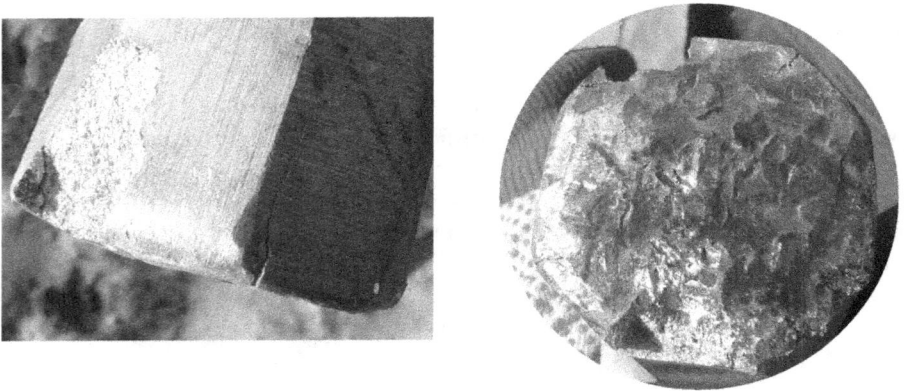

FIGURE 5.63   Brittle "Brass" Hammer.

5.15.8 Crossovers

The service company stock several options of crossovers and adapters to rig up and to connect to other service units. The working pressure of unions with male wings are easier to verify, size- and rating-wise, by the stamps on the wings, but with female options there are none. To compensate for the lack of information, all the details in regard to sizes and ratings of crossovers are stamped on the identification rings common to all pipework.

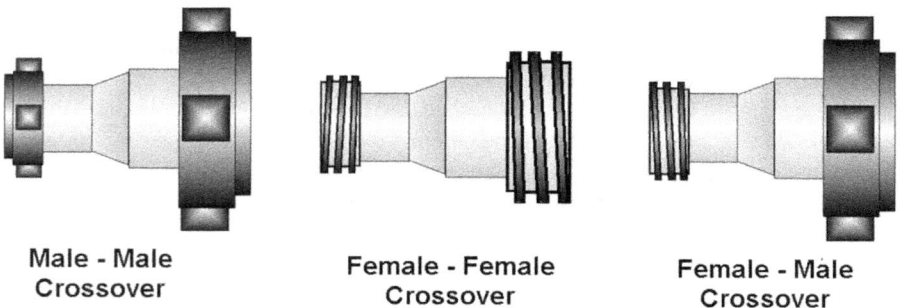

Male - Male Crossover   Female - Female Crossover   Female - Male Crossover

FIGURE 5.64   Pipework Crossovers.

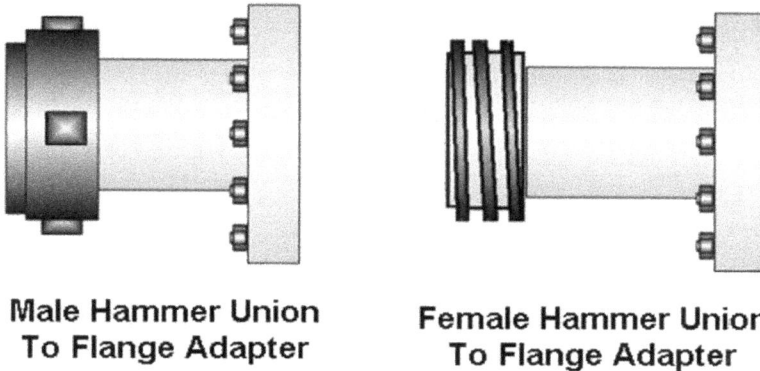

**Male Hammer Union
To Flange Adapter**

**Female Hammer Union
To Flange Adapter**

FIGURE 5.65 Hammer Union Flange Adapters.

NOTE: *The pressure rating of the assembly should be the lower of the two fittings on the crossover.*

5.15.9 DATA ON PIPEWORK TEST RING

There must be a welded steel data ring, or identification ring, on the pipework or crossover that is irremovable, and it should include the following basic information:

i. Identification serial number
ii. Working pressure
iii. Test pressure
iv. Service (sweet/sour)
v. Union types (figure number)
vi. Average pipe thickness
vii. Minimum allowable pipe thickness at maximum pressure
viii. Date of survey

NOTE: *Pipework without this type of identifier ring should not be used.*

The original obsolete data ring must be cut off and discarded before a new data ring with updated certification details is attached.

Detailed certification of the pipework should accompany the well testing package; when the pipe was numbered, inspected, and tested to the company standards by a third-party inspection company.

5.15.10 NPT-TYPE FITTINGS

It is now generally accepted that with pressures equal to and above 10,000 psi (10 kpsi) all NPT fittings should be replaced with autoclave fittings, both illustrated in Figure 5.66. This also means that an autoclave to NPT adapter cannot be used, and all gauges, liners, and fittings must be equipped with appropriate autoclave fittings.

Where NPT fittings are allowed to be used, Teflon or PTFE tape wrappings should be no more than three layers.

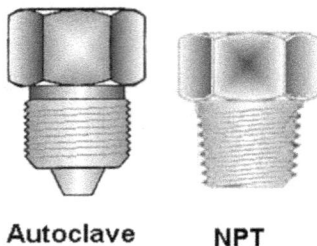

Autoclave   NPT

FIGURE 5.66   Standard Pipe Fittings.

Under no circumstances should Teflon tape be used on autoclave fittings. Furthermore, on many installations/locations the use of Teflon/PTFE tape has been banned in favour of a liquid thread sealant. This should be part of the TWOP and site visit discussions before equipment mobilisation.

All pressure fittings should be suitable for the purpose (H_2S etc.), sourced from acknowledged and professional manufacturers, of the correct material, and with their working pressure stamped on them. Brass, among other materials, must not be used.

All well test crew members are accountable for using the correct, undamaged, and correctly installed fittings.

5.16 BURNERS

The well test oil burners do not always have the best reputation, often mistaken for performance, when it is mainly due to the lack of correct or sufficient quantities of air, water and gases required for adequate atomisation and burner operation.

5.16.1 RADIANT HEAT

All types of well testing burners, ground flares, and flare stacks produce radiant heat which can seriously damage equipment, surroundings, buildings, offshore rigs, and personnel.

Therefore, it is necessary to assess the likelihood of radiant heat damage prior to burner installation. This is not a simple calculation and software have been developed to simulate and do the maths. One of the more comprehensive packages is Flarecheq from Triangla, with a simpler version called CATS:

www.linkedin.com/company/trianglia/about/

Having water shields reduces the amount of radiation that is transmitted; for this reason, effective water shielding is critical.

5.16.1.1 Personnel Exposure to Radiant Heat

When considering heat radiation exposure, the geographic location and weather should be accounted for. Exposure would be dramatically greater in the Gulf Cooperation Council (GCC) during the summer time than in Alaska. The effect of atmospheric heat combined with radiated heat increases the hazard to health and all necessary precautions must be taken.

Radiant heat factors within this context do not only apply to well test burners but to flares and ground flares as each becomes a dominant source of radiant heat.

The information presented here is a well testers compilation of more than 40 years of safety courses, lectures, seminars, and programs from service companies, specialist gas safety companies, and oil companies.

TABLE 5.5
Example Heat Radiation Levels Summary from Multiple Sources

Btu/hr/ft^2	W/m^2	Description
330	1,041	The maximum radiated solar heat at ground level.
440	1,388	The maximum limit for harmless exposure to human skin.
500	1,585	Maximum limit for heat radiation where personnel are continuously exposed.
1,500	4,732	The API RP 521 recommended upper working limit for a worker intermittently sheltered or sprayed with water.
2,000	6311	Maximum limit for heat radiation where emergency actions lasting up to 60 seconds can be undertaken by trained personnel with the appropriate protective clothing.
3,000	9,464	The upper limit for unprotected structures and equipment. If exposed to this level can avoid harm by leaving immediately.
4,000	12,618	Heats wood to 300 degrees F and causes ignition.
2,000	6,309	Exposure is human pain threshold reached in 8 seconds Skin blistering in fewer than 20 seconds.

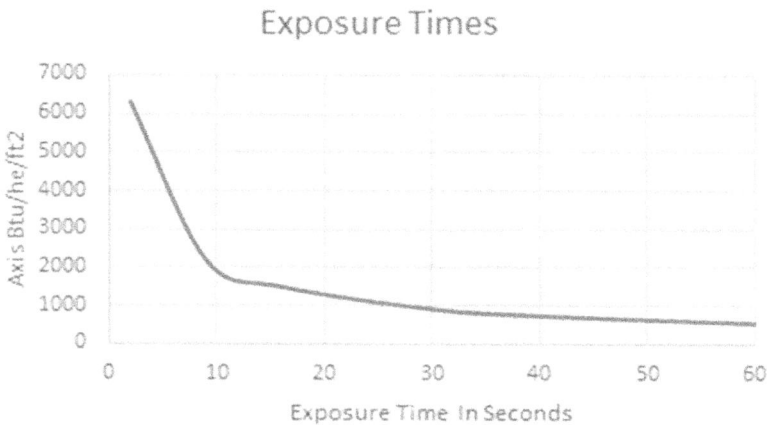

FIGURE 5.67 Derived from API RP 520: The Exposure Times to Reach Pain Threshold.

The radiated heat from a burner can be significantly lower than the total generated heat of the flame due to the effect of convection. The radiation heat release from the flare is calculated by the following equation:

$$Qr = FW\Delta h$$

This seems like a fairly simple equation, but the variables affecting the value of F are very complex and are dependent upon on the composition of the burnt fluids.

5.16.1.2   Heat Stress and Heat-Related Illness

Physical exertion leads to heat stress, which in locations with high ambient temperatures or extreme heat can result in heat-related disorders like heat stroke, which need quick action in the early stages to help undo the adverse effect on personnel.

5.16.1.2.1 Heat Exhaustion

Heat exhaustion can ensue if an individual becomes excessively hot, does not drink enough hydrating fluids, or experiences physical exhaustion. The most common and obvious symptoms of heat exhaustion are:

- Profuse sweating
- Vertigo and dizziness
- Feeling of sickness
- Muscle cramps, generally in the legs and abdomen

Medical attention may be needed if an individual is not responsive or is severely damaged from heat exhaustion. Professional medical assistance should be sought in the interim and as initial first aid may include:

- Removing the person from the work area to a shaded and preferably air-conditioned area
- Laying the person down on their side and making them comfortable
- DO NOT give the person ice
- Applying cool water, not iced, to the back of the neck and wrists
- Giving the victim small amounts of water every two to three minutes until the victim recovers
- If there is no improvement, then evacuate the person as soon as possible
- After recovering, the person should not return to work immediately until found medically and mentally stable

Muscular cramps are typically caused by salt deficiency from excessive sweating; an individual's minerals and salts should be replenished.

5.16.1.2.2 Heat Stroke

Heat stroke is less common than heat exhaustion but lethal if not recognised and treated soon. A heat stroke usually occurs after the onset heat exhaustion with the following symptoms:

- The person will have a high body temperature where the person is very hot to touch
- The person stops sweating, resulting in dry skin
- Their breathing will be more rapid than normal with fast, shallow breaths
- The person can become dizzy, lightheaded, and confused

This can quickly lead to unconsciousness and initial first aid response would include:

- Removing the person from the work area to a shaded and preferably air-conditioned area
- Laying the person down on their side and making them comfortable
- Removing all upper body clothing
- Calling for medical assistance immediately
- Placing packed ice or cold water around the victim's wrists, under the arms, and near the groin
- If at all possible, giving the victim tiny amounts of water every two to three minutes until the victim is feeling better
- Do not allow the person to swallow ice or iced water
- Do NOT give victim salt at this stage
- Stay with the person and keep them calm until medical assistance arrives

5.16.2 BURNER REQUIREMENTS

The specifications should be clarified and the burner manufacturer's specification document should be submitted before any equipment is shipped. The following parameters should be thoroughly investigated with regard to a burner:

i.	Air	• What volume and how is it supplied • If insufficient air is available then the oil cannot be fully atomised, resulting in smoke and burner fallout • All air compressors and supplies MUST have individual non return valves • Is a backup supplied?
ii.	Water	• Origin of water supply • The rate and pressure it is being supplied at. • The quality of water – the water has to be filtered especially if supplied from rig pits with contaminants and debris • There must be a switchable backup water supply that can be immediately switched in or the burner will be destroyed
iii.	Propane	• Depending upon the duration of the test or burn, sufficient industrial-sized full bottles of propane should be supplied in an adequate rack system • Regulators for all bottles • Manifold to allow propane bottles to be switched when necessary • If a propane rack is supplied then it should be full
iv.	Compressors	• Hydraulic and engine oil • Diesel (unless arranged by the rig) • Spare air compressor • Spark arresters and all necessary safety precautions
v.	Spares	• Burner spark plugs • Cabling • Seals • Basically, any component part of the burner that can be damaged by flame or heat
vi.	Manuals	• Manuals for burner with full specifications and requirements • Compressor manuals • Company operating procedures

5.16.3 BURNER FUNCTION

The majority of well test burners currently in use feature a design that effectively atomises the oil in a mixing chamber with supplied compressed air.

However, oil is seldom produced without other fluids, primarily water, and though only oil should flow out of a separator oil outlet, it is necessary to ensure that BS&W reaching the burners does not exceed an average of 20–25%, to be verified with burner model and design. A higher percentage of BS&W at the burner could affect the efficiency of oil burning and lead to environmental issues.

Therefore, if feasible, the separator should be operated in three-phase mode to remove the majority of water from the oil line, and only the oil flowline should be diverted to the burner. The segregated water should be flown to a tank or specified pit.

Burner functionality must be discussed at the TWOP in light of the local HSE and company regulations. The sequence of events in using a burner should also be discussed to ensure different circumstances are considered and taken into account.

When first opening a well for clean-up after well completion or stimulation, slug water and other minerals or fluids would flow through the surface equipment. Burner pilots should be constantly on at this time, regardless of the flow content, to ensure maximum burn of produced fluids. Pilot gas would be substantially consumed, and since it is a logistical challenge to replace flammable gas bottles, surplus/spare bottles should be anticipated and made available or on standby.

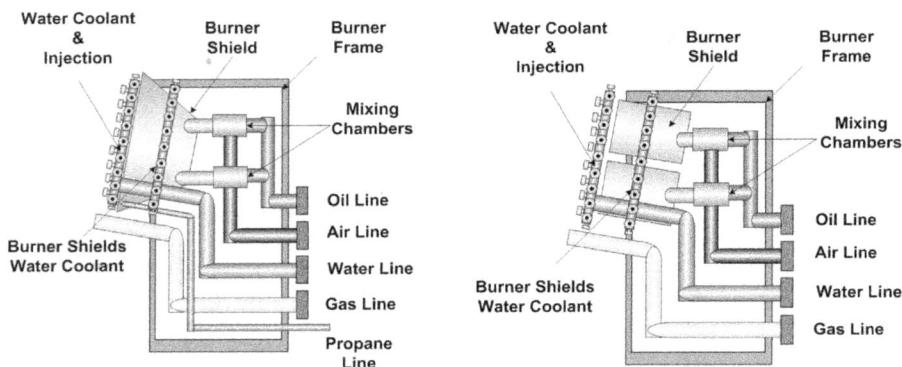

FIGURE 5.68 Example Burners.

TABLE 5.6
Example Specifications for a Burner

Maximum burner water cut	Up to 25%
Maximum oil pressure (separator max)	1440 psig
Maximum oil rate	15,000 bbl/d
Maximum air pressure	120 psig
Maximum air pressure	750 psig or higher
Minimum air flow rate	750 scfm @ 100 psi

The value of water in a clean burn is something that is underestimated. The black smoke that is frequently associated with oilfield burners will be reduced with the injection of water into the flame.

Uncombusted hydrocarbons dropping out of the flames in basic burner designs can be reduced by increasing the compressed air supplied, increasing atomisation. Proper chamber and mixing designs allow for the generation of smaller hydrocarbon droplets that can burn quicker. Supplied air is key to a successful burn.

It is obvious that this atomised oil spray will not ignite by itself. This is why the ignition systems and pilot flames are so important for igniting the hydrocarbon fluids and gases as well as for ensuring prompt reignition in the event the flame is extinguished.

NOTE: *As a safety issue; under no circumstances must gas from the separator be used for the gas pilot on burners.*

Entrained gas in the oil further assists in burning, and if the burner flames frequently extinguish, the GOR can be increased by increasing the separator pressure. As a result, the oil has more entrained gas, which should improve burn efficiency. This may not always be possible and the well response should be monitored.

Another strategy is to reduce the viscosity of the oil by connecting a steam exchanger to the surface equipment. The lower oil viscosity can be atomised more easily than a higher viscosity oil. This ought to be discussed during the TWOP and contractual stages.

5.16.4   BURNER PILOTS

The usual ignition source is a pilot system with remote "spark plugs." Either gas or an atomised liquid serves as the pilot medium, which must be installed or stored in a secure location far from the burners' heat radiation. In a desert or extreme solar radiated heat environment, any flammable items should be protected and shielded. The same principle applies with flares and ground flares.

When trying to ignite a burner pilot, extreme caution must be used. Ignition must be carried out using a system significantly isolated from any lingering gas and beyond the range of any coolant water spray. The ignition system should be waterproof and, if located in a hazardous area, explosion-proof. As a safety precaution, ignition systems should not be located in areas of standing water to prevent the risk of electrical shock. When working on a rig, the rig electrician should inspect the electrical ignition system and professionally connect it to the rig's electrical supply.

The majority of burner ignition systems employ a high voltage, so all safety inspections and precautions should be performed by electrically qualified personnel prior to mobilisation and during operations following all safety procedures. The burners should be earthed at a suitable point from the ignition box, which must be clean and using proper earthing lugs.

Under no circumstances should the ignition systems be worked on while power is applied.

In some locations a hot work permit is required to work and use these systems, applied for at the start of the job. The installation should always be checked and verified by the rig or location electricians. In the case of pilot gas, Propane is usually used with a 1/4" minimum sized stainless-steel line used for piping the propane to the system. Rubber or plastic hoses should never be used.

It is crucial to mount a water spray ring around the burner head – shown in Figure 5.69 – or heads in the case of multiple headed burners, designed to pump a water shield encompassing the flame for cooling and jetting fine water mist into the flame to help reduce black smoke, depicted in Figure 5.70.

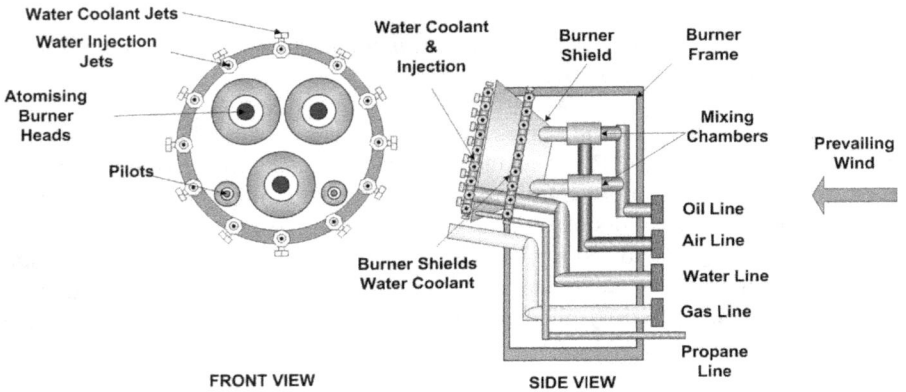

FIGURE 5.69   Burner Schematics.

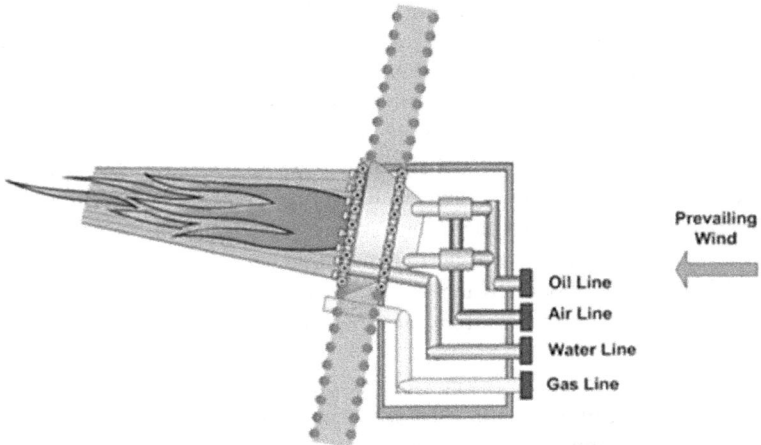

FIGURE 5.70   Pumped Shield and Coolant Water.

5.16.5 Positioning of the Burner

A wind direction indicator should be erected in a clear area that is viewable from the burner, the well testing area, and any other areas pertinent to the test so that any wind direction changes can be easily observed and recorded. The burner/flare/ground flare are mounted in line with the prevailing wind to ensure any smoke does not drift over adverse areas.

The burner is typically situated on the edge of a flare pit, unless it is an offshore setting where the burners are mounted on a boom. The flare pit is for burner operations only and not for disposing of rubbish or surplus liquids. The flare pit cannot contain any flammables, liquids, or solids to avoid uncontrollable fires, particulates, and black smoke. The operating company representatives, safety officer, or rig tool pusher must be made aware of any concerns regarding the physical state of the pit or its contents. If there are foreign flammable objects or liquids in the flare pit when the burner is ignited, the flames can drag smoke and particulates from the flare pit into the air, giving the impression that it's from burner performance, when in fact the burner may be performing quite well.

5.16.6 Supplied Water to the Burner

Figure 5.70 illustrates that the pumped water should effectively shroud the flame, helping to reduce smoke and improve the burn. The coolant water is sprayed in a circular pattern around the burner with a radius sufficient to act as a shield for the radiated heat from the burner flame. It also acts as a coolant for the burner assembly and frame. Without this water shield the metal burner components will deform and warp.

The water supply cannot be emphasised enough for the correct and safe operation of a burner. Clean water must be used that is completely solid free, otherwise the solids clog up the burner and prevent the burner operating properly. Often the water supplied by the rigs is full of solids, (rags, fittings, mud, etc.) It is no good complaining about the excessive smoke if the water supplied is not fit for purpose.

To counter this inadequate water supply, some companies have used *clean* frac tanks to store clean water that is then used for the burners. This can be water from the rig that has gone through a debris/junk catcher. In the case of a water supply system as shown in Figure 5.71 the tanks must be full of clean filtered water before burner operations commence. This usually helps eliminate any debris in the water supply, reducing multiple filter changes during operations.

If rig water is supplied to the burners through a filter line this can – and has – caused problems as the rags cause the filter to plug and restrict the water to the burner causing damage or smoke production. There are frequent cases of large objects like welder's hats and gloves getting stuck in the filters; filtering using a debris/junk catcher, shown in Figure 5.72, instead of a standard filter before the tanks as shown in Figure 5.71 will allow the filter to be cleaned without stopping water to the burner.

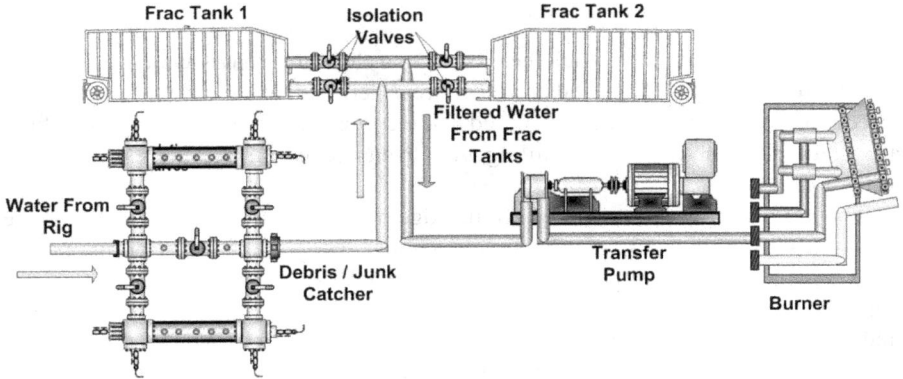

FIGURE 5.71   Clean Water Using Frac Tanks and Filters.

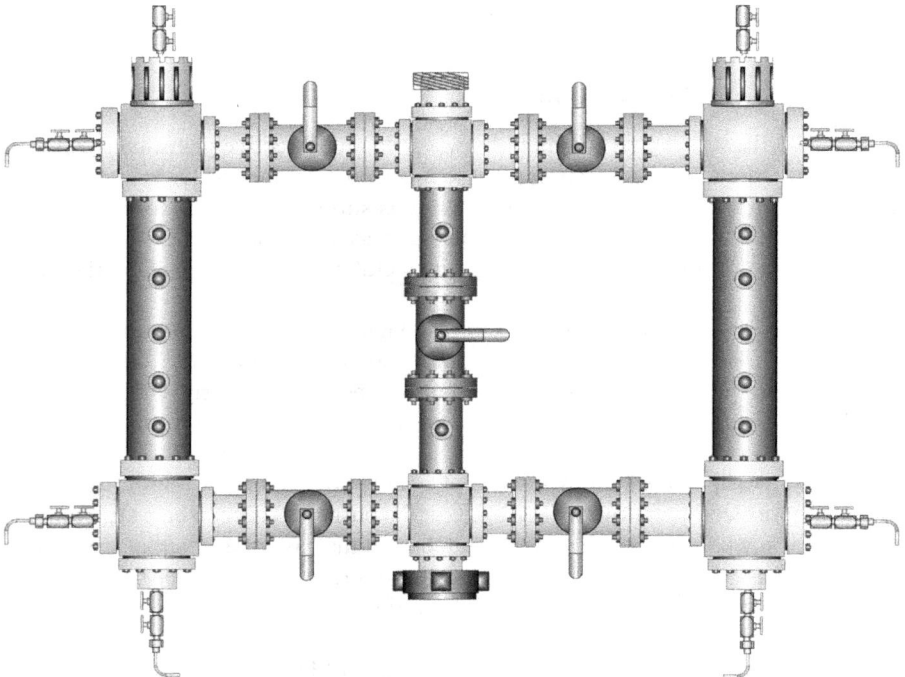

FIGURE 5.72   Low-Pressure Debris/Junk Catcher.

5.16.7   BURNER AIR SUPPLIES

It is generally accepted, even in offshore locations, that separate compressors are used to supply air to burners because of the potential dangers.

There are records of an accident on an offshore rig where the rig air compressors were used to supply air to the well test burners. This happened because **NO** non-return valves or gas detectors were in place. Gas from the burners travelled back along the airlines and into the rig compressors which resulted in an explosion with severe damages to the rig. From that moment onwards all well testing burners have their own air supplies. This helps demonstrate the importance of using non-return valves on air supply lines. Other problems with compressors occur due to lack of maintenance and attention.

The serious lack of air is probably the single biggest contributing factor that is responsible for a poor burn and in turn black smoke and particulates. This is probably the easiest to solve, with the burner specification sheets specifying the volume and pressure of air required.

Figure 5.73 is not to be taken as a true reference but shows the possible amount of air required and the size of compressors needed for a specific manufacture of burners. The graph is included to show the importance of an adequate supply of compressed air for optimum burner operation. A similar graph/data for the used burners should be supplied with the burners on site.

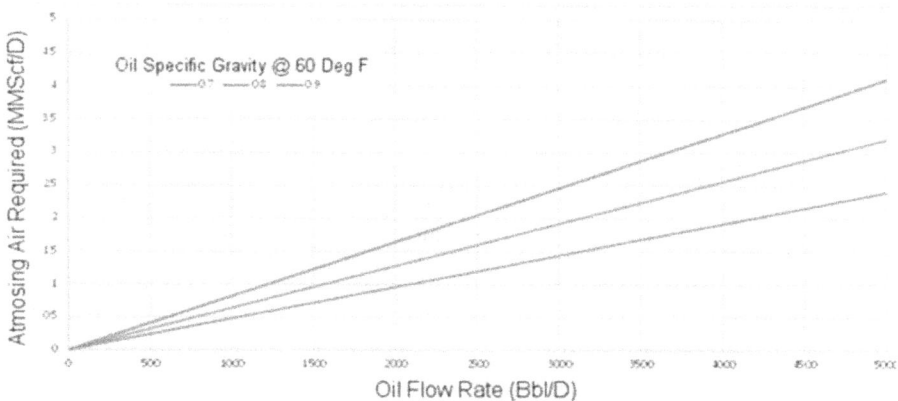

FIGURE 5.73 Hypothetical Graph of Supplied Burner Air Atomisation Estimation Graph.

This does not necessarily mean a single compressor of such capacity has to be used, but with a manifold and non-return valves, multiple smaller-capacity compressors can be used to provide the necessary air, as shown in Figure 5.74.

In onshore land jobs the compressor supplied is often far too small for the job, often being borrowed from other departments irrespective of capacity. Multiple compressors would be required if the duration of the burn was to exceed 12 hours, at least to account for the necessary shutting of the compressor to carry out maintenance and checks.

It is better to have a higher capacity compressor than a lower one. All compressors used for supplying air to a burner must be fitted with non-return valves, air reservoir, over-speed valves, spark arrestors, and safety controls. All controls, dials, and indicators should be fully functional and calibrated on all the compressors.

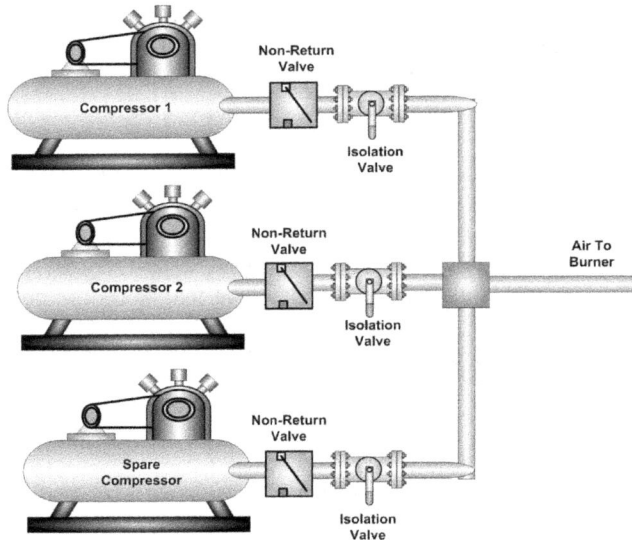

FIGURE 5.74   Compressors Used with Burner.

The positioning of all the compressors must be safe position *upwind* of the burner and away from any other potential sources of hydrocarbons, dusts, or gasses to ensure that combustible materials cannot be sucked into the compressor air intakes.

A good reference regarding burners is:

www.offshore-mag.com/business-briefs/equipment-engineering/article/16759268/well-testing-proper-oil-water-atomization-needed-for-burner-combustion

5.16.8   Burning Gas

Often, a straight forward pipe commonly suffices to flare off the produced gas when a well is solely producing gas. On an offshore setting, a burner boom would still be required to support the pipe length required to keep the flame a safe distance from the rig body. However, with the presence of liquid hydrocarbons it is becoming more common to utilise a suitable burner to prevent pollution.

The straight pipe for gas flaring would normally be equipped with a remote ignition and pilots that function in a similar manner to the more complicated burner configuration. The coolant and protective water sprays apply equally.

In a burner assembly the gas line outlet is strategically positioned underneath the configuration to assist in burning any flammable fallout from the burner, as can be seen in Figure 5.75.

A "drip tray" may be placed underneath the burner to try to catch any unburnt fallout hydrocarbon, which allows the dip tray content to ignite by the flames.

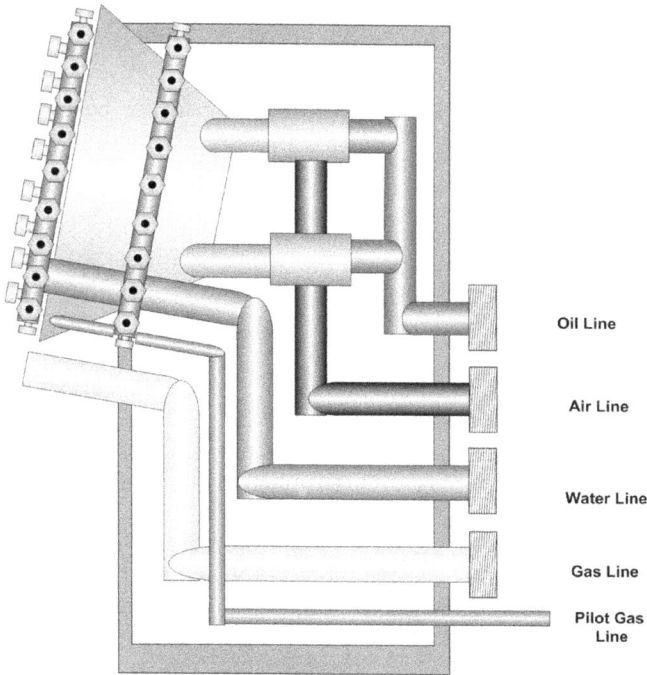

Oil Line

Air Line

Water Line

Gas Line

Pilot Gas
Line

FIGURE 5.75   Side View of a Burner Configuration with Gas Line beneath the Burner.

5.16.9   OFFSHORE BURNER

Onshore or offshore burners operate under the same principles.

In an offshore setting there are, as standard, two burners positioned on burner booms on either side of the rig, and the flow can be switched to either one using the oil and gas manifolds. The rig is positioned accommodating wind directions so that one of the burners can be used.

The lengths of the burner booms are determined primarily by the magnitude of generated heat radiation, with the purpose of protecting the rig from heat and possible damage. To further aid the control of heat radiation or to compensate for the inadequate length of the boom, water sprays of different volumes and combinations are employed, to an extent of completely encompassing the burner. The cooling water spray shield is distinguished from the water supply for the burning of intended fluid.

A specialised contractor with the necessary expertise is charged with mounting, commissioning, and certifying the booms, to avoid historical failures that have led to fatalities. The connected pipework will also help to reinforce the boom's stability.

Air supply is critical to combustion and the source of air and supply setup must be independent and unassociated with any other purposes. Previously, air supply was shared and it led to catastrophes. A drilling rig supplied multi-purpose air to the burner and gas was pressured back into the compressors, leading to an explosion and significant damage. Since then, it has become a forbidden practice to share air supply with a burner. The use of non-return valves is critical especially with multiple compressors.

5.16.10 ONSHORE BURNER

In general, onshore burners have gained the reputation of the most polluting types, largely air pollution with the appearance of black smoke. However, this is often a fault of insufficient air supply to the burner, creating incomplete combustion due to the insufficient air needed to atomise and, in turn, burn the hydrocarbons completely.

The flare pit has a significant role in well testing; its size, dimensions and location(s) depend on the expected flow rates and consequently the flare size. It should be common practice to calculate an adequate flare pit size prior to mobilisation, and understanding of its location should be predetermined, as well as the positioning of the burner to the pit.

In exploration phases or in remote locations the burner is usually mounted at the entrance to the "flare pit" or "burning pit." In many circumstances, the spotting of the well testing equipment is reverse engineered from the location of the flare pit, to ensure the safety of personnel and equipment.

Flare pits must only be used with the burner. The operating company should make it clear that use of the "flare" pit for other uses is forbidden. Liquids or solids, especially flammables, are prohibited from being placed in the pit before or after the burner is operated. Otherwise, the flare pit contents may ignite and cause uncontrollable fires, particulates, and black smoke. The operating company representatives and the safety officer or rig tool pusher should be made aware of any concerns about the physical state of the pit or its contents.

The presence of foreign bodies in the pit often causes the black smoke polluting the air and is misconstrued as coming from the burner.

5.16.10.1   Flare Pit Construction

On permanent production stations or gathering stations, there may be a pre-constructed pit capable of handling the maximum production rates of the oilfield. While in an exploration well testing phase, the pit should be designed and constructed in consideration of the following:

i.   Maximum Flow Rate

The flame length will be a result of the maximum flow rate, so it is important that the maximum size flame is deduced.

ii.   Pit Length

The pit length should exceed the maximum expected flame size.

iii.   Fluid Type

The length of the flare pit should reflect the fact that gas typically has a higher escape velocity and produces a flame discharge that may be longer.

iv. *Flare Impacting on the Rear Berm*

With high velocity flares there is the potential ability of eroding the rear berm and forcing the material into the air. The Berm's structural integrity is a safety concern and environmentally crucial. When it is compromised flow rates must be reduced below the tolerance level until the flare pit can be reinforced to accommodate higher flow rates.

v. *Flooding*

If in regions where heavy rainfall is expected or where it frequently rains, metal drainage pipes connected to pumps should be placed in the pit to lower water levels.

vi. *Warning Signs*

Warning signs should be put around the entire pit at a safe distance.

vii. *H_2S*

Where the gas contains dangerous levels of H_2S, signs, detectors, and breathing apparatus should be placed at strategic areas, including around the pit.

When H_2S is burnt, sulphur dioxide (SO_2), another toxic gas, is produced, therefore an isolation zone should be set up with the necessary signs downstream of the flare pit.

Deciding the location of a flare pit takes into account a number of factors and requirements:

   i. In accordance with any governmental, local, or company regulations
   ii. Dug in accordance with any prevailing wind direction
   iii. Beyond the normal activity of any work or processes
   iv. Where there are flames, smoke, particulates, and any resultant gas, such as sulphur dioxide in the case of burning hydrogen sulphide, the following points need to be considered:
      a. Personnel
      b. Animals both wild and domestic
      c. Residential areas, schools, playgrounds, or walkways. Minimum suggested distances in excess of 300–400 metres (1,000–1,300 feet)
      d. Flaring interfering with any form of air transport
      e. Roads and road traffic; recommended distances are at least 100 metres (350 ft)
      f. Parking lots and surrounding areas may pose a hazard if flaring operations produce particle fallout
      g. Areas with dry plants or woodlands could be susceptible to fumes, ignited from heat radiation or fallout particles
      h. Water courses that become contaminated by fallout particles or from the flare pit's location

The design and construction of a flare pit requires consideration such as the following:

      a. The pit should be dug on a level section of land
      b. There should be no drainage or utilities crossing through or near to the pit

c. In the event of any ground water the pit location will have to be changed or a heat resistant lining will need to be added to the pit

d. Water being important to a well testing's burner performance, there should be as little water seepage into the ground as possible, to prevent environmental contamination. This is especially important when it comes to vegetation or nearby natural water sources

e. The likelihood of the pit flooding should be taken into account, especially during periods of heavy rainfall, as this could contaminate any nearby water stream or water source

The pit bottom should slope in the direction of the burner so that the water volumes will be separated into the lower section of the pit and act as a barrier to prevent any seepage of hydrocarbons into the ground. Also, the closer the floating hydrocarbons are to the flare the easier it is for them to burn. Figure 5.76 illustrates this pit design and the gravitational separation of water and oil

FIGURE 5.76   Cross-Section of Flare Pit Showing Liquid Levels.

f. The burner mounting position must be higher than the pit's maximum allowable fluid level. Draining facilities must be provided to lower flare pit fluid levels if there is any potential of flooding

g. If it all possible, the burner should be positioned at a lower level than the process equipment so after any flow periods the lines will naturally drain under gravity towards the pit. This makes flushing and rigging down of equipment a much easier task

Berms are ridges or slopes that surround the entire flare pit area, with a gap for the burner, and are an important part of the pit design with the following basic functions:

i. Act as a visual identifier to the area defined as the flare pit

ii. Assist in stopping or reducing the volumes of gas, vapours, and particulates from being blown out of the designated flare pit area

iii. Help to reduce radiant heat from the burner flame

Several attributes define the effectiveness of a berm:

a. The inner berm wall's slope shouldn't be vertical. The vertical structure hinders air entering or circulating into the pit and lowers burning efficiency

b. The outer walls should also be angled to make it easy and risk-free for any personnel to access

c. The berm walls should be sufficiently thick to contain the force of flames, especially in high velocity gas wells

d. In desert terrains, the berm is usually constructed with sand, which makes scaling the berm wall challenging. To facilitate easier access, aluminium ladders are embedded into the berm

e. There should be wind direction indicators, visible from different angles, that are out of potential combustion areas

5.16.11 MOUNTING THE BURNER AT FLARE PIT

On land-based operations, the burner should be self-supporting or mounted on a sturdy skid that is not relying on the connected piping to support it. Positioning a burner isn't as simple as spotting it on the ground; it requires certain preparations and considerations:

i. Tubing joints: often used in the line from the separator to the burner; these should be inspected and pressure tested for suitability

ii. Crossovers: Connecting selected tubing connections back to hammer unions must be available for each line of tubing used

iii. Securing flowlines: The flowlines to the burner should be secured to the ground, even closer to the burner, to help avoid equipment lift with flow and flaring. Weights placed on the flowlines and small diameter stakes driven in to the ground on top of the pipes are often used to secure flowlines in place, as shown in Figure 5.77

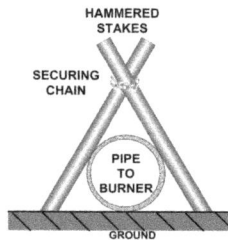

FIGURE 5.77 Securing Pipework to the Ground with Stakes.

iv. Flowing the well: This not only generates lift, but cooling using water sprays can cause the side of the pit to erode. This can be avoided by constructing a secure platform, with the burner mounted on a suitable skid for stability by weight distribution, as shown in Figure 5.78

v. Flushing should be performed before the burner is positioned or connected to flowlines and secured on location, initially with air, as these flowlines (especially rig pipes) may contain shale, rust, or mud particles that are damaging to the burner

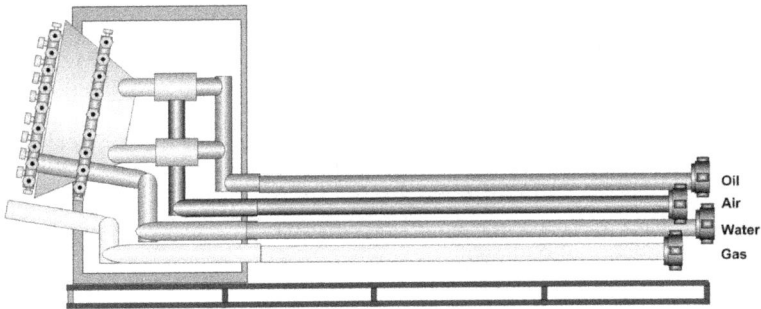

Burner Mounted On Skid

FIGURE 5.78   Burner Mounted on Skid.

> vi. It is recommended that the ignition system is set up and tested with the pilot gas prior to any flowing operations

5.16.12   OFFSHORE BURNERS OPERATION

On an offshore rig, two burner booms are typically directed, taking into account rig positioning, to allow their use with dominant winds. Each burner boom's design, structure, and assembly must be approved and certified by a third-party certifying company and installed by professional installation companies, then certified for use.

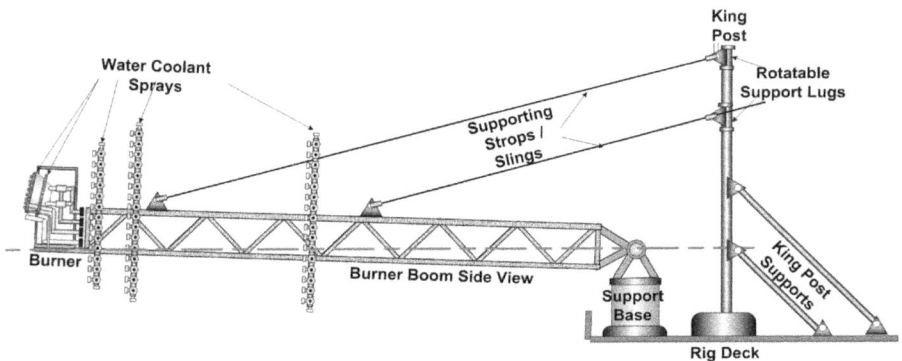

FIGURE 5.79   Example of Offshore Burner Mounted on Boom.

In Figure 5.79 note the cooling water jets/sprays mounted on the boom despite there being a water line connected to the burner. The volume of cooling water sprays depends on the generated radiant heat from the burner. If the cooling spray proves insufficient to their purpose, then auxiliary hoses can be additionally used.

Offshore boom designs and installations must consider these parameters:

i. All offshore booms should be designed to withstand all stresses and strains from weight, weather conditions, and rig movement
ii. Length of offshore burner booms vary and are selected to reduce the radiated heat towards the rig, nominally between 50 and 100 ft in length
iii. The burner booms have to be certified as fit for purpose by a third-party certifier
iv. Burner booms are modular and are normally shipped to the location in sections
v. There should be sufficient space allocated on the location for the assembly and mounting of the booms
vi. When hanging booms, they are predominately inclined upwards
   Booms are generally manufactured in three supporting designs. In the case of the box and triangular section booms, illustrated in Figure 5.80, the walkway access is on the top of the boom but in the U section boom it is part of the boom structure

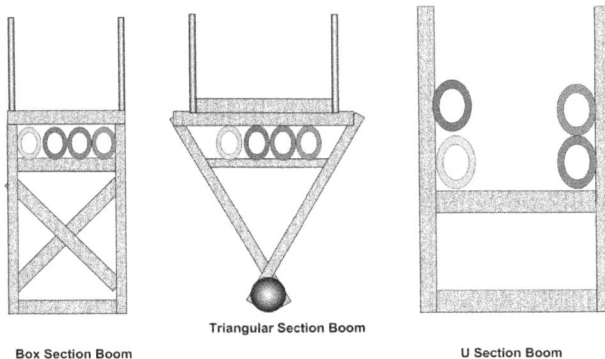

Box Section Boom   Triangular Section Boom   U Section Boom

FIGURE 5.80   Offshore Boom Structure Support Designs.

vii. The pipework is integral to all booms
viii. The King Post, illustrated in Figure 5.79, is mechanically supported and will also be used to mount the burner vertically during rig moves. The rig structure, normally around the legs in a jack up, can be used instead of a king post. Regardless of which mechanism is used, they all have to be mechanically inspected and certified
ix. Safety meetings shall be held before any boom operations that include all rig personnel, especially crane drivers
x. Assembly of the boom sections is carried out under the guidance and control of certified riggers. Work permits should be issued

xi. Hanging the booms off the platform will only be performed by certified riggers

xii. Weather conditions are critical to the safety of hanging burner booms. All decisions shall be made by the certified riggers in whether it is safe to proceed under new work permits

xiii. Booms should not be hung at night or in low visibility conditions

xiv. The burner should be disconnected from the flowlines and air blown through the flowlines independently; when sufficiently cleared from any debris, rust, etc. the burner should be reconnected and the flushing with air process repeated

xv. Before any attempts to "hang" the burners the tool pusher/barge master/captain must be informed and standby vessels moved to the side of the rig where booms are to be hung

xvi. Potential helicopter landing timings and schedule should be taken into account to avoid downdraft effects and crane difficulties

xvii. Upon completion of hanging the booms, certificates of compliance should be issued, with copies held by the rig and well test supervisors.

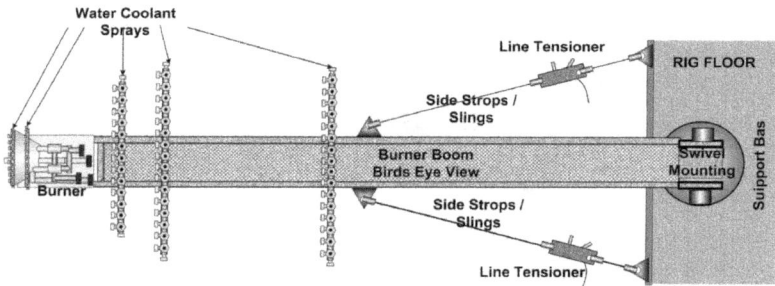

FIGURE 5.81   Bird's-Eye View of Burner and Boom Example.

The burner is usually held in place by strops mounted either side of the burner boom. These side strops are tightened with heavy-duty line tensioners, often referred to as come-alongs, to prevent movement from wave or wind effects. The side strops are secured to fixed, normally welded, points on the rig body.

Side strops can be used to adjust the boom's angle to accommodate the wind's direction. The burner should not be in use when doing this; the flow is diverted to the other boom, and then line tensioners can be adjusted. Most burner booms are equipped with a rotating base plate where the boom is connected to the rig, allowing for these angle adjustments in the clockwise and anti-clockwise directions but not a complete rotation.

Walkways on the booms are normally made from steel grating to avoid any liquid pooling and to ensure a good grip.

5.16.12.1 Burner Operations

When using a burner for the first time it is necessary to be aware of the following general points:

i. It is a generally held safety issue that the well cannot be opened for the first time during the hours of darkness

ii. Ensure there is sufficient pilot gas available on location and that it is possible to swap or switch gas containers with no loss of flame on the burner pilots

iii. Inform the tool pusher that well testing operations are about to begin; flow will be diverted to the well testing tanks until the volume of fluid in the tubing has been returned. Tubing volume calculations are relatively easy, but it is recognised practice to check with the drillers; there could be lost circulation or lost fluids in the formation

iv. A safety meeting must be conducted by the well test supervisor with all the involved personnel; operating company representatives, rig, service companies, and well test crew. The work permits should be checked for validity

v. If helicopter operations are imminent or crew changes, then opening the well should be delayed

vi. If not already covered by the current work permit, a new one should be processed for burner operations in addition to flowing the well

vii. All personnel involved in the flowing of the well should have an intrinsically safe radio, tuned to the same frequency, and individuals should be committed to proper communication

viii. The tool pusher gives a general announcement about flowing the well

ix. Standby boats are to be informed and moved to the side of the rig situating themselves at a safe distance upstream of the burner

x. Burner pilots should be started and their functionality verified. They should be kept on for the duration of the clean-up phases until clean oil is established

xi. Coolant liquid pumps should be ready for operation

xii. Throughout all burner operations, a watch should be made available for time recording and a radio, while monitoring radiant heat impact on the rig, with care to fallouts and pollution

xiii. If the radiant heat is causing concern the cooling water should be increased or, if all else fails, the flow rate reduced to safe levels

xiv. If radiant heat is deemed excessive and may cause a risk on the pilot gas bottles, then they should be relocated to a more sheltered spot or a cooling agent can be used

xv. Never encroach on the boom while burner operations are active

Accessing the burner/boom follows a strict procedure to ensure safety. Weather conditions primarily determine if it is safe to access a boom/burner, and preferably during daylight hours. Before any attempt to access the burner/boom the tool pusher must be informed and a work permit for the purpose has to be issued. A safety meeting with all on-duty personnel should be held by the well test supervisor emphasising the wear of PPE, safety clothing, and equipment necessary to the job such as work life vest. Safety lanyards should be used to secure the crew to the boom and prevent them from falling into the sea. The standby boat must move into position close to the boom but not directly underneath it, to avoid the risk of falling objects or personnel, and they should be informed of the people on the boom. Each tool or fitting taken onto the boom should have a lanyard/rope securing them to the equipment in case it slips or falls making the loss irretrievable.

The burner should never be worked on while it's still hot or while coolant is being sprayed.

5.16.13   BURNER PERSONNEL

There should be dedicated technicians that are trained and fully conversant with the burner operation, maintenance, and procedures on each shift. The responsibility for running the burners should *not* be the job of the supervisor but a trained technician per shift.

These personnel should also keep a separate manual readings sheet of the compressor settings, wind direction, and performance as well as the burner condition and burn condition. Regular physical checks and records on the following conditions are mandatory.

i. Burner Flame
ii. Smoke
iii. Fallout
iv. Coolant
v. Flow rates

These records will help future burner configurations on existing reservoirs.

If possible, the burner technician should be at the burner when any significant changes to the well are made in case it disturbs or extinguishes the flame. It is also necessary that all the well test crew are aware that burner configuration changes are imminent before proceeding to the burner location.

5.16.14 BURNER FLOWCHART

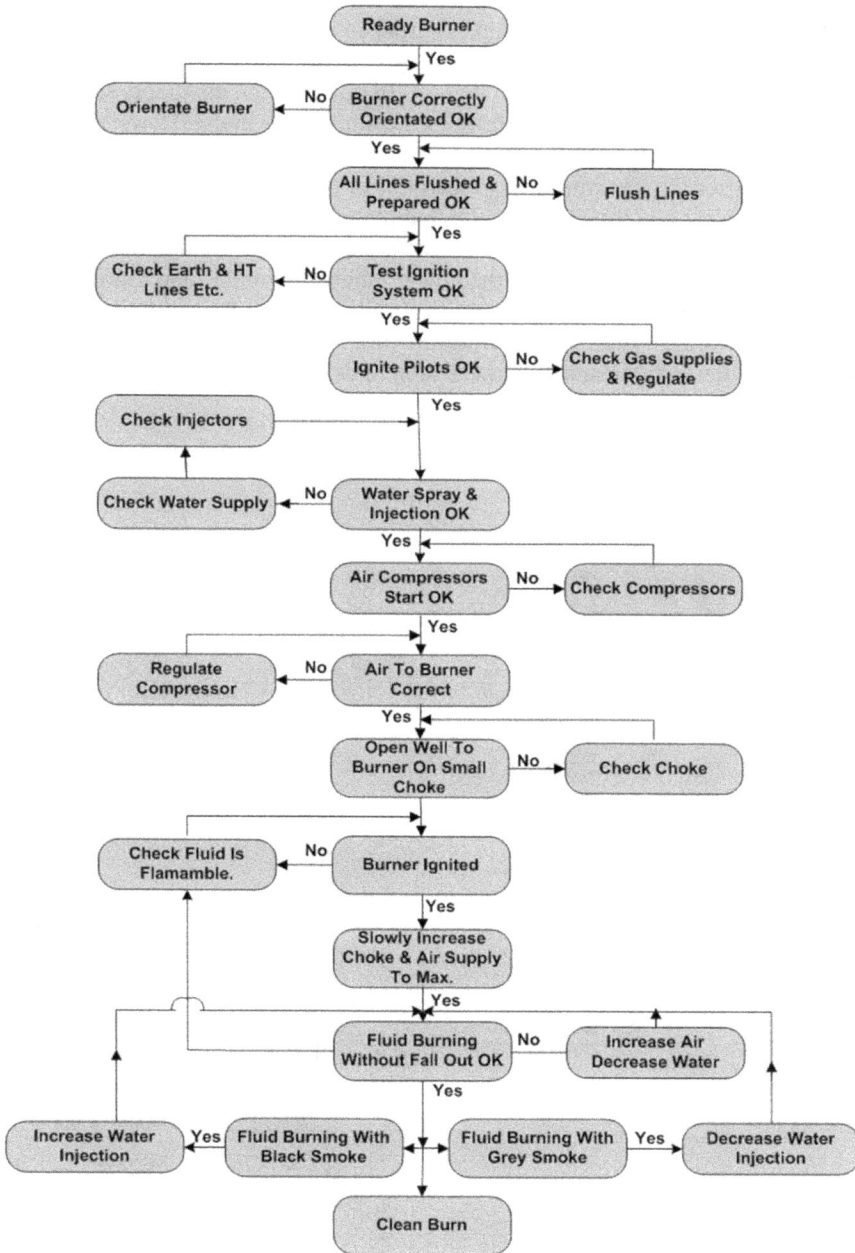

FIGURE 5.82 Burner Operation Flowchart.

5.16.15   Mud and Heavy Oil Burners

Not all burners are suitable for heavy oils and the burner specification sheets should be checked and verified before attempting this.

With oil-based mud flowed through a burner in an attempt to flare it off, unfortunately, the main components of the mud, the solids, are generally incombustible, and only the oil component is flared off, if at all. In this case the air supply should be as high as possible in an attempt to atomise the flow to its maximum. However, the presence of the mud particles can be erosive and cause permanent damage to the burners.

In other times the gas entrained in the oil – or the majority of the constituents – are not flammable and this impedes the flammability of the fluid. For this purpose, some burners have been modified to include a diesel injection ring, inside the water cones, which will assist in keeping the flow of liquid flammable and minimise dropouts. For better efficiency the lower burner head should be fed with diesel which is then atomised and provides a more efficient burn. Any fallout then will pass through the atomised diesel flame and be ignited where possible.

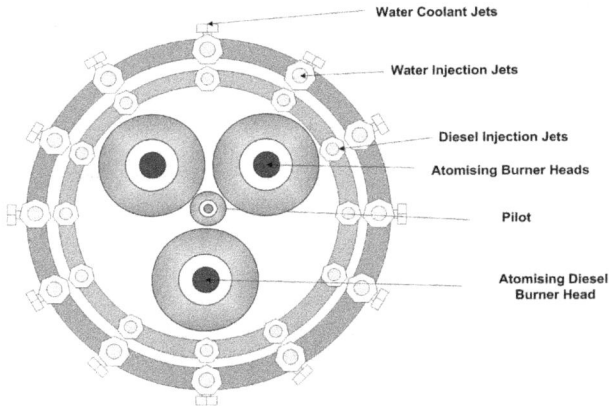

FIGURE 5.83   Modified Burner Head.

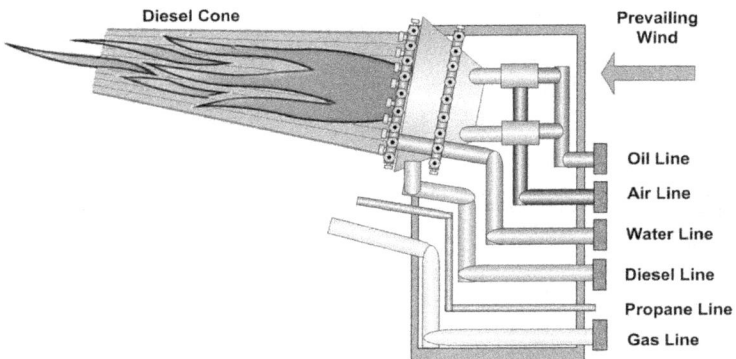

FIGURE 5.84   Side View of Modified Burner Head.

If the modified burner approach is used then there should be a permanent technician monitoring the performance at all times and adjusting where necessary.

5.16.16 Monitoring Burner Performance

If the well testers are using a data acquisition system (DAS), then critical parameters from the burner can be monitored. This should have been discussed and agreed within the TWOP meeting and within the contractual agreement.

The easiest of them to monitor is

i. Supplied Water to Burner

The flow of water to the burner can be monitored by installing a turbine meter in the water flowline near to the burner. An alarm can be programmed to warn personnel if water flow rate drops or stops, before the heat can cause damage to the burner. Fluid rates can also be monitored and correlated for coolant purposes

A bypass manifold should be in place and used if the turbine meter gets blocked with debris

ii. Supplied Air to Burner

Similarly, a gas turbine meter can be installed in the air supply line close to the burner to measure airflow to the burner. Alarms on the DAS can be set to notify personnel if the airflow stops or falls beyond a set threshold to avoid any burner operations without air that lead to pollution

A pitot tube is another instrument that can be used to monitor air flow but requires additional programming to the data acquisition package

More detailed monitoring can be:

iii. Flare Flame

Certified infrared monitors are available but have a long lead time on delivery from manufacturers. The DAS would receive signals of the flame-out on a burner and alarm personnel through a simple alarm setting

iv. Smoke

Certified infrared monitors are also equipped to notify the DAS with any issue regarding smoke on the burner. Using this instrument in combination with the supplied air monitoring, would prove informative. An algorithm on the DAS or a direct alarm can be programmed for either input to alert personnel on duty

v. Wind

Wind direction and speed measuring devices are available but tend to be used on longer production tests such as on early production systems

5.16.17 Smoke Measurement

Configuring, mounting, and using a well testing burner is the main aim to cause minimum damage to the environment. The fallout and liquids are caught in the burner pit, but the smoke is not caught and escapes into the atmosphere.

Dark smoke consists of partially burned particles of fuel that are the remnants of incomplete combustion. The smoke can be hazardous to health as the small particles can be absorbed into the lungs.

White smoke consists of mainly minuscule water droplets that are produced when water vapour released during the combustion processes condenses.

There are multiple ways of measuring the level of smoke ranging from filtering to optical systems but because of the positioning of a well testing burner it is not always possible to have the access to smoke. Therefore, a simple cheap observation system is a good indicator.

Smoke from a burner is measured in terms of the density in relation to a fixed grey scale, known as the Ringelmann scale. Full details and how to use it are on

www.soliftec.com/Ringelmann%20Smoke%20Chart.pdf

The Ringelmann Scale is a scale to measure the density of smoke. It was developed by Maximilien Ringelmann of the station d'Essais of machines in Paris in 1888. The Ringelmann Scale has five levels of density inferred from a network of black lines on a white surface, which, blend together in shades of grey, shown in Figure 5.85.

- 0 smoke level is represented by the white
- The levels of 1 to 4 are 10 mm square grids drawn with 1 mm, 2.3 mm, 3.7 mm, and 5.5 mm in width of lines respectively
- Level 5 is all black

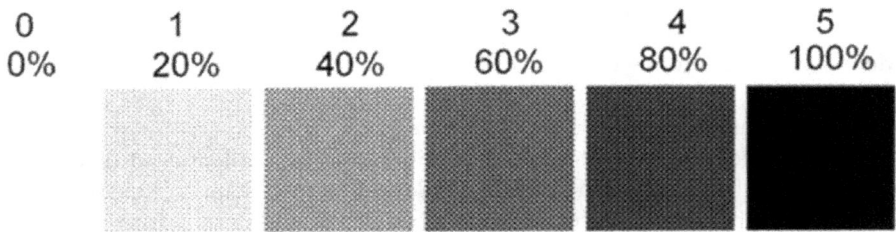

FIGURE 5.85 Chart Example of Basic Ringelmann Scale.

This is now a reference for the burner smoke intensity observed and recorded on regular intervals. Observing and recording of smoke intensity should be on an hourly basis or more frequently with the availability of dedicated personnel assigned to monitor the burner's performance.

5.16.18 MACRAE FLARE SMOKE EVALUATOR

An easier method to use with oilfield burners had been derived and used in Oman shown in Figure 5.86.

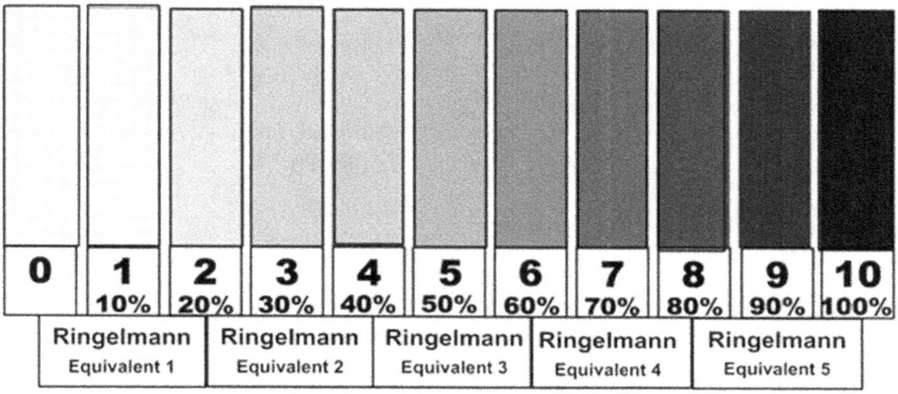

FIGURE 5.86   Macrae Flare Smoke Evaluator.

Source: Adapted by Mr Alex Macrae; used with his kind permission.

Instructions:

- To use the chart in Figure 5.86, hold it at arm's length towards the smoke from the flare
- Every ten seconds for a period of three minutes, record the smoke number which corresponds most closely with the smoke chart evaluator on the log sheet, example sheet shown in Table 5.7
- Repeat every 30 minutes
- Ensure all readings are taken from same smoke area

If these record sheets are compiled then a record of the functional settings for the burners can be determined by comparison and deductions; eventually with trial and error, it may be possible to perfectly set the burner parameters before the burn starts.

TABLE 5.7
Example Burner Record Sheet

Date	Time	Oil Rate	Water Rate	Gas Rate	Number of Compressors	Supplied Air Rate	Supplied Air Pressure	Oil Gravity	Gas Gravity	GOR	BS&W	Average Ringlemann Number	Fallout

6 Well Test Separators

This section is intended to explain the *basics and fundamentals* associated with the well testing separator rather than a guide to its configuration. The separator is the most crucial piece of equipment in the well test setup; all the other equipment installed are designed to condition the well flow to pass it through the separator and should be setup or configured by experienced well testers.

6.1 WHAT IS A SEPARATOR?

A separator, as its name implies, is a vessel that is designed to physically separate the components of a well flow into its three major components.

These are:

   i. Oil
  ii. Gas
 iii. Water

This is called three-phase separation since the phases are separated and measured independently. However, wells exist that only have two phases, gas and water or oil and gas, or the fluid phases of oil and water cannot be easily separated then we resort to two-phase separation.

In underbalanced drilling operations a fourth phase is introduced, solids, the cuttings or sand. This requires a four-phase separator which differs from a three-phase separator design and construction.

To summarise difference in design: a four-phase separator has internally installed water jets and exits to allow the cuttings to be washed out of the separator, whereas a three-phase separator has to be depressurised and the inspection hatch opened for the cuttings to be washed out manually. A three-phase separator normally has restrictions regarding flowing four-phases through it; the standard operating procedures of the service company, the operating company, and the manufacture specifications of the separator dictate this for the most part.

6.2 PRE-JOB

   i. Before the separator is mobilised to location, it should have been maintained and pressure tested
  ii. The separator must be accompanied with a document package detailing the operational data, maintenance, pressure tests, and calibrations of all instruments on the separator

DOI: 10.1201/9781032623689-6

iii. Prior to ANY live hydrocarbons being put through the separator and ANY of its associated equipment, it must be pressure tested using water or water with glycol if hydrates are an issue. Water is used as a pressure testing medium as it is a relatively non-compressible liquid; this is a fundamental safety requirement

There may be an issue pressure testing with water if safety valves are mounted on the separator as these should only be tested using a non-flammable gas, usually nitrogen

There is a habit amongst the mobile test units to use the well pressure as a pressure test source. This is potentially unsafe and should not happen

iv. The well test separator is normally rated as 1,440 psig and if the maximum wellhead pressure to be tested is lower than the operating pressure of the separator, then testing using tanks should be considered as an alternative

Other common separators are pressure rated to 720 psig and 2,160 psig – good knowledge of the separator and equipment used is a must before well testing.

6.3   WELL TESTING SEPARATOR BASIC RULES

Before a well can be put through a separator there are set conditions that must be met, otherwise the separator may not operate properly or may operate in a hazardous state.

6.3.1   CHOKE CHANGES

As a matter of safety, the choke must not be increased while the well is flowing through the separator with the Orifice Plate in the lower chamber or measuring position. Any major choke changes can cause a surge in both flow and pressure that effectively "blow" the orifice plate out of place and into the pressure control valves. This can cause a pressure increase in the vessel and trigger the safety valves.

If the plate is "blown" into the control valves it is a hazardous condition; it will put the separator out of operation until repaired or a replacement separator delivered.

Well test personnel will either bypass the separator or "raise" the orifice plate before any choke changes.

This is one of the reasons why non-well test staff should not touch or adjust the well test equipment when they are not aware of any ramifications to the adjustment.

6.3.2   CLEAN UP

The separator is not intended to be used as a clean-up device. Fluid should only be put through the separator after the clean-up phase. Flowing the well through the separator before it is fully clean will damage the separator internals and metering system and cause future issues with vessel capacity and therefore separation functionality.

If the well is not fully clean and a rate is essential then a surge tank could be used to obtain a fluid rate. Gas rate is usually measured from a pitot tube fitted on the gas export or vent line of the tank.

6.3.3   INLET PRESSURE

The inlet pressure must be below the maximum operating pressure of the separator. Just assuming the choke manifold will reduce the pressure automatically can cause accidents.

Most well test separators have a maximum operating pressure of 1,440 psig but there are other models of separators with lower/higher operating pressure; this needs to be verified. There must be a plate on the separator detailing all of its operating parameters. This should be free of paint and easily accessed and viewed. If the separator does not have a plate detailing its operating pressure and other physical parameters then it will not have passed an inspection and should not be used.

6.3.4   PRESSURE RELIEF LINES

The pressure relief lines must be run to a safe area separately from the separator export lines; otherwise, back pressure on a combined vent line can stop the separator venting properly. This can result in a catastrophic pressure rupture of the separator.

If the separator pressure relief lines are not run properly or of a sufficient size, then the separator should not be used.

6.3.5   SURGING PRESSURES

If the fluid flow into the separator is surging or widely fluctuating then it should not be put through it, since this can lead to carry over and instability within the separator.

6.3.6   HIGH BASIC SEDIMENTS AND WATER (BS&W)

The separator is a three-phase device and the introduction of the sediment into the separator is effectively a fourth phase which will settle in the bottom of the separator and accumulate. Unless the separator is designed for use with under balanced or pressure managed drilling, it will not have the capability of removing or "jetting" the accumulated sediment out of the vessel.

Unless it's the four-phase separator design, the separator inspection hatch will need to be opened and the internals manually cleaned. All seals will then have to be replaced and the separator fully pressure tested and re-certified for use. The inspection hatch seals are one of the largest and spares are not normally carried on units.

With larger flows and high sediment rates then the velocity combined with the solids can cause erosion/damage especially on the baffle plates, which should be avoided.

As a general rule, the downstream choke sediment rate should be below 1% of the total flow or better.

6.3.7 ACID

High concentrations of acid should not be put through the separator as it may affect and damage seals along with measuring instruments. Mild concentrations of acid may be put through the separator as it can help to clean the insides of it.

Remember distilled vinegar typically has a pH of around 2.5 and we put that on our food, so before just saying it "can't be put through the separator," verify the effects first if unsure.

Unless it is direct from the pumping unit, the returned acid from the well is called "spent acid" and has been pumped downhole or into the reservoir where it has reacted with the rock. Spent acid is defined as – "acid weakened by use." Acidity checks should be made with litmus papers and recorded on the manual readings sheet.

6.4 PHASES OPERATION

The technicality or practicality of this will be better understood when covering further parts of this chapter. It is mentioned here so the reader can notice this choice of functionality while expanding on the separator details.

A separator can normally be run in either of two setups or phases.

6.4.1 TWO-PHASE

Two-phase separation is where the fluid stream is split into two constituents, gas and liquid. There is discussion about whether or not it is truly two phases when the liquid component is in two parts, oil and water, but termed so when these two liquid phases cannot be effectively split from each other by the separator and considered collectively as a single phase. Hence the term two-phase.

- Gas
- Liquids

If oil and water can be split and measured independently, then the separator should be run in three phases. In two-phase separation the liquid phases are not separated – or split – and are flowed out through the oil line for their total volume to be measured.

This flow makes a surge tank more favourable with only two-phase measurement, similar to Figure 6.1.

To re-iterate, a two-phase well test happens if:

i. There is a gas phase and only a single liquid phase produced
ii. Both or one of the liquid phases, oil/water, are too low in flow rate (volume) to be measured independently

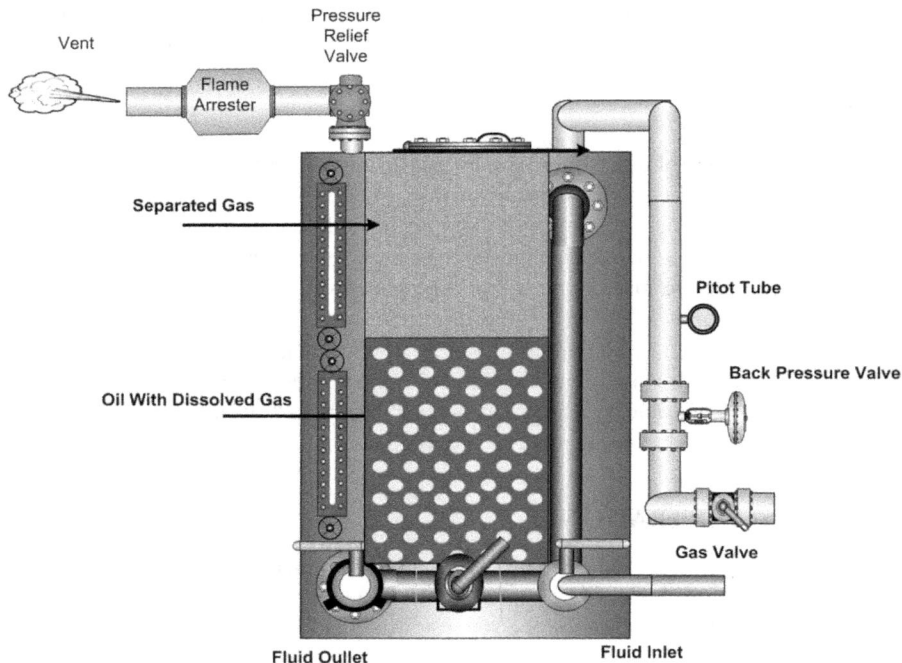

FIGURE 6.1   Vertical Two-Phase Separation.

Any leftover liquid volume at the bottom of the separator, when run in two phases, should continue in the two phases of flow rate and cumulative calculations, regardless of whether the liquid present is both oil and water or one of them. Any separated liquid within the separator can be used for sampling, etc., but not for flows or cumulative calculations.

6.4.2   THREE-PHASE

Three-phase separation is where the fluid stream is split into three constituents:

   i. Gas
  ii. Oil
 iii. Water

The liquid in a three-phase configuration has the produced liquid separated or split into two completely different phases (Figure 6.2) and is flowed out through the oil and water lines respectively.

The internals of a test separator do not differ for either two- or three-phase operation; it is only the way the separator is operated.

As a generality if there is measurable produced water then the separator should be run in three-phase operation.

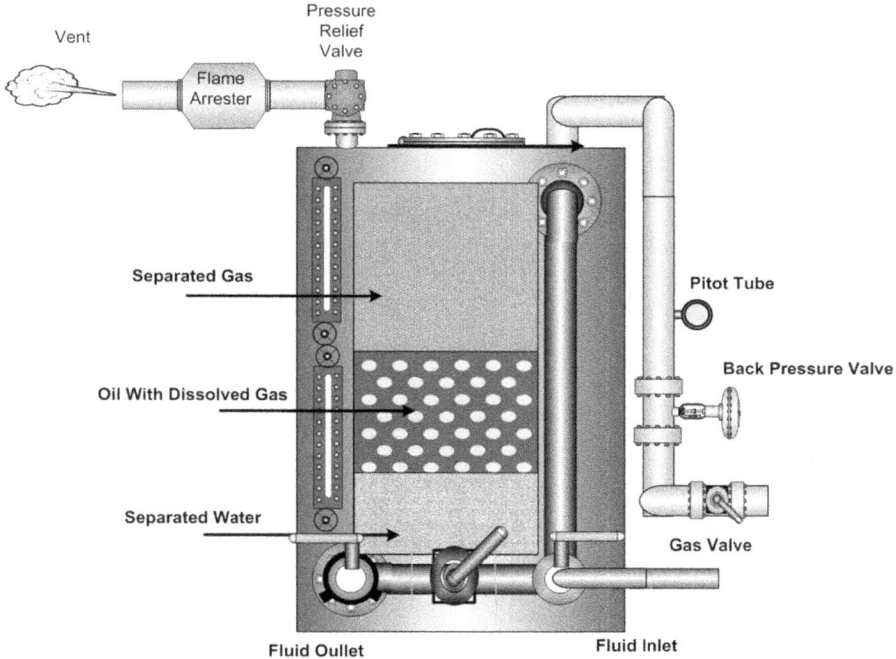

FIGURE 6.2 Vertical Three-Phase Separation.

6.5 WELL TEST SEPARATOR COMPONENTS

6.5.1 SEPARATOR MANIFOLD

On the inlet side of the separator there is a piping manifold that allows the flow to bypass the separator vessel and commingle flow streams. Figure 6.4 illustrates such a manifold configuration. In some instances, an independent and dedicated clean up line is used until the well is clean enough to be flowed through the separator; even with a dedicated clean up line aspects of the separator manifold are in use.

The flowline arrangement and number of valves on the manifold depends on the separator's engineering design. Vertical separators have a similar manifold system with the same functionality but arranged differently, due to the physical footprint.

Often, the separator manifold is missed from the pressure test sequence. The valves used in the separator manifold are not high-pressure valves and pressure testing should not exceed the pressure rating of the separator. Separator manifold valves are often unidirectional valves and should only be pressure tested in the direction of the flow.

Valve operation always follow the same rule; a valve must be open before closing the other, a two-man job.

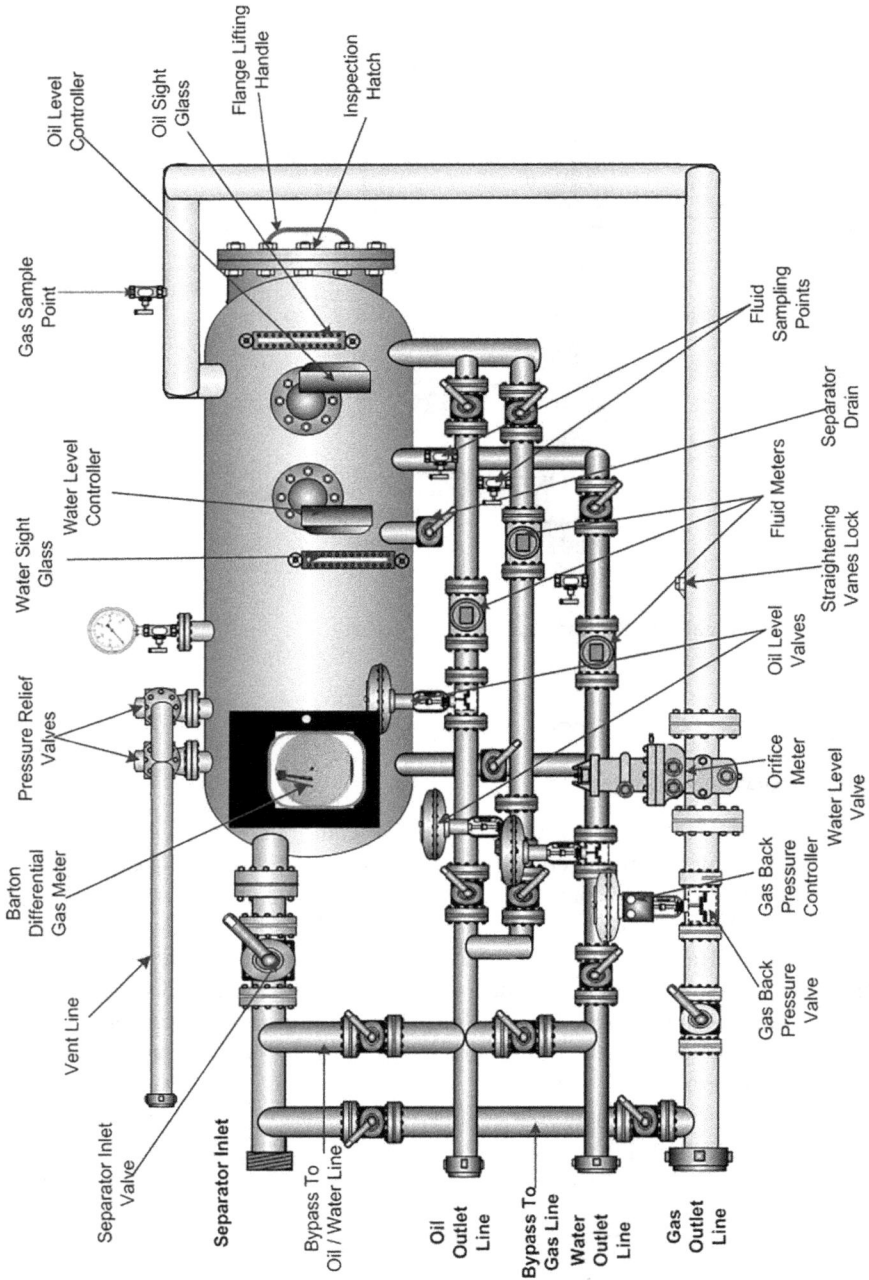

FIGURE 6.3 Horizontal Three-Phase Separator Major Parts.

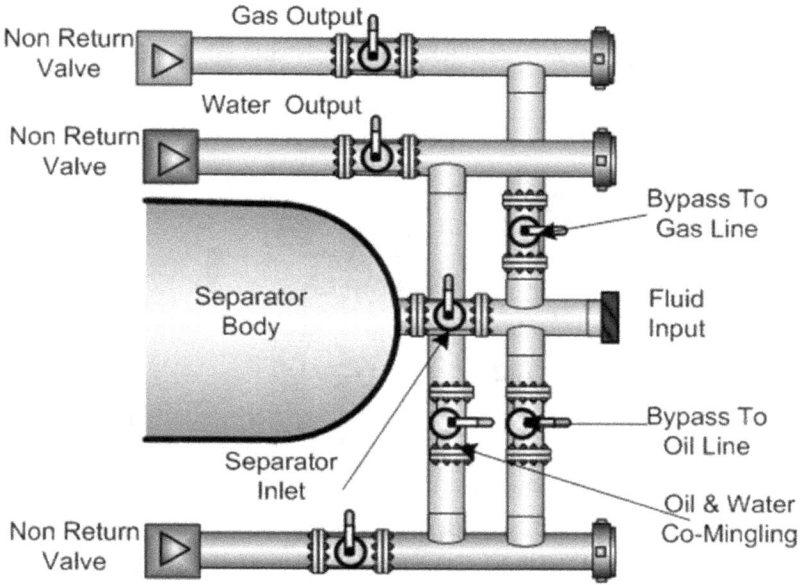

FIGURE 6.4   Separator Piping Manifold Example.

The symbols shown in Figure 6.5 are used to specify the valve status in the following figures and diagrams.

Shut Valve   Open Valve   Non-Return Valve

FIGURE 6.5   Example of Symbols Used.

During initial stages of opening a well, clean up, or when the flow is considered unsuitable for flowing through the separator, high BS&W, acid etc., the flow is switched to the separator bypass flowline. The configuration of the separator manifold allows flow diversion to either the oil or gas lines depending on the expected flow returns or regime.

Where an oil well or a predominately oil flow regime is involved, the flow is initially bypassed the separator and flowed to either a tank or the burner's oil line, Figure 6.6 illustrates the separators' manifold arrangement. The output from the oil line on the separator will be directly connected to a fluid manifold that in turn diverts the flow to either tanks or burners.

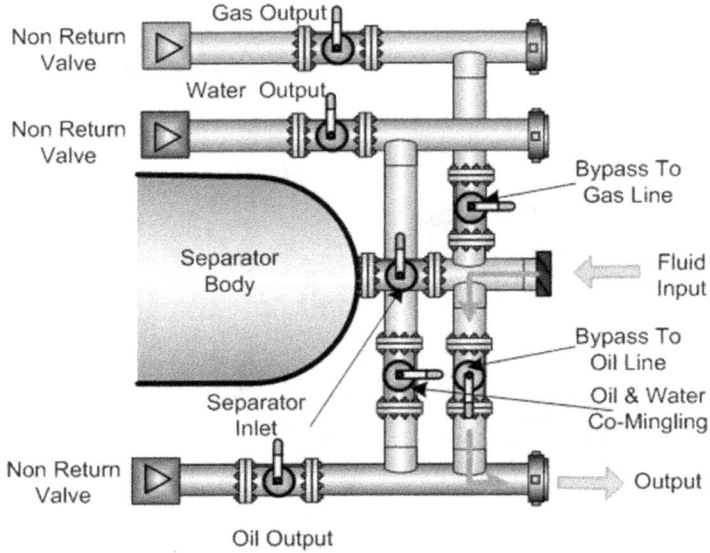

FIGURE 6.6 Bypass to Oil Lines.

If cleaning up a gas well or a water zone/well, then the flow can often be switched to the gas line on the separator manifold as shown in Figure 6.7. The bypass to the gas line is used for gas flows or if unloading a non-polluting water/toxic water cushion. Polluting returns should always be flowed to a tank or lined storage pit.

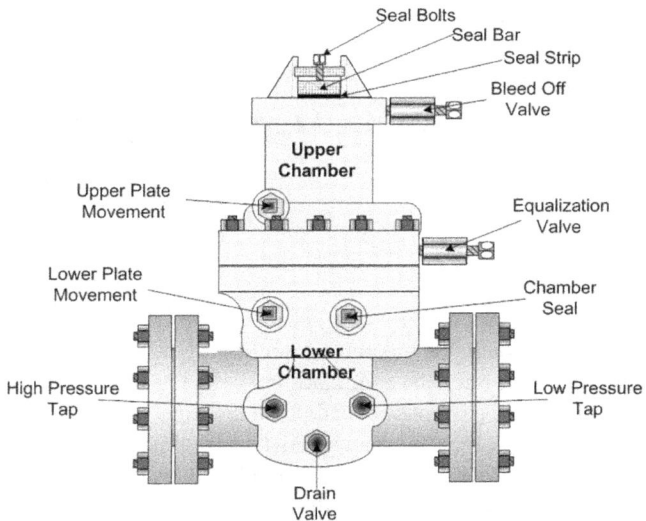

FIGURE 6.7 Bypass to Gas Lines.

Occasionally the flow from the water line is joined – or commingled – with the oil flow, as shown in Figure 6.8. This is usually carried out if there is oil or other pollutants in the water flow so they are collectively sent to the burner for safe disposal – burning. In this case the water line is usually plugged off (Figure 6.8) or isolated from a fluid manifold.

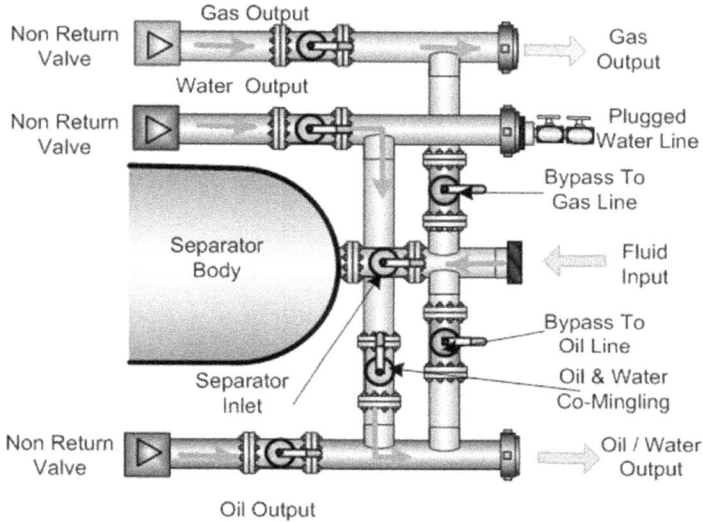

FIGURE 6.8 Co-Mingling Water and Oil Flows.

Where the oil phase of the flow does not burn easily or consistently then the gas component can be recombined with the oil flow to provide a more flammable flow (Figure 6.9). This is not always an option on some separators.

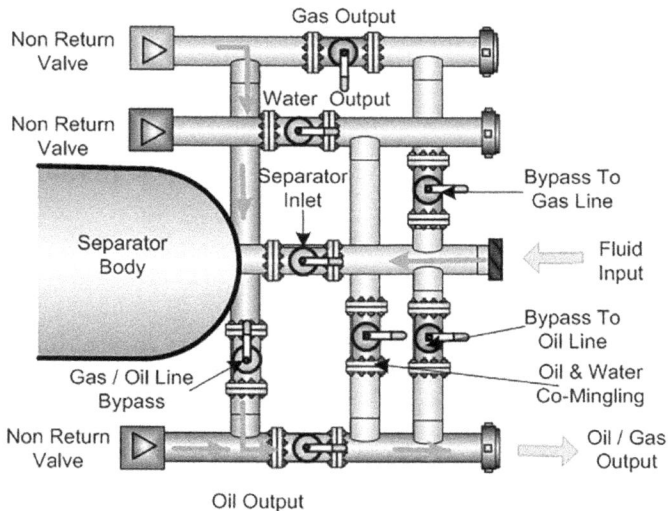

FIGURE 6.9 Co-Mingling Gas and Oil Flows.

6.5.2   NON-RETURN VALVE

A non-return valve (NRV) needs no manual intervention that allows the fluid to flow through it in one direction but not in the other. These non-return valves are often a flapper valve, preventing any backflow into the separator. A direction of flow arrow on the body of the valve indicates the allowed flow path, inlet to outlet.

In Figure 6.10 a weighted seal assembly seals against the inlet. The seal assembly is weighted so that it defaults to the closed position. When resultant fluid flows in the allowed flow path direction, it pushes the seal assembly to the open position allowing flow to pass through the valve. These types of valves would have a removable flange opening on the top to allow internal access for any maintenance and checking.

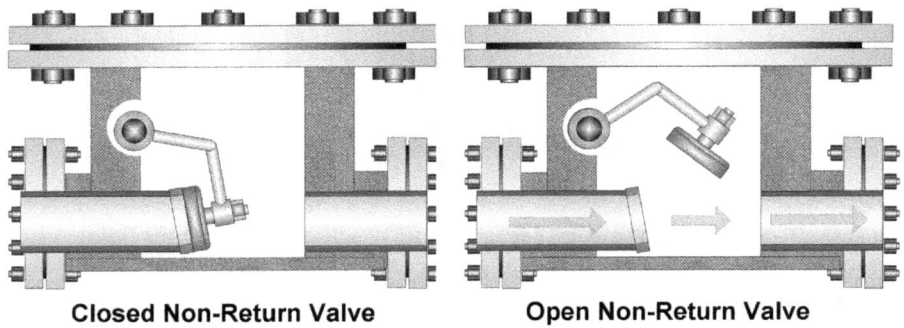

Closed Non-Return Valve   **Open Non-Return Valve**

FIGURE 6.10   Example Non-Return Valve Assemblies.

6.5.3   WELL TEST SEPARATOR INTERNALS

6.5.3.1   Basic Three-Phase Well Test Separator

Figure 6.11 shows the basic internals of a three-phase separator. The flow path within a separator follows the sequence:

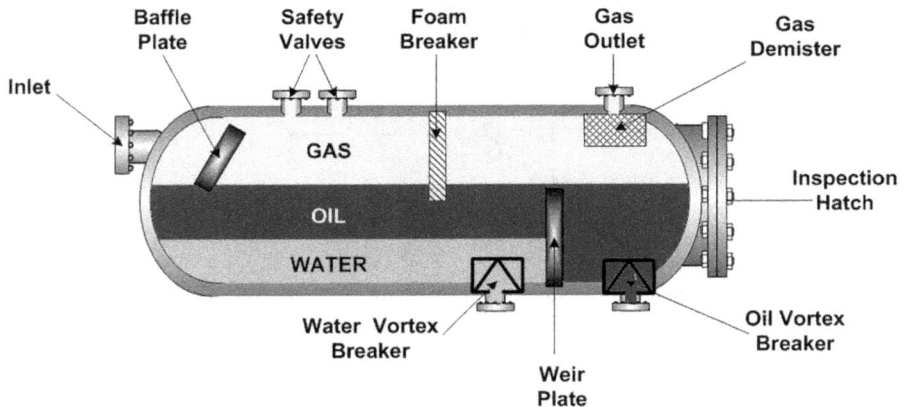

FIGURE 6.11   Three-Phase Separator.

i. The un-separated flow enters the separator at the inlet
ii. The flow immediately hits a plate; designs vary from separator to separator. This plate has many different names, baffle plate, inlet diverter, slug catcher, etc. but the function is the same

Separation begins to occur due to the following conditions:

- The change of flow direction due to the impact of the flow on the baffle plate which also agitates the fluid and causes the fluid to begin to separate
  - The liquid falls to the bottom of the separator, high specific gravity
  - The gases tend rise to the top end of the separator, low specific gravity
- Reduction of Velocity
  - The velocity of the flow is significantly reduced because of the change in volume from the smaller pipe to the larger separator body
- Reduction in Pressure
  - The pressure is now determined by the function of the control valves which is normally considerably lower than the pressure at the inlet of the separator

iii. The gas will tend to rise to the top end of the separator and flow towards the gas outlet. This initially separated gas will not be a "pure" gas but will contain liquid in droplet or mist form
The liquid droplets will tend to fall to the bottom half of the separator by the effect of gravity. The rate the liquid droplets fall out of the gas stream is dependent upon
  a. The physical size of the droplets, greater pull of gravity;
     - This is determined by the composition of the fluid stream
     - Smaller droplets usually remain in the gas stream in the form of mist
  b. The velocity of the gas stream. The flow velocity is dependent on the physical size of the separator and the well flow rates
  c. Turbulence within the gas stream, normally from the effect of the baffle plates
     - Reduction of turbulence is reduced by the inclusion of straightening vanes within the body of the separator, shown in Figure 6.12. These are basically a series of plates mounted parallel to each other so the action of flowing through them causes the turbulent flow to be all in the same direction
iv. As the separator pressure is lower than the wellhead pressure, the gas begins to segregate from the fluid flow in the pipe work and combines with the free gas in the separator
The longer the fluid remains in the separator the better the separation. This is known as retention time

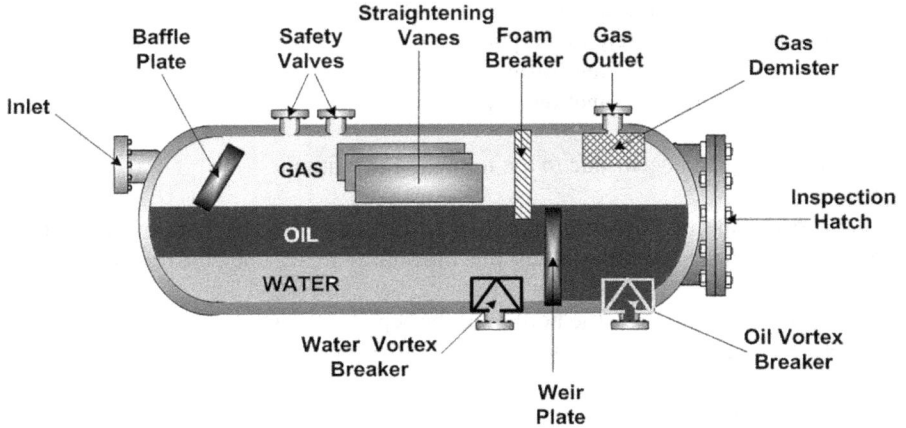

FIGURE 6.12   Separator with Straightening Vanes.

v. As the gas then passes through the separator it will exit the separator via a demister which helps knock out any small fluid particles trapped in the gas phase not yet separated under the influence of gravity, so the exported gas is as dry as possible

Different technologies are used in the mist extraction section; however, the most common is a compressed or knitted wire mesh that the gas flows through. The mesh demister functions by fluid particles impacting on the mesh where they are separated and fall to the bottom section of the separator

Sometimes liquid coalescers are used as well, to improve the quality of the exported gas. A coalescer is effectively an evaporation plate which assists in the separation of the fluids. Think of steam on a pane of glass where the coalesced liquid (water) runs to the bottom section of the glass pane. Similar action occurs in the separator where the liquid runs to the bottom of the separator

vi. The liquids dropped to the bottom of the separator will segregate under the effects of gravity and density into the oil and water phases mainly within the first section of the separator

As entrained gas exits the liquid solution it can form a foam which is broken back into a liquid by the foam breaker. Again, the longer the liquid remains in the vessel the better the separation–retention time

vii. Water and oil do not easily mix, therefore if the separator is large enough to retain the fluid long enough, gravity will separate the two liquid phases

As the oil phase of the flow is lighter than the water phase it automatically rises to the top of the liquid level; the water being the heavier phase falls to the bottom. There is no physical agitation; the process is under the influence of gravity alone, where lighter densities rise and heavier densities fall

viii. Within the liquid the bottom section of the separator where the liquid accumulates there is a weir plate, dividing the lower end of the separator into chambers. If the oil and water interface is kept below the top of the weir plate, it will only allow the oil to flow over it (skim) and prevents the water from entering the oil chamber

In order to configure the level control so that only the oil flows over the weir plate in three-phase operation; there must be a physical mark on the separator, reflecting the height of the weir plate. This is usually a metal indicator attached to the separator sight glass as shown in Figure 6.13, so levels can be compared against the weir plate level indicator.

Without knowing the height of the weir plate, it is often not possible to correctly configure the oil and water levels.

NOTE: There are usually two type of sight glasses used, one that can be seen through and a reflective option.

ix. Control valves regulate the pressure and fluid levels within the vessel
x. Fluid production is measured by individual meters on the gas, oil, and water outlets of the separator

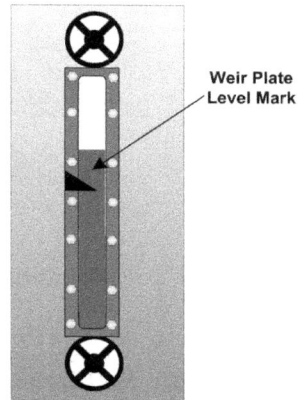

FIGURE 6.13 Separator Sight Glass Illustration.

Often the oil and water phases are so commingled that the well test separator cannot fully separate the oil and water into their separate phases. This is referred to as a two-phase operation; refer to Section 6.4 Phases Operation.

With two-phase flow the unseparated flow is passed through the oil line, no flow through the water line or it's considered three-phase. Water production then is calculated from the BS&W water measurement taken from the *wellhead* or before the separator inlet, if possible, preferably after a heater, if used.

This is one reason why BS&W measurement is so important in well testing; an erroneous BS&W will result in erroneous oil and water rates.

This also applies to the separation of gas from the oil phase in instances where the operating pressure of the separator prevents all the gas from fully escaping from the oil. This is measured and corrected using shrinkage and GOR2 factors discussed in further detail in later sections.

6.5.3.2 Two-Phase Well Testing Separator

Within a separator running in two phase all the components and functions are still in place, as illustrated in Figure 6.14. The gas is separated as with the three-phase separator operation, but in this case, there is no separation between the two fluid phases and it is allowed to flow over the weir plate and all fluid is passed through the oil lines.

Water production is calculated by the BS&W reading taken from **upstream** of the separator. The chapter on BS&W will elaborate on this.

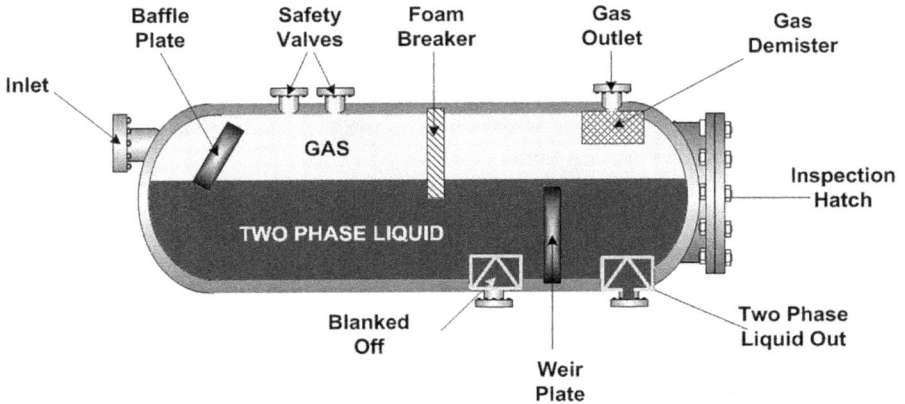

FIGURE 6.14   Two-Phase Separator.

IMPORTANT NOTE: *BS&W sampling must **NOT** be taken from the separator when in two-phase operation; this often causes confusion in some spheres but separator BS&W is limited to three-phase operation only.*

Where the separator is run in three-phase operation the BS&W taken at the choke manifold becomes irrelevant and need not be taken unless requested by the customer as a general reference to verify any sand or debris production. With the separator in three-phase operation, the BS&W results from the **separator oil line**, then mathematically corrects the oil production and calculates the amount of water entrained in the oil stream.

Again, the importance of proper and accurate BS&W measurement techniques cannot be stressed enough. Otherwise, the errors are passed on to the oil, water, GOR, and GOR2 rates and cumulative calculations.

6.5.4   Separator Liquid Levels

The only guaranteed way that the fluid levels within the separator can be determined is by using the standard see-through sight glasses, unless using an ultrasonic or radar-type level detector, which rarely work properly in a well testing separator because they need the internals of the separator mapped for level correction. Guided beam level detectors seem to get better results. These devices work well in a fixed/non-moveable vessel, but with the vibration and manhandling of a portable system these do not survive the transportation very well.

Also, the act of gasses leaving the oil phase can result in foam being generated. This can lead to the appearance of false levels and readings from the electronic sensors, even with a series of plates (commonly called foam breakers) are installed to help minimise this.

Foaming can also be reduced by increasing the separator pressure; this has the knock-on effect of increasing the GOR and GOR2 values.

Many well test operators often do not understand the importance of knowing the level of the fluids and interfaces within the separator and hence the sight glasses are often left uncleaned or not maintained regularly, losing their intended functionality.

If no digital instruments are used then the reliance is on the built-in sight glasses to determine the interface and oil level, as can be seen in Figure 6.15. Even with

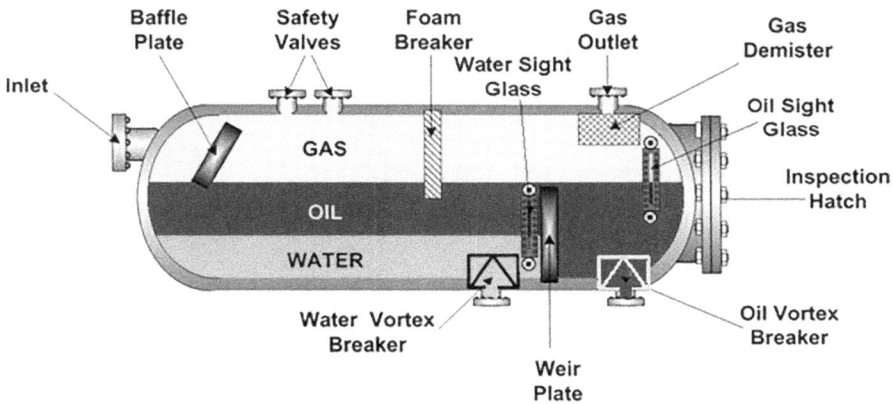

FIGURE 6.15   Separator with Sight Glasses.

digital level sensors the sight glasses are used as a reference and for calibration. This means that the sight glasses *must* be kept clean and the fluids easily visible through the sight glass. This often necessitates cleaning sight glasses throughout the period of the test while the separator is in use.

The height of the Weir Plate *must* be marked on the water sight glass so that the oil water interface does not exceed the height of the weir plate. This should not be a marker line but a permanent indication.

The oil sight glass must also be kept clean so the top of the fluid can be monitored to avoid overfilling the separator.

The liquid used to clean the sight glass must not have a low flashpoint such as solvents; diesel is preferred. Check company operating procedures for safety information and safety sheets.

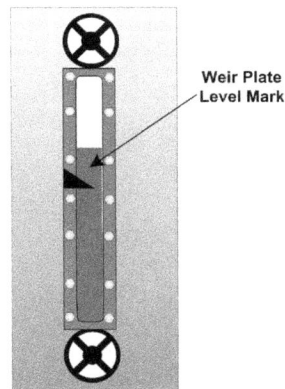

FIGURE 6.16   Separator Sight Glass with Weir Plate Marking Illustration.

The use of a well-known make of cola, with lemon, has proved to be effective in cleaning sight glasses.

NOTE: *Petrol and gasoline are banned as cleaning agents in all well testing operations due to flammability.*

Whatever cleaning liquid is used it must be fully drained before opening the isolation valves to the separator vessel.

A separator sight glass must use 1/2" (or greater) ball valves at the top and bottom of the sight glass, to enable a bottle brush to be passed through it. Needle valves should not be installed as a brush cannot pass through them.

From the bottle brush, Figure 6.17, note the length of the handle; this should be long enough to pass through the entire sight glass length. The brush handle can be extended easily using an aluminium (or other suitable metal) rod or a slickline sheer rod. Do not use a stick or other breakable materials as they can cause an obstruction in the sight glass that can mean complete disassembly and repressure testing the vessel.

FIGURE 6.17   Bottle Brush.

Where electronic level detection devices are used, illustrated in Figure 6.18, it is often the case that there are two used in tandem to jointly determine the interface of the two fluids and the highest level of the liquid (oil level).

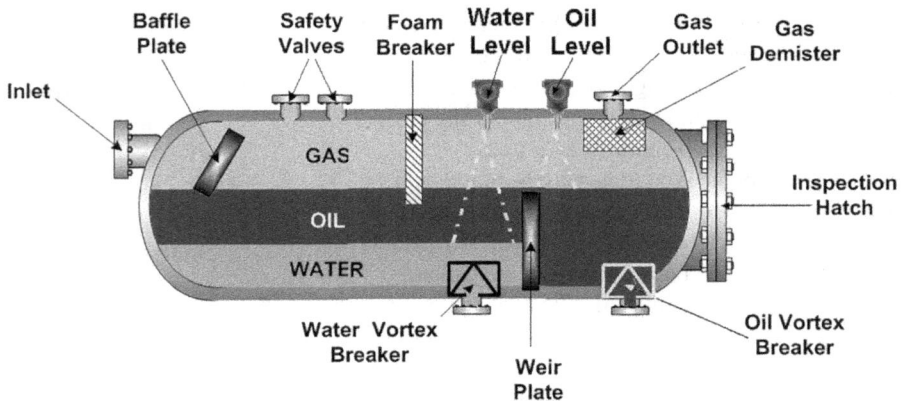

FIGURE 6.18   Separator with Level Sensors.

These devices work using various methods – radar, ultrasonic, etc. – and are chosen to suit the application as often foaming oil or intense mist causes false level readings. This is where the importance of fully clean and functional sight glasses is relied on. An advantage of electronic type of level measuring devices is their possible integration with emergency shut-down systems and SCADA data acquisition systems.

6.5.5   LIQUID LEVEL CONTROLLERS

To maintain a constant level within the separator there are control valves linked to a level sensing device mounted through the body of the well test separator. The controllers consist of a "float" which has its density selected so that it floats on the fluid media it is controlling, whether water or oil.

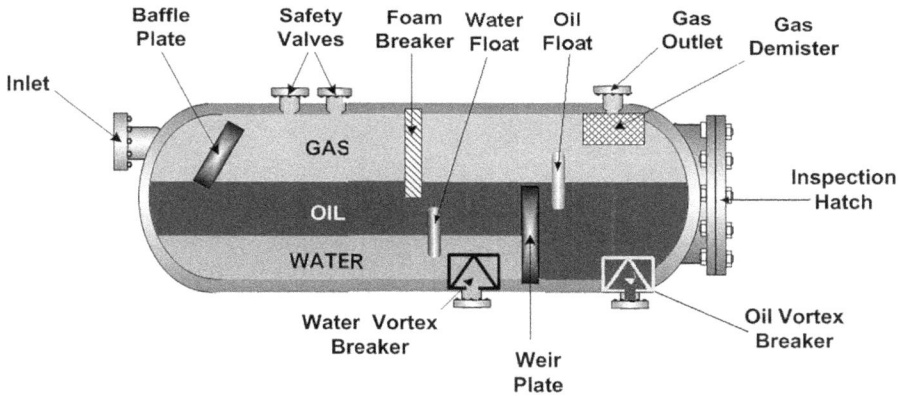

FIGURE 6.19   Separator with Liquid Level Floats.

The water float is manufactured using a high-density polymer or equivalent, which is dense enough that the oil will not provide sufficient buoyancy to make it float, therefore it will sink through the oil level until it encounters water, offering it sufficient buoyancy to float, whereas the oil float has a lower density so it can be supported by the oil.

The individual float design and weight is geared towards the density of the oil and water. Therefore, there are two different densities used.

i.   Oil Interface Float

The float density is based on the range of crude oil densities.

It is placed in the oil chamber of the separator after the weir plate. These are designed and weighted to float in the interface between the oil and gas.

ii. Water Interface Float

The float density is based on water or brine densities.

It is placed in the main or water chamber of the separator. These are designed
and weighted to float in the interface between the water and oil.

These two floats are connected by a mechanical linkage to an independent pneumatic system that is connected to the respective control valves on the liquid's flowline, depicted in Figure 6.20.

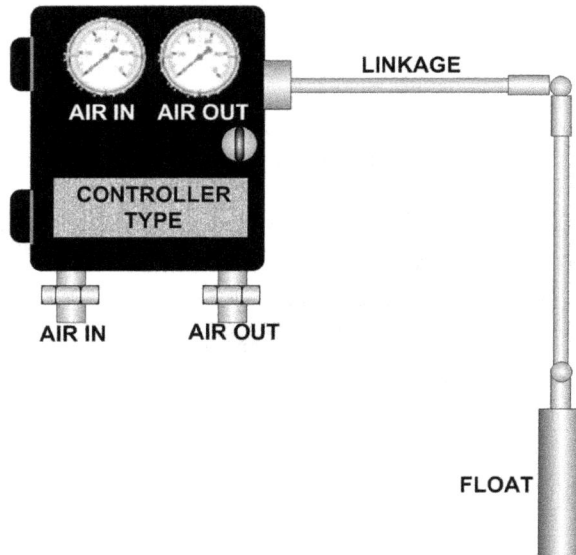

FIGURE 6.20   Liquid Level Float and Controller.

As the float is raised or lowered by its respective liquid level, it actuates the pneumatic pressure on the liquid control valve. The control valves on the separator liquid
flowlines are normally CLOSED valves as a safety issue. If the system malfunctions
(loss of air) then the valves close to prevent any spillage.

With pneumatic pressure required to open the valve, lowering the pressure closes the valve, and increasing the applied pneumatic pressure opens it.
Pneumatic pressure is applied at the bottom of the valve diaphragm, lifting it up
to open. This function is a safety consideration. If the air pressure to the separator is lost and there is no pressure on the liquid valves they fully close; this way
the liquid is trapped in the separator and not discharged unintentionally with
severe HSE consequences.

The fluid level in the separator is maintained by the position of the individual floats for both oil and water. In some separators the floats are mounted in separate sections, open at the top and bottom, which makes access to the floats easier.

The action of the valves on the fluid lines are usually a proportional type, which means that the valve maintains a constant level in the separator instead of fully opening and closing with high and low levels. Instead of the standard proportional level controllers, a snap-acting valve system is used on older separators which "dump" the level from the high set level to the low set level. A comparison graph illustrates this in Figure 6.21. This means that the level in the separator is allowed to increase to a set high level. Once the level has been reached the valve is then fully opened until the level reaches a set low level, when the valve then closes and the cycle restarts.

This applies to both oil and water controllers.

The proportional system is the far more accurate system and is more compatible with measurement readings in general and data acquisition systems. However, in unusual circumstances it is not always easy to achieve the desired proportional configuration.

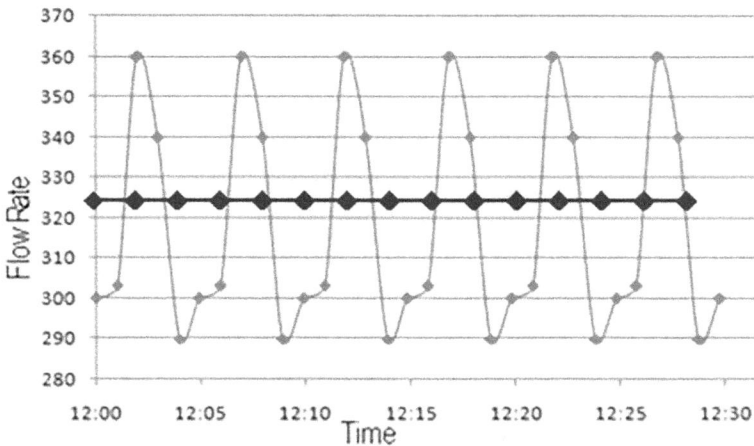

FIGURE 6.21   Fluid Output.

Thin line ◆ – Snap Acting Controller
Thick line ◆ – Proportional Controller

6.5.6   Vortex Breakers

When high flows of liquid are exiting the separator body at both oil and water lines, a vortex can occur at the bottom of the separator. This is similar to the effect when the plug is pulled from a washbasin; a cone shaped whirlpool can be formed.

The vortex can then drag the gas down to the liquid outlets, which can seriously affect the operation of the separator and the measurements. Therefore, vortex breakers are installed immediately above the fluid outlet holes in the separator to prevent this action.

6.5.7   WAVE BREAKERS AND SURGE PLATES

Where a separator is used on a floating rig or platform, the action of the sea waves on the rig can cause artificial waves within the body of the separator which can result in separator instability and carry over into the separator outlets.

At first look at Figure 6.22, the wave breakers and surge plates seem similar. But, looking at the same separator with a top view and cross sections, Figure 6.23 and Figure 6.24 respectively, the construction becomes clear.

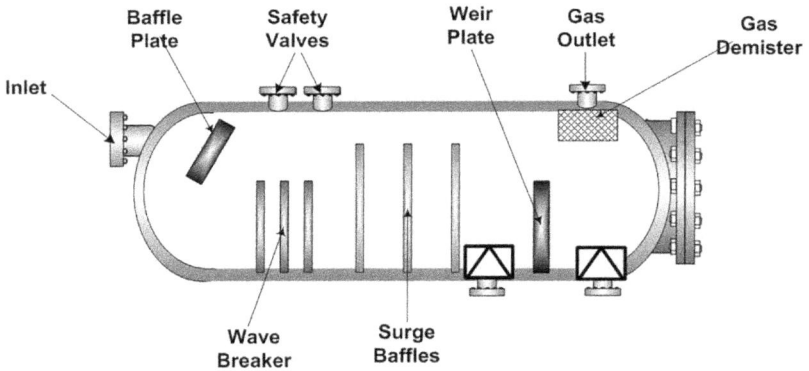

FIGURE 6.22   Wave Breakers and Surge Baffles Seen from the Side.

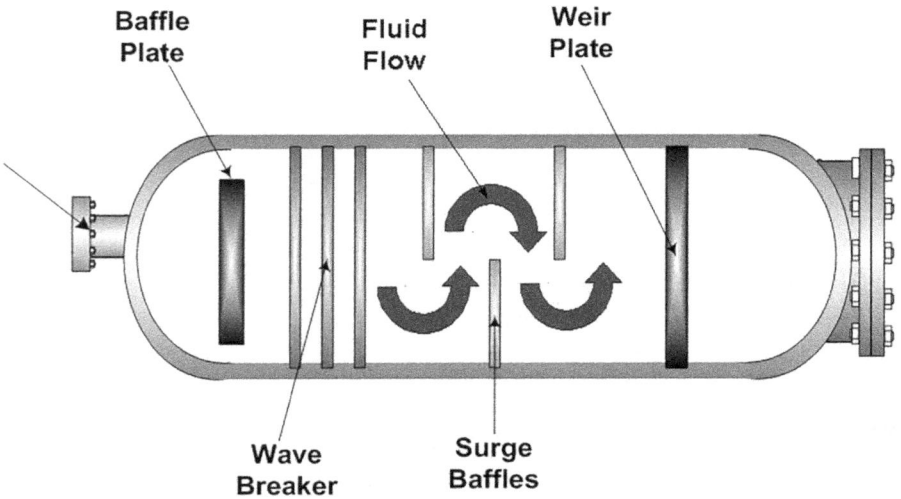

FIGURE 6.23   Wave Breakers and Surge Baffles Seen from Above.

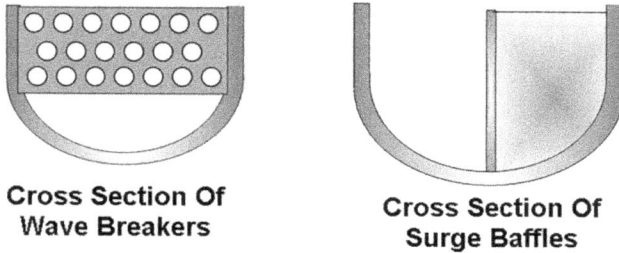

**Cross Section Of
Wave Breakers**

**Cross Section Of
Surge Baffles**

FIGURE 6.24 Separator Body Cross-Sections.

The wave breaker works by permitting the fluid to move freely in the bottom section of the separator but, by using plates with holes in them, they restrict the flow in the upper section. This limits the kinetic energy of the fluid flowing through the holes and therefore "breaks the wave," resulting in a constant liquid level.

The surge baffles work by directing the flow around the baffles, thus reducing the kinetic energy in the flow for a more stable liquid level.

6.5.8 SEPARATOR GAS LINE

The separator gas pressure is regulated by a valve which maintains the pressure in the separator by a tapping from the gas line into a Bourdon tube which in turn regulates the amount of pressure applied to the control valve.

Unlike the liquid flowline control valves, the control valves on the separator gas flowline are normally OPEN valves. With pneumatic pressure required to close the valve, lowering the pressure opens the valve, and increasing the applied pneumatic pressure closes it. Pneumatic pressure is applied at the top of the valve diaphragm, pushing it down to close. The applied pneumatic pressure causes the valve to close proportionately.

This is also for operational safety reasons; if the applied pressure reduces or is lost then the valve opens to reduce the separator pressure. The loss of pressure scenario is to ensure no dangerous pressure build-up in the separator if control pressure is lost.

6.5.9 SEPARATOR INSTRUMENT AIR

As detailed previously the well test package should have a dedicated air supply that cannot be disconnected, isolated, or used for other purposes while the test is in progress.

It should be re-iterated that separators may often be set up to use the separated gas, through suitable regulators etc., in place of instrument air.

This is a dangerous practice and has been discontinued from all major service companies due to risk of fire and explosion.

The internals of separator controllers are not H_2S certified and will corrode if H_2S is present, which adds to why separator gas should not be used. An increased risk is the venting of gas as it is released to the working environment from the controllers.

This needs to be verified as some newer separators are still, unfortunately, manufactured with the gas instrument supply option.

6.5.10   SEPARATOR METERING

The separator gas metering is carried out before the control valves, which control pressure and flow rates.

6.5.10.1   Gas Line

The measurement of gas flow is normally carried out with an orifice plate meter, although it can be replaced with a Coriolis meter or even a gas turbine meter. What has to be considered is that the orifice meter has the advantage of having what is called rangeability whereas other meters such as a Coriolis meter have fixed upper and lower ranges.

The orifice system is considered rangeable because the measuring orifice plates can be selected/interchanged to match the gas flow rates measured without having to bleed off the separator pressure.

Daniel Orifice Box

There are several different models of orifice boxes, but the most common in well testing is the Daniel Senior orifice box which facilitates the changing of orifice plates under pressure.

The Daniel box is a dual chamber device.

  a. **Top chamber** – insertion and removal of the orifice plate and its carrier assembly
  b. **Bottom chamber** – measurement chamber with the orifice in the gas flow creating a pressure differential across the plate

The two chambers are isolated by an internal gate which can be opened and closed to allow the plate assembly to be inserted or removed.

The Daniel box, Figure 6.25, should be in good condition and regularly inspected for damage and corrosion especially on the gear wheels and internal gate.

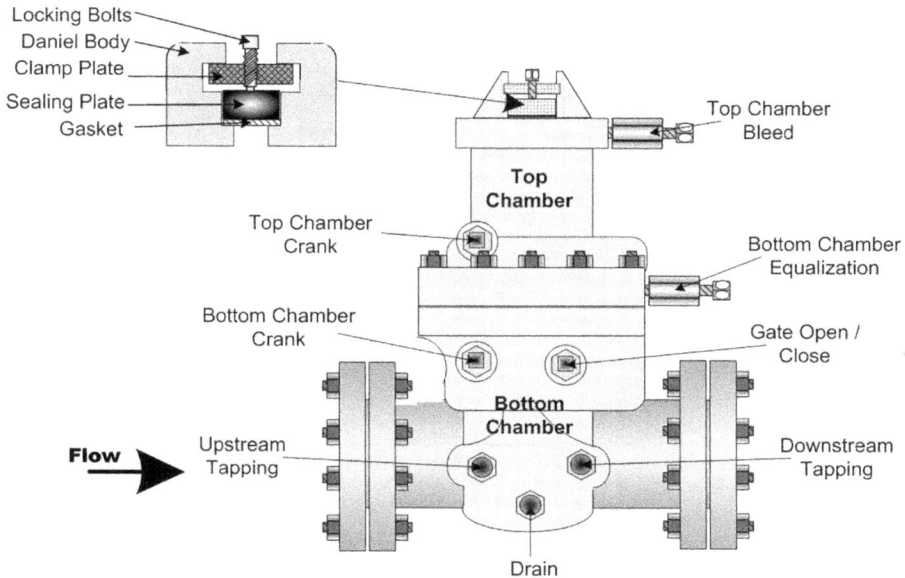

FIGURE 6.25   Daniel Senior Orifice Box Details.

Many Daniel boxes are used in high level H_2S wells, especially prevalent in production monitoring tests, while they are not fully H_2S material certified and will corrode after time which will result in:

   i. Difficulty lowering and raising the plate and holder
  ii. Internal gate holding pressure and sealing
 iii. Orifice plate not engaging or sticking in the Daniel box

If H_2S is present, especially in high concentrations, it can corrode the internals of the Daniel box as well as the moving parts. This is not often picked up during a pressure test. Therefore, it is critical that the Daniel boxes are disassembled during maintenance and any damage is corrected.

Corroded and damaged Daniel box gear wheels (Figure 6.26) can prevent the orifice plate carrier from being inserted and removed, posing a source of a potential hazard. These gear wheels must be checked during yearly inspections as a minimum and, if necessary, replaced.

FIGURE 6.26   Corroded Daniel Gear Wheel.

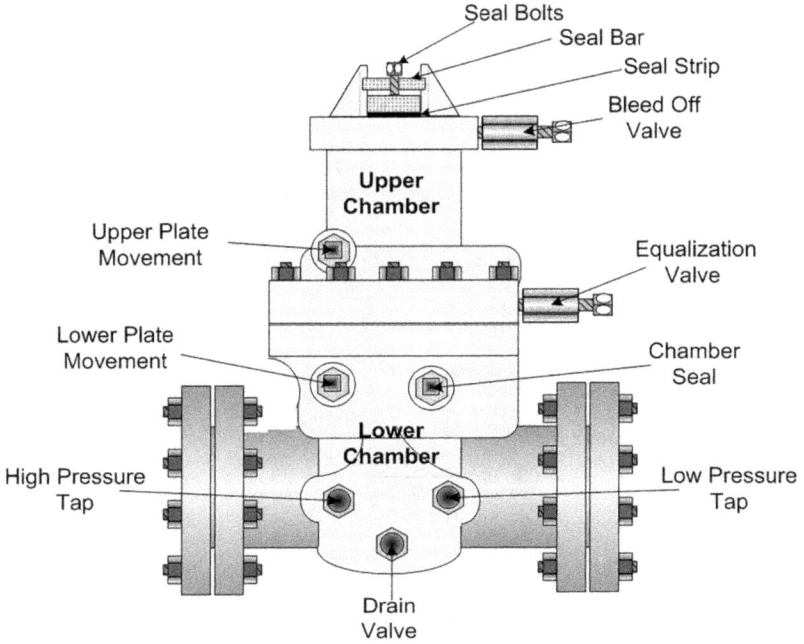

FIGURE 6.27   Daniel Senior Orifice Box.

The Daniel box and associated components should have a separate inspection record that is included in the yearly inspections.

There are two types of orifice meter in use (Figure 6.28):

  i. Flange tap
 ii. Pipe tap

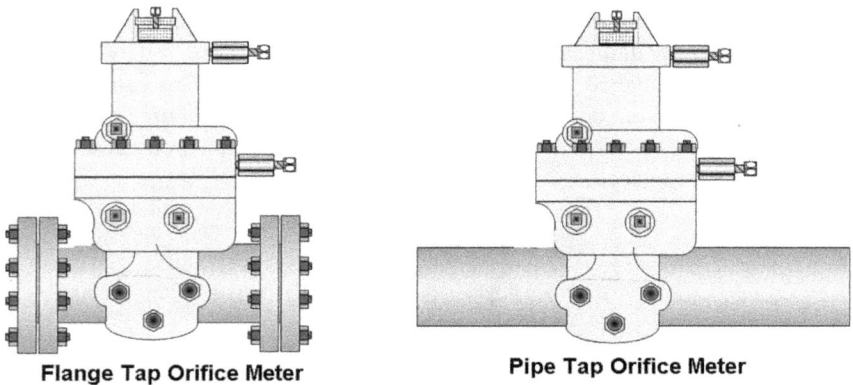

Flange Tap Orifice Meter   **Pipe Tap Orifice Meter**

FIGURE 6.28   Orifice Meter Types.

There is no difference in operation between the two models, however, the most common form in use in well testing is the flange tap option. There are other variants of the Daniel box, but in well testing the majority of them are the Daniel senior series.

A simple method of determination of the meter type is if there are flanges attached to the orifice meter then it is a Flange Tap model.

Although the operation of the orifice meters is the same, the flow calculations for each are different and it is critical that the correct version of calculation is used. This difference is important when a DAS system is used.

The gas flow through an orifice has to be laminar, which is when all the gas flow is in the same direction. The gas from the separator will exit at the top of the vessel and go through a series of bends in order to get to the gas line. The changes in flow direction in the gas line will contribute to the gas flow being turbulent.

As the gas line pipe runs short before the Daniel box on mobile separators, it is usually insufficient to achieve "natural" laminar flow because of length limitations upstream of the meter, specified by American Gas Association (AGA); it is therefore common practice to have a bundle of straitening tubes in the gas line upstream of the meter to condition and force the gas into laminar flow as shown in Figures 6.29 and 6.30.

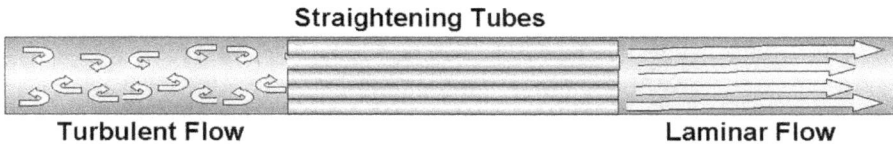

Straightening Tubes

Turbulent Flow   **Laminar Flow**

FIGURE 6.29   Effect of Straightening Tubes on Turbulent Flow.

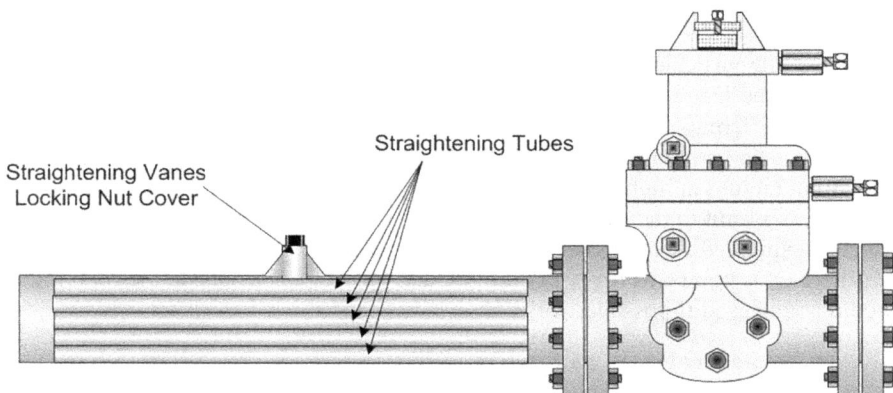

Straightening Tubes

Straightening Vanes
Locking Nut Cover

FIGURE 6.30   Gas Flow Straightening for Laminar Flow.

If a cross-section of the gas line up to the Daniel box is examined it will give a better image of how the straightening is carried out, as in Figure 6.31.

A critical part of the straightening tubes is the locking nut which must be in place otherwise the tube bundle can be blown across the orifice box and into the control valve.

The locking bolt is inside a pipe weldolet which is positioned approximately in the centre of the gas line. Many junior operators think this is a pressure tapping and put a pressure gauge in it. This should not register pressure.

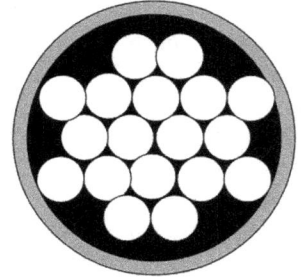

FIGURE 6.31   Cross-Section of Gas Line.

NOTE: *As a safety aspect none of the fittings on the locking nut assembly should be touched while there is pressure in the separator. In fact, these should never have to be undone unless replacing the straightening bundle which is a job carried out in the workshop.*

If the locking nut cover is removed and the inside viewed from above, the actual locking bolt will be shown.

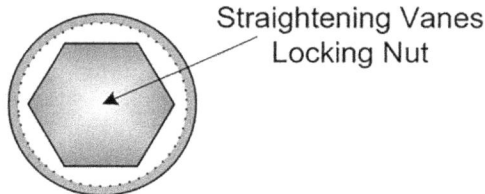

Straightening Vanes
Locking Nut

FIGURE 6.32   Straightening Bundle Locking Bolt.

If the bolt is not there then the separator is potentially in a dangerous state and this should be rectified immediately.

Details of the gas calculations are dealt with in a dedicated chapter. However, the purpose of the Daniel box is to house an orifice plate. The orifice plate is used to generate a differential pressure across it that is dependent upon the flow of the gas through it. The output from the Daniel box is a differential pressure. The Daniel box has output tapings at high pressure and low pressure respectively.

There is a limit to the size of the orifice plate that can be installed in the gas line, which is expressed as a ratio between the orifice plate size and the line bore, denoted as d/D. This is known as the beta ratio (ß), and orifice plates outside the ratio limits should not be used if possible.

AGA 3 states that:

For meters with flange taps ß (beta) shall be between 0.15 and 0.70.
For meters with pipe taps ß (beta) shall be between 0.20 and 0.67.

The majority of all well test separators use a 6-inch Daniel box, with the most common bore of 5.761" although this is not always the case. Gas calculation errors have been made because the well testers have assumed that the line bore is 5.761" and in fact it was 6.065". The line bore is normally detailed on a metal plate riveted to the top of the Daniel box (Figure 6.33). This should be checked as it is the only verifiable source without disconnecting the Daniel box. The plate should be completely legible and must be a documented part of the annual equipment inspection.

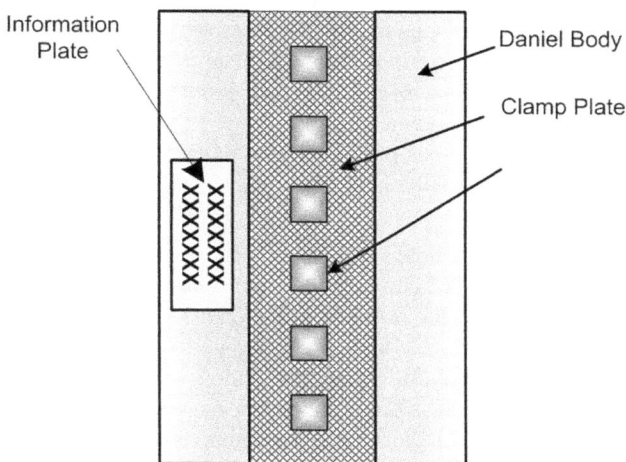

FIGURE 6.33 Daniel Box, Top View, Information Plate Location.

Most data acquisition software uses a configuration file to record the fixed variables for the separator being used. The software often resides with the data acquisition engineers and as such the configuration files are rarely updated or assumed constant. This can be a source of error with the calculated flow rates. Therefore, it is essential that all separator fixed variables are verified before the well is flowed through the separator, otherwise calculation errors will result with all fluids.

Published line bores for the 6-inch Daniel box and their orifice plate limits according to beta errors are listed in Table 6.1. The table gives the maximum and minimum size orifice plate for the corresponding Daniel box line bore.

TABLE 6.1
Bore Line Size and Orifice Sizes

Minimum Orifice Size	Line Bore	Maximum Orifice Size
0.73	**4.8970**	3.43
0.78	**5.1870**	3.63
0.86	**5.7610**	4.03
0.91	**6.0650**	4.25

Source: American Gas Association (AGA) Orifice Metering of Natural Gas Report Number 3.

However, the ß ratio restrictions are only normally applied to fiscal metering options.

There are other factors that cause higher levels of error.

Opening of Daniel Box General Rules:

  i. Full PPE must be used at all times
 ii. Untrained or junior operators should not operate the Daniel box unsupervised
iii. An operator must never lean over the orifice box when removing or installing plates as the pressure could cause the plate to be ejected at speed
 iv. Regular lubrication and sealing compound should take place using the correct compounds and lubricants
  v. When opening a Daniel box the probability of gas release/venting is high so all HSE precautions must be in place

It is critical that no mobile phones are in the vicinity as a mobile phone has the potential to cause a spark which can lead to an explosion. This should be enforced as a disciplinary offense even if the phone is switched off.

The separator would normally have been shipped with the orifice plate holder inside the Daniel box this should be verified as soon as possible when the equipment is on location as without the plate holder no gas measurement is possible. The Orifice Plate Seal and range of Orifice plates should also be verified.

The Daniel box should always be left in a safe condition with all the bleed off valves closed and the top seal plate in a closed state with the gasket strip in place. Even if the pressure gauge on the gas line shows zero pressure, it is a mistake to assume that there is no pressure trapped inside the Daniel box. It is possible for pressure to be trapped and with it H_2S therefore all safety and PPE precautions should be taken before opening, eye protection especially.

Operations with the Daniel box related to orifice plates should be done using the Daniel box spanner depicted in Figure 6.34.

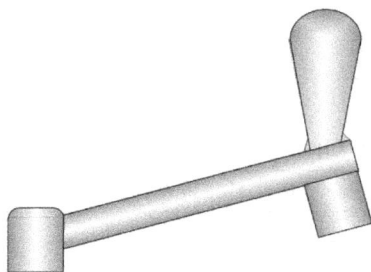

FIGURE 6.34   Daniel Box Spanner.

Steps Determining Plate Location

i. The Daniel box is fully closed and sealed
ii. Ensure that the bleed off on the top chamber is closed
iii. Using the Daniel box spanner, put it on the bottom chamber crank and slowly turn the handle clockwise to lower the plate. Do not use force
   • If the handle rotates without any resistance, then the plate is not in the bottom chamber
iv. Put the spanner on the top chamber crank and slowly turn the handle anti-clockwise to raise the plate
   • If the handle rotates without any resistance, then the plate is not in the top chamber

Removal of Plate from Bottom Chamber

All actions and problems must be recorded on the manual readings sheet using the Daniel box spanner and appropriate PPE:

a. Ensure that the top seal is in place and tight
b. Ensure that the top chamber bleed valve is closed
c. Slowly open the equalisation valve to equalise the pressures in the top and bottom chambers
d. When the pressure is equalised, slowly open the gate valve so the plate can be lifted
   • This should be done slowly in cases of higher pressure still in the lower chamber, which can force the plate carrier to lift
e. Only when the gate is fully open, place the spanner on the bottom crank and slowly rotate it anti-clockwise to lift the plate. Again, without using force
f. The action of lifting the gate will suddenly cease as the gear wheel disengages
g. The top crank should start to move, indicating the plate has entered the top chamber and beginning to mesh in the top chamber gearwheel/crank
h. Transfer the spanner to the top crank and slowly continue to lift the plate assembly to the top of the upper chamber
i. Transfer the spanner to the gate valve and fully close the gate
   • On some of the newer orifice boxes, the gate indicator plate, indicated in Figure 6.35, is still in place. These tend to get lost with time
   • The use of the gate indicator is a good safety option as the position of the gate valve can be determined without any mechanical operation. This can be considered a safety option that is easily implemented/manufactured
j. Close the equalisation valve before proceeding
k. Transfer the spanner to the top bleed off and slowly bleed off pressure to zero

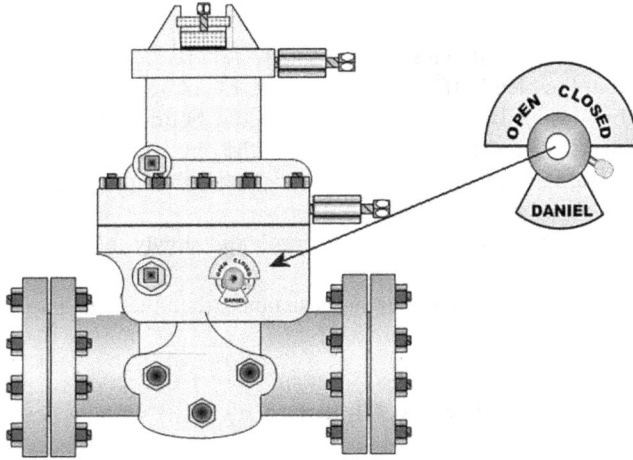

FIGURE 6.35   Daniel Box with Gate Indicator.

l. When the pressure is fully released, using the spanner, loosen the locking bolts located at the top of the Daniel box, shown in Figure 6.36

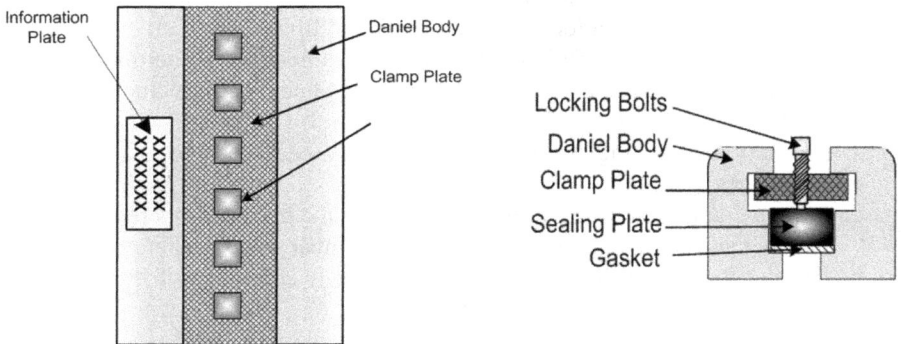

FIGURE 6.36   Daniel Box Top Seal Assembly.

m. Slowly loosen the first bolt slightly and progress to the next in the series/row

All bolts should be present

- If while loosening there is an escape of gas then the gate is either not fully closed or is leaking
- Check the gate for full closure. Sealant can be pumped in to assist seal

- Only if the gate is secure and there are no leaks should the remaining bolts be loosened
- If the sealing plate is forced up towards the Daniel body it indicates trapped pressure in the chamber and this should be released using the bleed off
- There is a potential risk of the sealing plate or trapped gas blowing out of the Daniel box so ensure that no personnel are leaning over the Daniel box

n. Release the screws so the clamp plate can slide out
  - Under no circumstances should the locking bolts be removed
  - If it is difficult to slide out the seal bar, it is an indication that there is still pressure in the top chamber
  - Be aware that when the clamp plate is removed there is the potential of the sealing plate and also the orifice plate holder to be blown out

No leaning over the Daniel box

o. With the clamp plate removed, the sealing plate and gasket can be removed
p. To release the seal, the orifice plate holder is lifted using the spanner. Place the spanner on the top crank and slowly rotate clockwise
q. Remove the sealing plate and gasket and place in a clean area with the clamp plate
r. Finish lifting the orifice plate holder from the Daniel box
s. Inspect and clean the orifice plate.
t. Verify the plate removed is the one specified in the report/manual readings sheet

Draining the Orifice Box

While the orifice plate holder is not in the orifice box it is recommended that the gas line and orifice box are drained of all liquids. This also applies to the sensor lines connected to the Barton recorder and DAS sensors.

Selection of Orifice Plate Size

Before the introduction of well testing data acquisition systems (DAS), the size of orifice plate used to be set to read mid-range of the Barton differential recorder. This was due to the non-linearity response in the lower (< 25%) and upper (> 75%) ranges of the Barton chart recorder.

DAS differential sensors are useable throughout the full range.

However, the Barton chart recorder is often still used for its mechanical recording charts, for reports, and for manual readings, so the orifice size would ideally still need to be within the linear mid-range of the Barton recorder. This ensures both the electronic and mechanical recording systems are functional and verification checks can be carried out.

If the flow rate is known – or a good estimate – the approximate orifice plate size can be estimated from:

Orifice Size Estimation

$$Q \approx \sqrt{\left(H_w \times P_f \right)} \times F_b$$

Where
   Q = estimated gas glow rate in MMSCF/D
   H_w = gas differential in inches water gauge (50% of Barton diff cell)
   P_f = gas static pressure in psia (gas line pressure)
   F_b = orifice plate factor
   ≈ almost equal to

The equation then becomes

$$F_b \approx Q \div \sqrt{\left(H_w \times P_f \right)}$$

Once F_b has been determined then the corresponding orifice size can be looked up in the AGA calculations book. Alternatively, there are proprietary charts that can be used for this calculation.

The graph in Figure 6.37 can be used to estimate the size of orifice plate from the given equation. Table 6.2 data is based on a 6" Daniel orifice box with a 5.761" inch Line Bore with Flange Taps. As an estimation the graph can be used for all 6" bore orifice meters.

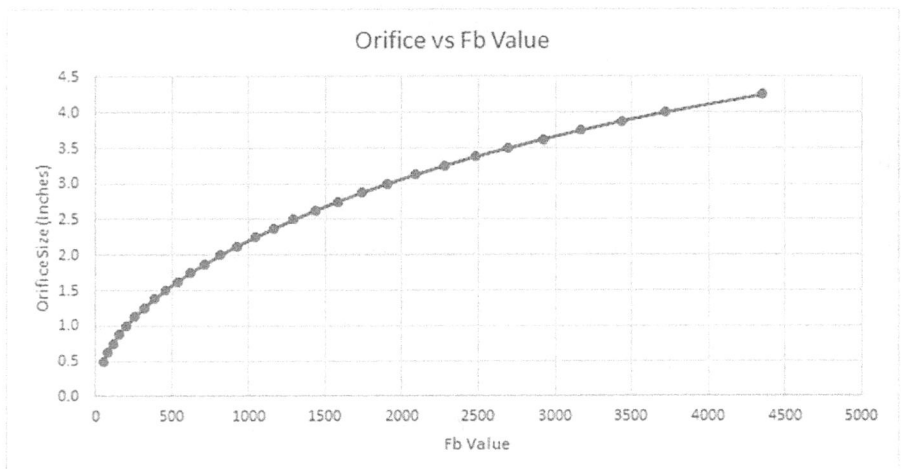

FIGURE 6.37   Orifice Plate and Corresponding Fb Number for Flange Taps For 5.761" Line Bore.

TABLE 6.2
Sample of Fb Results for Different Line Bores

Orifice Size	Line Bore Size			
	4.897"	5.189"	5.761"	6.065"
	F_b Results			
1.000	201.34	201.19	200.96	200.85
2.000	823.99	820.68	816.13	814.41
3.000	1986.60	1952.40	1907.80	1891.90
3.625	3180.80	3065.30	2925.70	2876.00

Source: American Gas Association (AGA) Orifice Metering of Natural Gas Report Number 3 Basic Orifice Factors Fb – Flange Taps.

Insertion of Plate

All sequence of events, actions, and problems to be recorded on the manual readings sheet. The correct PPE must be used with emphasis on eye protection, and breathing apparatus is essential if any H_2S is present. Using the Daniel box spanner:

i. Preselect the required plate, clean and inspect it for any defects
ii. Install it in the relevant orifice plate seal and carrier, illustrated in Figure 6.38

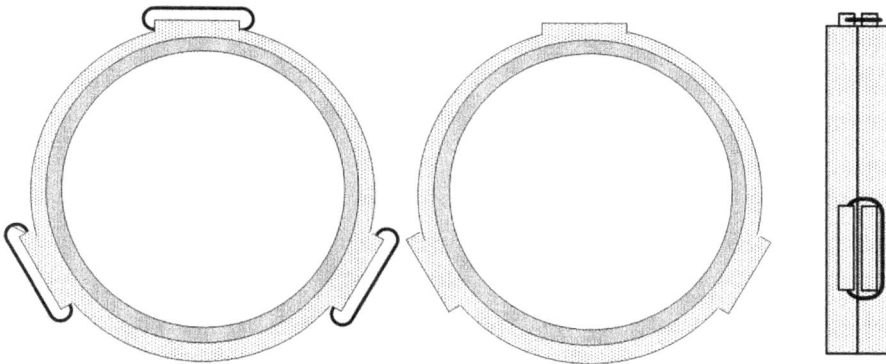

FIGURE 6.38 Orifice Plate Seal Assembly.

The orifice plate is installed between the two halves of the seal assembly, mounting checked and verified, the two halves clipped together, and assembly verified before inserting into the orifice plate holder
iii. The choice of size of orifice plate can be critical; if too small for the gas flow it can be blown down the gas line into the control valve causing a hazardous condition. Always err on the large size of orifice plate as if it is too large the plate can easily be changed for a smaller one

iv. Have the selected plate size, orientation, and mounting witnessed by another well test member crew and documented on the manual readings sheet. Mandatory!
  • Ensure the bevel is in the direction of the flow
v. Verify the Daniel box is in a safe condition, equalisation valve closed, and bleed off valve open
vi. Mount the orifice assembly in the Daniel box and engage/mesh the drive teeth using the spanner on the top chamber crank, as in Figure 6.39

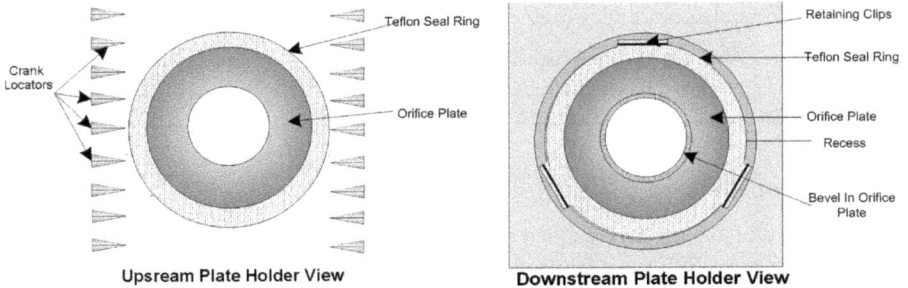

Upsream Plate Holder View   **Downstream Plate Holder View**

FIGURE 6.39   Orifice Plate, Orifice Seal, and Holder Side View.

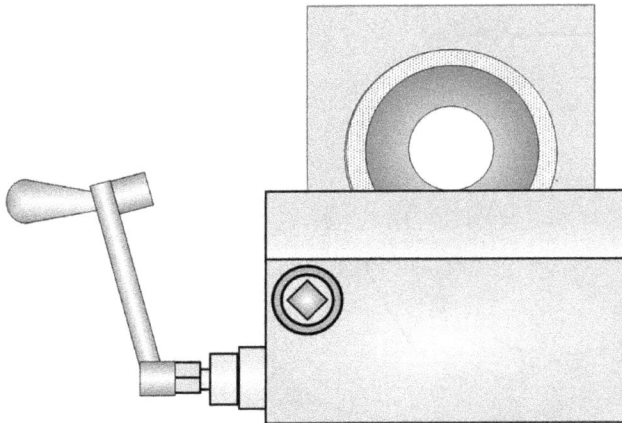

FIGURE 6.40   Orifice Plate Holder in Daniel Box.

vii. Slowly lower the orifice assembly into the top chamber
  • Make sure that the assembly enters the Daniel box square not on an angle
  • Verify smooth operation of the crank system
viii. Slowly lower the orifice assembly till it is just below the level of the Daniel box

ix. Place the basket to cover the opening

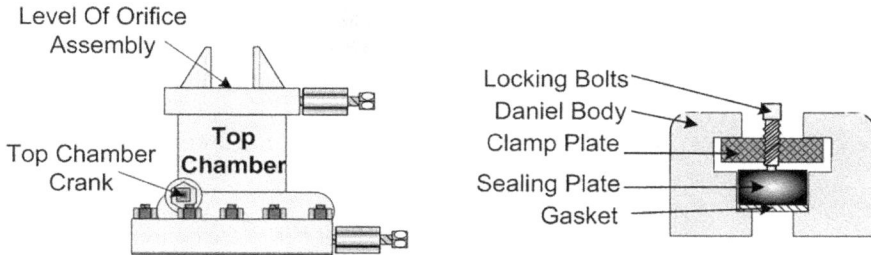

FIGURE 6.41   Top of Daniel Box.

- In some cases, the well test crew will not use the gasket as they consider it unnecessary. Without it if the separator is running at a low pressure it may function normally, but at higher operating pressures the lack of a gasket can cause a leak. This is a safety concern
- If the manual is checked the gasket is an integral part of the design and should be present. It becomes a safety issue as the gasket must be present
- The gasket can be replaced by a handmade one using suitable gasket material if missing

x. Put the sealing plate on top of the gasket with the indentations/marks made by the locking bolts upwards, Figure 6.42. It is good practice so that a smooth face is always presented to the gasket

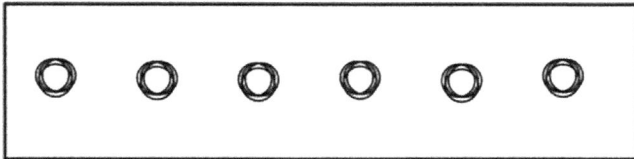

FIGURE 6.42   Seal Plate Top Face with Bolt Indentations.

xi. Slide in the clamp plate
xii. Tighten the bolts in sequence using the orifice box spanner as if tightening a flange
- After initial tightening double check the bolts
xiii. Close the top chamber bleed off valve
xiv. With the orifice plate box sealed and verified, slowly open the equalisation valve
- Check for any leaks. If found, close equalisation valve and bleed off pressure
- Check for source of leak and start again
- Use of supplied sealant may be necessary to rectify leak
xv. Only when the two chambers have fully equalised, slowly open the gate valve

xiv. When ready, do not rush; slowly lower the plate into the lower chamber and the gas flow
  - This is a potentially hazardous operation as the orifice plate can be blown down the gas line and into the gas back pressure valve
  - This can result from wrongly inserted orifice or orifice plate too small for the gas flow
xvii. Ensure the orifice plate holder is in the bottom of the lower chamber and begin to close the gate
xviii. All plate operations should be smooth
  - If the gate meets a resistance, then check that it is in the bottom and retry
  - Open the gate and move the orifice plate holder up into the top chamber, and slowly run the plate assembly straight down into the bottom chamber
  - Failure to seat the orifice plate holder and close the gate can result in a Daniel box failure and remedial action needs to take place
xix. With the Orifice holder firmly seated in the bottom chamber, close the orifice gate
xx. Close the equalisation valve
xxi. Observe the resultant differential gas reading on the Barton and DAS systems and alter the size of the orifice if and as required

Any leaks on any of the Daniel orifice valves should have the separator recorded and marked for maintenance.

With the plate correctly positioned in the gas stream, the gas rate then is calculated using the following:

Parameter Measurement

  - Average gas line pressure (psig)
  - Average gas line temperature (degrees F)
  - Average differential pressure across the orifice plate ("wg (Inches of Water))
  - Gas gravity at time of measurement
  - Gas contaminants (H_2S, CO_2, N_2 etc.) at time of measurement

Physical Measurement

  - Gas line bore (inches)
  - Orifice plate size (inches)
  - Pressure tapping type

From the aforementioned physical measurement parameters, the line bore and pressure tappings are a fixed part of the separator, and the orifice plate size is selected to suit the gas flow.

The gas gravity and contaminants in the equation can also make considerable differences in the calculation. This is why there must be a consistent method of measuring all gas properties. Full details in the flow rate chapters.

Orifice Plate Errors

From Table 6.3 it becomes apparent that the care and inspection of the orifice plates is critical to the derived results. Liquid in the meter is also a cause of errors (~11%); this reinforces the need for orifice plate and lines to be bled down periodically, draining accumulated condensate and other liquids. Often the bleed off line at the bottom of the Daniel box is plugged, restricting accumulated liquid bleed down; this is a design violation to the purpose and all areas of the Daniel box should be accessible for bleed off.

https://www.emerson.com/documents/automation/product-data-sheet-orifice-plates-plate-sealing-units-en-43806.pdf

Before inserting an orifice plate the following should be carried out to avoid errors or operational issues:

   i. The plate size should be confirmed and witnessed
   ii. The plate should be cleaned
  iii. The plate should be checked for any warps
  iv. The direction of the plate should be determined

TABLE 6.3
Table of Orifice Plate Errors %

Conditions	Error	
Orifice plate installed backwards	17–30%	Low
Dirt upstream of plate	6–11%	Low
Dirt downstream of plate	2%	Low
Dirt both sides of plate	3%	High
Liquid in meter tube	11%	Low
Grease and dirt in tube	11%	Low
Dirt between pipe and face of plate	23%	Low
Dirt upstream of face of plate	4–27%	Low
Valve lubricant on upstream face	4%	Low
Valve lubricant on downstream face	3%	High
Rounded and scratched plate	11%	Low
Dull plate	0.5%	Low
Upstream edge bevelled	2–13.5%	Low
Upstream edge rounded	0–9.3%	Low
Warped plate	2–10%	High
Methanol injection upstream	6%	Low

Source: Orifice plate errors and their relative magnitude based on work by Batchelder (1985), Burgin (1971), Hoch (1983), and Schepers (1981) in *Oilfield Processing of Petroleum*. Oilfield Processing of Petroleum. Author: Francis S. Manning, Richard E. Thompson. Published by PennWell Books. ISBN 0-87814-342-2.

6.5.10.2 Oil and Water Lines

Normally, turbine or positive displacement meters are used to measure liquid flow rates, oil, and water flowlines. With advancements in the technology these can be replaced with a Coriolis meter.

The oil line normally has two outlet oil lines, a low rate and high rate, with differing pipe size, and consequently differing size meters are used.

A separate water meter line using similar instruments is considered as a separate measurement. This is normally standardised as a two-inch meter.

Figure 6.43 is a generic cross section of the turbine flow meters; designs and models vary but they all use the flow to turn the turbine blades or rotors. The speed of rotation is a function of the flow rate; the greater the velocity of the flow, the faster the rotation of the turbine.

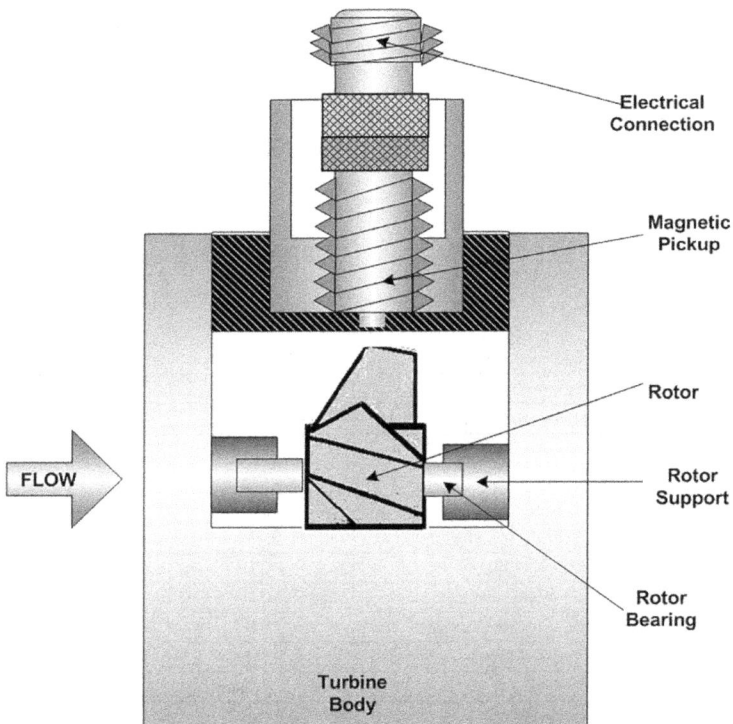

FIGURE 6.43 Generic Turbine Liquid Flow Meter.

However, the cross-section area of the pipe is a major function, so the flow rate is a function of the turbine speed and the area of the pipe. This is all combined into a factor that is known as the K factor which is specific to each individual flow meter size, and because the area of the pipe limits the flow in the pipe, each meter has a flow range.

Turbine Flow Meters encompass a fixed envelope or range of flow rates, with an upper and lower limit dependant on the size of the meter. If the flow through the meter is outside of this range, then significant flow errors can take place. Table 6.4 gives an example listing of the standard turbine flow meter sizes used on the separator fluid lines and their corresponding flow ranges.

TABLE 6.4

Generic Turbine Flow Meter Ranges Used on Separator

Meter Size Inches	Lower Limit Bbl/D	Upper Limit Bbl/D
1	170	1700
2	1300	13000
3	2750	27500
4	3400	41000

From Table 6.4 it is evident that the meters overlap in the lower limit flow range and it is possible to use different Flow Meters for the same flow rate.

As detailed earlier the meter factor will play a major part in calculating an accurate fluid flow rate. Meter factors can also be referred to as "meter slippage." Meter slippage occurs because the turbine blades are not an exact fit to the diameter of the pipework. They cannot ever be an exact fit otherwise it will impede the free rotation of the blades. Because there are gaps all the way round the turbine blades it gives a potential path for fluid to flow round the blades without causing rotation. This is compensated for in the meter factor measurement as long as the flow rate is within the limits set by the manufacturer.

All the flow meter details are in the manufacturers manuals which should be on the location for easy reference.

The output from these types of turbines is a magnetic pulse, created between the turbine blades and the magnetic pick-up as each blade passes the magnet. The turbine has multiple blades and one full rotation gives a set number of pulse counts.

NOTE: *Like all electrical devices used in a hazardous area the magnetic pickup must have a valid electrical safety certification if it is to be used on a separator.*

Knowing the internal size of the turbine, the manufacturers calculate a basic calibration referred to as the meter k-factor, corresponding to the number of full blade rotations per volume flowed through it, translating into pulses. This measured k-factor is in pulses per unit, which for oilfield purposes is either pulses per barrel or pulses per m^3. The operators will need to know this factor for every turbine as it is required to calculate flow rates for the digital readouts, SCADA, and data acquisition systems.

Differences between data monitoring results can often be attributed to errors in the pulse per unit being different.

NOTE: *Each turbine meter usually comes with a specified pulses per unit volume. This figure is normally based on the flow of water which unlike oil is not a compressible fluid.*

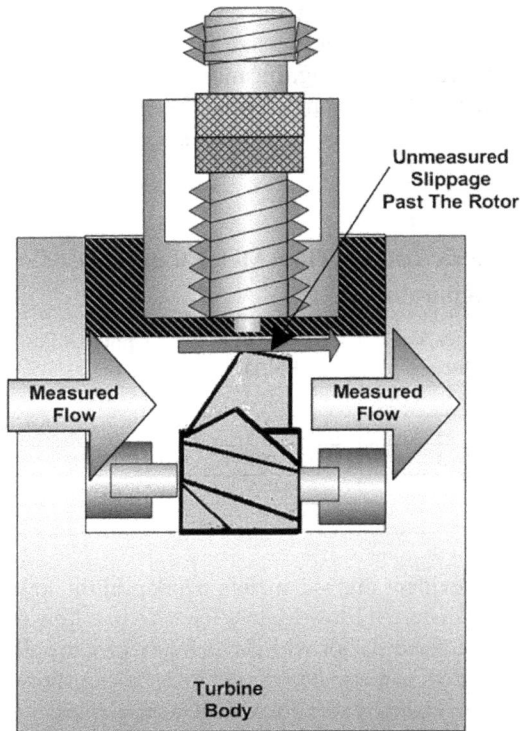

FIGURE 6.44   Examples of Flow Meter Slippage.

Meter Factor

The meter factor should be carried out using a pressurised surge tank. A pressurised surge tank can also be used to determine the shrinkage factor. The "true" meter factor is determined using the fluid from the separator at the pressure and as close to the temperature of the separator. This is not always possible if a surge tank is not included in the well test package and alternatively a non-pressurised tank – or stock tank – is used.

Without a pressurised tank, the liquid level will shrink when the entrained gas escapes from the solution. This has to be taken into account so the result of this measurement is known as a combined meter factor, a meter and shrinkage factor combined.

Meter factors supplied from the manufacturer in routine maintenance and meter checks are generated using water, and all they prove is meter functionality. In order for the meter factor to be valid for flow calculations it has to be carried out at the pressure, temperature, and with the same fluid flowing through the turbine so it takes into consideration the viscosity, gravities, and, if present, the amount of entrained gas, which vary with each flow rate.

Water is a non-compressible fluid, whereas the fluid from the separator comprises oil and entrained gas, both of which are compressible and do not behave like water.

6.6   WELL TEST MANUAL READINGS

The importance of manual readings has been stressed earlier. These are a direct comparison to the readings generated by the electronic data acquisition systems and it is critical that these readings and measurements are taken and recorded from independant transducers, rather than those used on the DAS system.

These readings *must not* be from the displays of the transducers used for the data acquisition/SCADA systems.

Seperator manual pressure readings should be taken from the barton chart recorder or pressure gauge on the gas line; a pressure gauge on the separator body is not the same reading! Temperature readings must be taken from the oil and gas lines. Oil temperature is not the same as gas temperature; they must be taken seperately.

Standard manual readings:

Description	Location	Instrument
WHP	Upstream of choke manifold (*)	Deadweight tester
WHT	Upstream of choke manifold (*)	Thermometer or dial
Downstream Pressure	Downstream of choke manifold	Dial gauge
Separator pressure	Separator gas line	Barton recorder
Separator temperature	Separator gas line	Barton recorder (if fitted)
		Thermometer
Separator differential	Orifice plate holder	Barton recorder
Oil temperature	Separator oil line	Thermometer or dial
Oil flow	Separator oil lines	Liquid flow/mass meter Digital display
Water flow	Separator water line	Liquid flow/mass meter Digital display

(*) Except in HPHT wells.

All these readings MUST be recorded, with their associated time, on the manual readings sheet with set intervals not greater than one hour. Failure to do so must be considered a disciplinary offence.

All manual readings MUST be checked against the DAS readings regularly. It is also recommended that the DAS operator verifies these readings at random times.

If there are any doubts regarding the electronic data acquisition systems then a manual readings calculation should be carried out. There are several options on these, mainly in Excel, but the version used should be a version approved by the operating and/or service companies.

With systems like Excel, well test operators can change the original spreadsheet or create their own. This has happened several times in the past resulting in errors in the calculations.

These software packages must remain unchanged.

It is recommended that the operating company representative inspects the manual readings sheets and the Barton charts when visiting the location or site. These results should be compared to the digital data. More often the company representative will just look at digital displays on the electronic meters and not verify them against DAS and dial gauges.

It is critical that random samples of manual readings be taken from the well testers to ensure that the correct procedures and methods are being employed. Failure to maintain and update a manual readings sheet should be considered a disciplinary offence, for both operators and the service company.

7 Hazards

7.1 HAZARDOUS GAS WELL TESTING

7.1.1 HYDROGEN SULPHIDE (H₂S)

All well testing jobs are considered hazardous; the term testing infers uncertainty, and with that risks are involved. The flowing nature of fluids from a reservoir and related well activities are various and pose threats with:

- Hazardous Gases
- Pressures
- Injected Chemicals
- Unconsolidated Formations

Even with known and older wells, with historical data, hazards are compounded by deterioration of completions, well integrity, and changes whether natural or intervened by operations.

With the growing global demand for hydrocarbons and advances in technology, oil and gas reserves that were once challenging are being explored and developed. With new territory comes higher risks and hazards.

Although hydrogen sulphide (H_2S) is the most common poisonous gas encountered in the industry, well testers and well testing equipment will be exposed to greater H_2S risks. Reservoirs can contain levels of H_2S as low as ten parts per million (ppm) to much higher percentages.

Equipment and Guidelines for H₂S

The American Petroleum Institute, more commonly referred to as API, sets standards and publishes guidelines for equipment and risks of substances in the oilfield, including H_2S. The following publishing is relevant to the H_2S operations.

- API Recommended Practice 49 (API RP 49) "Recommended Practice for Drilling and Well Servicing Operations Involving Hydrogen Sulfide"
- API 16C Specifications for Choke and Kill Systems
- API 6A Specification for Wellhead and Tree Equipment
- API Q1 Specification for Quality Programs for the Petroleum and Natural Gas Industry

The API papers are periodically updated and it is important that ALL operations involving H_2S (and other operations) follow these guidelines unless other regulatory body rules are applicable.

DOI: 10.1201/9781032623689-7

Despite the training, qualifications, and competence of staff, H_2S annually claims several lives, being an exceedingly deadly substance. Naturally existing in some reservoirs, H_2S can also be produced in the reservoir by foreign substances, such as injecting water or other fluids, as part of the improved and enhanced recovery processes. Associated microbes and other bacteria unintentionally introduced would produce H_2S. It can also be produced as a by-product of acid washing.

Therefore, it is a mistake to assume H_2S is not present now or is not in one part of the reservoir since it didn't exist then in any other part of the reservoir or wells. Individual wells should be taken on their own merit and may can contain H_2S. Its presence should always be checked for. Even in processing plants and refineries that handle raw hydrocarbons, H_2S can also be present.

Operating companies normally enforce an HSE Training Passport for the oil and gas field, including H_2S. The qualifications differ between territories and within oil and gas fields, and many well services companies run internal safety training systems for working with H_2S. When a job is at hand with known concentrations of H_2S then only a qualified crew may be selected for the task.

The information presented here is a compilation of more than 40 years of safety courses, lectures, seminars, and programs from service companies, specialist gas safety companies, HSE divisions, and oil companies.

However, regulations and rules vary from country to country and from one region to another, and individuals MUST get intimately familiar with the regulations at their place of work. Before beginning any operations, any concerns or questions must be raised with the supervisor or operating company safety personnel.

H_2S Facts

- H_2S is a colourless gas and odourless at very low concentrations
- Over 50% of people killed by H_2S exposure were in fact part of a rescue attempt to save others
- H_2S has a high specific gravity of 1.189 (air = 1.000) and consequently H_2S will accumulate in any low-lying areas such as well cellars, sumps, tanks, and ditches
- H_2S is soluble in both water and liquid hydrocarbons
- H_2S is corrosive to all electrochemical series metals
- H_2S is a skin irritant
- The main hazard with H_2S is that it can cause immediate damage to a person's health and even death at exposure. High concentrations of H_2S "numb" a person's ability to recognise the presence of the gas
- H_2S is also known as sulfane, sulfur hydride, sour gas, sulfurated hydrogen, hydrosulfuric acid, sewer gas, and stink damp
- H_2S is six times more lethal than carbon monoxide and half as lethal as hydrogen cyanide
- In concentrations of 1 ppm H_2S has a smell described as that of rotten eggs. However, as the concentration of H_2S increases so does its odour, until H_2S concentration reaches a 100 ppm and above; then anyone exposed to H_2S

will have a temporary paralysis of the olfactory nerves in the nose, leading to an odourless perception. In anyone exposed to high concentrations of H_2S, the ability to detect or smell H_2S is lost instantaneously

- At concentrations of 200 ppm and greater, anyone exposed to H_2S will sufferer irritation to the eyes, throat, and respiratory system
- H_2S will form an explosive mixture in air with concentrations of between 4.3 and 4.6 percent. It is a very flammable gas as it has a low ignition temperature of 232 Deg. C, (450 degrees F)
- H_2S burns with a blue flame, and burning H_2S results in a chemical reaction that produces sulphur dioxide (SO_2), another highly toxic gas. With a strong scent and very irritating to the eyes and lungs, SO_2 is only slightly less toxic than H_2S. Extreme caution must be exercised when setting burners for H_2S contaminated gas

7.1.1.1 Safety Limits

H_2S is released from crude oil when containing pressure is reduced, which is why we find the highest concentrations of H_2S at the choke manifold.

The effects of H_2S are slightly different from one person to another, based on their physical attributes, but as a set standard all the following are generally accepted H_2S exposure limits:

Concentration	Limits
10 ppm	This is a time weighted average (TWA).
	Average concentration of exposure must not exceed eight hours at a time.
	The short-term exposure limit (STEL).
15 ppm	Personnel must not be exposed to this concentration for periods longer than 15 minutes at a time.
100 ppm	This is the level of concentration which any exposure to any personnel will pose an extreme danger to life and health (IDLH).

Other exposure limits have been published as:

Concentration	Effect
0.03 ppm	Rotten eggs odour. Safe for eight hours exposure.
4 ppm	May cause eye irritation.
	Exposure may damage individual's metabolism, therefore breathing masks are compulsory.
10 ppm	Maximum exposure ten minutes. Kills sense of smell in 3–15 minutes. Causes eye irritation (*soreness, light sensitivity, seeing "rainbows" around bright lights, or a gritty pain with a spasm of the eye lids*) and throat injury.
	Can react with some mercury-based tooth fillings.
20 ppm	Exposure for more than one minute may result in damage to eye nerves.
30 ppm	Exposure will result in the loss of smell plus potential injury to the blood-brain barrier through olfactory nerves.

(Continued)

Concentration	Effect
100 ppm	Exposure will cause loss of consciousness within 15 minutes.
	Respiratory paralysis begins in 30–45 minutes and prompt artificial resuscitation would be necessary.
200 ppm	The eyes and throat are irritated and sting.
	Potential severe eye injuries and permanent damage to eye nerves.
300 ppm	Commencement of losing sense of reasoning and balance.
	Continued exposure leads to respiratory paralysis in 30–45 minutes
500 ppm	Suffocating conditions.
	Requires immediate evacuation and artificial resuscitation.
	If not evacuated will lead to unconsciousness within three to five minutes.
700 ppm	Exposure will result in immediate unconsciousness and probably permanent brain damage.
	Unless rescued immediately breathing will cease and death will be the result.

Threshold Limit Value (TLV) for H_2S is 25 ppm. Which infers a maximum exposure not exceeding a period of eight hours, without the use of any respirational equipment. The published lethal H_2S concentration in air for personnel:

- 600 ppm/30 minutes
- 800 ppm/5 minutes

7.1.1.2 Zoning

According to API RP 49 the following zones must clearly be marked on location to indicate likely occurrence of hazardous gases:

   i. No hazard area (accommodation and muster points)
  ii. Condition 1 area – low hazard < 10 ppm
 iii. Condition 2 area – medium hazard > 10ppm < 30 ppm
 iv. Condition 3 area – high hazard > 30 ppm

7.1.1.3 Personal Gas Monitors

Earlier on in the industry personal H_2S detection relied on reading the reaction on a lead acetate paper sensor or using reaction tube detection, both requiring individual involvement. With modern technology automatic sensing and warnings of H_2S is possible, aiding the prevention of H_2S harm. Personal gas monitors are small, compact, and explosion proof, making them easy to wear by personnel for H_2S detection.

Basic Features

- Available in hydrogen sulphide, carbon monoxide, and oxygen configurations
- Two-year operating life, no charging or calibration required

- Three-point alarm system including audible, visual, and tactile alarms
- Customisable LCD display with real-time gas readings and remaining operating life
- Intrinsically safe certification

Figure 7.1 is an example of a personal gas detector/monitor, with generic dimensions. They all feature an easy-to-use mounting clip; other versions are powered continuously and are disposable, while the better versions have a replaceable sensor shield. Regardless of the version the intent is the same – to save life.

Personnel need to remember H_2S is heavier than air, with a specific gravity of 1.189, when placing their sensors on themselves. Specialised courses have recommended wearing the personal detectors at waist level, bringing the sensor closer to the heavier gas. It is a malpractice to have detectors clipped to the hard hats with the intentions of keeping it cleaner from the hazards of the job, as that position defeats its purpose.

If these sensors are damaged or deemed dysfunctional it must be reported to the job supervisor and immediately replaced.

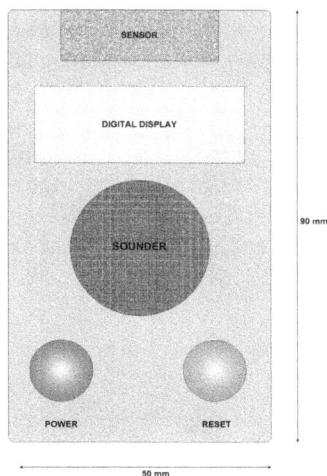

FIGURE 7.1 Example of Personal Gas Detector.

7.1.1.4 Permanent H_2S Sensors

A well test unit may have its permanently installed sensors throughout the location, at positions where gas release is probable, such as the choke manifold, the separator particularly close to the sampling points, and the orifice plate gas meter. These sensors would be integrated into an alarm system, audible and visual, relative to detected gas levels.

If no fixed H_2S detectors are installed on the separator then the supplied DAS system should be expanded to include H_2S detectors at points where gas can be released, choke manifold, orifice plate holder, sampling points, etc.

If possible, the DAS system should be linked to other fixed systems usually through a pair of open contacts that complies with safety standards.

On rigged locations or offshore platforms, gas sensors, and alarms would be part of the setup and, despite being for the same purpose, there would be technical challenges to integrate them into the well testing DAS. Extrinsic and intrinsic safety certifications must be considered carefully with any integrations between both systems. The existence of a gas detection and alarm system on a rig or platform, although an advantage, should rarely compensate for a dedicated well testing system that covers particular locations such as the orifice plate and sampling points.

7.1.1.5 Portable Detectors

Where permanent gas detectors are not installed, portable gas detectors can be used, with the capability of detecting and measuring multiple gas components. Similar to the personal detectors they are battery operated devices, often rechargeable, and require frequent calibration. Calibration certificates are usually issued at least annually for these devices, but company or regionals regulations may differ; manufacturer recommendations should also be adhered to.

The audible alarms from portable gas detectors are often limited, so they should be positioned to be heard over the ambient noise levels and seen.

Confined spaces within the unit, such as tanks or the cellar, would contain heavier gases. Portable gas detector inlet hoses are often lowered into these spaces to check for potential gas hazards. The detector should be switched on and stabilised at topside of any enclosed area or confined space to make sure the surroundings are safe before lowering, typically with a rubber type hose.

7.1.1.6 Wind Direction

Wind direction is critical in well tests in consideration of H_2S and other toxic gasses. Knowing the direction of the wind blows is a safety measure. In the event of an alarm, personnel, after donning a SCUBA set, muster into the direction the wind. Dedicated muster points would have been set up on site.

7.1.1.7 Breathing Apparatus

It is mandatory for personnel, in hazardous working environments, to be completely trained with the breathing apparatus. With several different manufacturers of breathing apparatus, personnel should be made familiar with apparatuses used at location, often with a familiarisation course when arriving on site. There are different types of breathing apparatus:

- Short-term work self-contained breathing pack, with five to ten minutes' capacity, often used as an escape pack
- Thirty-minute work self-contained breathing pack
- Cascade system breathing apparatus

In a cascaded system the breathing harness is connected by hose pipes to a reservoir of air cylinders. The harness will contain a short-term work pack cylinder, effectively acting as an escape pack if there is failure in the air supply, assisting personnel to escape. In any case the harness is worn when the users exit the dedicated muster point, as a precaution, even after a safe environment is declared.

Facial hair may be an issue with breathing apparatus masks. Some facial hair restrict the facemask seal, and a clean shave is recommended to ensure a good seal. The use of positive air pressure ensures no toxic gasses can seep into the facemask.

Under some operating companies in different regions, hoods and head covering alternatives are allowed instead of facemasks to accommodate facial hair.

7.1.1.8   Job Preparation with Toxic Gasses

If toxic gasses are suspected, all personnel must be wearing their breathing apparatus when the well is first opened, until gas tests have been conducted and the all clear is given. As soon as gas surfaces, a reaction tube test is conducted, while personal are wearing breathing apparatus with facemasks. Regardless of the presence or absence of toxic gases, it is mandatory to frequently perform these tests throughout the well test.

When no toxic gasses are detected at the opening of the well, the job may continue without the need for personnel to use breathing apparatus, but they should be made ready for use and located throughout the working site. However, all personnel must continue to wear gas monitors at all times.

NOTE: *In some areas the breathing apparatus are left unassembled and in sealed bags until they are needed. This strategy has the benefit of cost reduction, but it is a serious breach of standard HSE guidelines, leaving the crew exposed to the risks for the time it takes them to assemble and wear the apparatus.*

7.1.1.9   Working with H_2S

In any working environment a proper and fit-for-purpose PPE is mandatory; this is even more so as hazards are higher risk, such as with H_2S and the requirement of a breathing apparatus.

The escalated risks and hazards within these dangerous areas and environments calls for a watcher/observer situated outside the danger zones and keeping watch over the site and work. Referred to as "The Buddy System" in the industry, it is adopted in the case of H_2S as well as in other potentially dangerous circumstances.

The watcher will be equipped with an assembled and ready for use breathing apparatus, without the facemask or air on, until he is required to interfere in case of an emergency. Radio contact must be maintained between the watcher and the rest of the well test crew or other personnel involved in the operations.

No rescue attempts are allowed under any circumstances without full PPE and breathing apparatus on. Failure to do this will multiply casualties.

7.1.1.10   The Effects of H_2S on Equipment

H_2S chemically attacks steels and selection of equipment materials must be to the NACE standard MR0175, or MR-01–75. All materials and equipment subject to use and exposure to different levels of H_2S must comply with this standard. The National Association of Engineers, commonly referred to as NACE, is an international standards association that is considered the leading authority for corrosion control.

H_2S is highly corrosive to steel and can cause metal embrittlement, often referred to as hydrogen embrittlement, and it can happen in a very short time depending upon the grade of steel. The equipment and materials used and exposed to flow with potential H_2S contamination must satisfy the NACE specifications. Specifications and compliances should be declared on equipment specification sheets.

Despite evidence from material specification sheets, it is a mandatory requirement that all materials and equipment are fully inspected, including welds, on a regular basis, at least annually. The parts of the unit are tagged with unique serial numbers or identification codes to ensure they are tracked and identified as fit for purpose.

Other materials besides steel that are present on a well testing unit, such as elastomers used in the sealing of joints, are deteriorated by H_2S attacks and would need to be resilient. Hammer union seals are particularly important; they should be inspected and replaced to suit each job requirement if and as required.

Where Coflexip type high pressure hoses are intended for use, the inner liner should be selected to ensure it can withstand the percentages of H_2S potentially present in the flow, otherwise the hose can fail.

7.1.1.11   Gas Sampling with H_2S Content

Gas sampling is a task frequently performed on surface well testing, generally categorised for their objective:

- Content of H_2S
- Gas gravity

In both accounts the gas sampling is taken for the purpose of gas flow rates, which the content of H_2S is a variable of. H_2S content is normally taken with engineered chemical reaction tubes; the gas measurement section of the book covers the sampling and measuring process.

Sampling should involve:

i. An operator wearing breathing apparatus while sampling
ii. An observer, buddy system, also wearing breathing apparatus
iii. Both the technician and observer should be positioned upwind of the sampling point
iv. A double needle valve arrangement, contingency, should be standard but even more important with H_2S wells. The needle valve directly connected to the equipment should be left open and the second, outer, needle valve operated with. This needle arrangement and operating procedure is a safety measure to allow the changing of the second, outer, needle valve in case of leaks or failure
v. The sample should be initially measured first with a high range tube, 0–100% range, unless actual readings have been conducted earlier. With an initial reading indicating a content range for H_2S, a second reading is taken with a more specific scaled tube
vi. The gas should not be left to vent between samples

vii. In the case of taking a sample for the Ranarex Gravitometer the balloon – or vessel – should be filled

viii. The operator and observer should take the sample to the Ranarex to measure gas-specific gravity. Extreme care should also be taken while operating the Ranarex; the apparatus exhaust will vent the gas sample, so it must be vented to a safe place, outside the lab

Surface sampling for PVT analysis of the fluids is performed during most surface well tests. These are normally performed by specialised professionals, but nevertheless, the task involves risks and should be performed under the supervision of the well test supervisor.

7.1.2 Sulphur Dioxide (SO_2)

The hazards of H_2S are more common and operators are trained in the safety measures. Unfortunately, the same level of training is not given to the risks of sulphur dioxide (SO_2), which is not a normal constituent of well fluids but a by-product from burning of hydrocarbons entailing H_2S.

SO_2 Characteristics:

- SO_2 is a colourless gas
- SO_2 has a powerful, irritating odour, similar to the smell from a just-struck match
- SO_2 is a heavy gas with the specific gravity of 2.212 (Air = 1)
- SO_2 is not flammable or does not create an explosive mixture with air
- SO_2 can react and combine with water to produce sulphurous acid
- The safe limit for an eight-hour exposure to SO_2 is five ppm. This is known as the lower threshold limit value (TLV), a level to which it is believed a worker can be exposed without adverse effects

Sulphurous Acid

Well test burners utilise water which makes the sulphurous acid production probable where water chemically reacts with sulphur dioxide produced by the burner.

The concentrations of sulphurous acid are difficult to determine, which creates a potential health hazard to personnel. It is necessary that all personnel are aware of these potential hazards to work around the burners, even after flaring operations are complete, to account for residual sulphurous acid.

7.1.3 Carbon Dioxide (CO_2)

CO_2 Facts

- CO_2 has a high specific gravity of 1.5189 (Air = 1.000) and would naturally settle and accumulate in lower parts of the atmosphere such as well cellars, slumps, tanks, and ditches

- CO_2 is a colourless odourless gas
- CO_2 is not burnt during flaring operations
- CO_2 can leave a sharp taste in the mouth
- CO_2 is soluble with water to form carbonic acid, considered a weak acid

Usually with the hazards of H_2S overshadow the dangers of Carbon Dioxide (CO_2) that can be produced with well flow. It is possible that CO_2 can constitute a significant percentage of the reservoir's gas composition and therefore needs to be measured.

The real hazard to human health with CO_2 is induced oxygen deficiency. This is normally associated with risks involved at high concentrations of CO_2 but may occur in all concentrations of CO_2. Oxygen deficiency is thought to happen when a person inhales air with less than 20% oxygen.

Personnel with oxygen deficiency are recognised, early on, by exhibiting a lack of coordination; a person's lips may exhibit a "blue like" colour. If oxygen in air is reduced to 8–10%, personnel will become unconscious and a 6–8% oxygen concentration would lead to death in a matter of minutes.

Carbonic Acid

Carbonic acid can be formed in wells where both water and carbon dioxide exist, which is more probable with gas wells. Carbon dioxide (CO_2) dissolves in water (H_2O) to form carbonic acid H_2CO_3.

Carbonic acid is known for causing corrosion, rust, and pitting, affecting certain materials such as steels more than others. Indications of carbonic acid attacks would be part of the well testing equipment and materials check, although the generally used stainless steels resist corrosion caused by carbonic acid.

7.1.3.1   Precautions with CO_2

- Check for CO_2 in the produced fluids whether on newly explored wells or fields or developments
- Allow proper ventilation around the well testing equipment
- As is the case with H_2S, wind direction is important so wind direction indicators are always essential
- The corrosive properties and reactions with elastomer seals from CO_2 within the produced fluids and flow. CO_2 and carbonic acids can react with the elastomers used in pressure and fluid seals, such as a hammer union seal, therefore elastomers must be chosen that resist all the chemical compounds that they are exposed to

7.2   CORROSION

Throughout the oilfield, corrosion is a silent destroyer of equipment, causing unnoticeable damage to failure and can be responsible for costly economic and human losses. Corrosion is generally defined as the weakening or degradation of a material as a result of chemical or electromechanical reactions between metals and their environment.

The first recorded events of H_2S corrosion were in the 1890s in the USA.

The diverse operations of well testing and the use of the same equipment and pipework as a "one stop" solution see them facing operational environments ranging from no contaminants to very high levels of contaminants. For this purpose, the well testing equipment undergo necessary regular and periodic inspections, especially after highly corrosive or erosive operations.

Fluids flowed from a well are constituted of various substances, with different pressures and temperatures, that interact with the surface equipment used in well testing. The materials – and their specifications – used in the well testing equipment should suit the fluids – and often solids – flowed through them. Materials in this context not only refer to the metals, types, and specifications that equipment are fabricated from but any material that would directly interact with the flow. Ceramic and elastomer seals are an example; they must be installed to suit their purpose and capable of resisting functional degradation from reacting with flowing fluids.

7.2.1 H₂S

H_2S is a major contributor to corrosion and specifications from National Association of Corrosion Engineers (NACE) define materials as safe for use in different H_2S environments, concentrations, and purposes.

H_2S and H_2O can combine to form a weak (sulfuric) acid. H_2S concentrations of 5,000 ppm or less are generally considered safe for normal steel compositions. However, with higher H_2S concentration levels, the corrosive nature of the mix will "attack" the steel. A "black scale" on surfaces in contact with the fluid, such as the inside of the piping, are a sign of this.

The "black scale" is frequently regarded as a protective layer or shield that hinders further H_2S corrosion. Although this can be viewed as self-preservation of material, the high velocity fluid flows, turbulent flow, and existence of solid particles within the flow, such as sand, often chip away and erode this protective layer, resurfacing the base material to corrosive conditions of H_2S.

A change in flow direction can catalyse the erosion and corrosion of materials. Within well testing setups these are bends in the flow, and they should be regularly monitored with thickness testers. These bends of flow most often would show the first sign of deterioration, with either corrosion or erosion.

Elastomers can also react with H_2S. Proper choices of seals and O-rings must be considered to resist H_2S presence and concentrations. This is particularly critical in use with high-pressure multiple layer hoses (10, 15, and 20 kpsi etc.), as the inner layer is exposed to the well fluids and susceptible to failure. The damage induced internally is not visible until fluids start to leak.

Refer to the section on hazardous gas testing for operational aspects regarding H_2S.

7.2.2 CO_2

CO_2 corrosion, also referred to as "Sweet Corrosion," is one of the main contributors to corrosion. NACE address material specifications as safe for use with CO_2 environments, concentrations, and purposes.

Free water and CO_2 concentrations greater than 1,000 ppm in the well fluid can chemically combine to form carbonic acid (H_2CO_3). This produces sharp pits, appearing as sharp-edged holes, which apparently look like the Mesa Mountains and are therefore often referred to as "Mesa-corrosion."

CO_2 also attacks elastomers, so installed seals and O-ring materials would be chosen to resist attack by CO_2. Similarly an inner liner for high-pressure hoses is also critical.

The numerous "fizzy drinks" that are offered around the world use CO_2 because of its solubility in water, and the presence of a little carbonic acid can give the beverages the "kick" associated with fizzy drinks.

Refer to the section on hazardous gas testing for operational aspects regarding CO_2.

7.2.3 H_2S AND CO_2

Numerous wells contain both H_2S and CO_2 at various concentrations. This combination can be extremely hazardous and challenging to monitor. Selection of personal monitors must take into account the presence of both gasses.

With increased flow rates, higher pressures, and temperatures with induced flow turbulence through the well testing package, the risk of corrosion is relatively amplified. Monitoring becomes even more necessary as well as a heightened attention to equipment deterioration.

7.2.4 WATER (H_2O)

The most common corrosion is due to water, as a chemical reaction that converts metals into a more chemically stable oxide, commonly known as rust.

Many metals corrode merely from exposure to moist air. Generally, increased temperature catalyses the corrosion rate.

Within the enclosed system of the well testing package this is rarely an issue, with equipment internals, with the absence of oxygen from the produced fluids, until rigging down. Following a well test job when the equipment are disconnected and being rigged down for mobilisation, the internal parts of the equipment are exposed to the atmosphere. The presence of residual water in the equipment and oxygen from the air introduced allows for rust.

The speed of rust in such an environment is dependent on the weather conditions and congregation of water on equipment.

Offshore well testing equipment is frequently flushed and washed with sea water. Salty sea water corrodes metal up to five times faster than fresh water. To extend the life of the equipment it is best to flush them with fresh water, including a rust inhibitor to help with safeguarding equipment against rust.

Beside the apparent internal metal surfaces and even outer surfaces, threads and threadlets are particularly vulnerable to corrosion and often unnoticed. They are left open to the atmosphere, and their thread designs make it easier to trap water.

In 2001 an incident occurred which was recorded and referenced to emphasise the dangers of corrosion, more precisely with threads. A well testing choke manifold's 3/4" threadlet was corroded, and to compensate for this mechanical change, the installed thermowell threads were wrapped in excess Teflon (PTFE) over the recommended wraps to secure the fitting into the corroded threads. This setup failed under pressure, releasing the thermowell from its position and propelling it through the air at a velocity close to 100 mph, resulting in serious injury. The report is published on:

https://www.pdo.co.om/hseforcontractors/LibraryDocuments/200609020
    83254.pdf

Incidents such as this emphasise the importance of properly checking each piece of equipment and tapings for damage. Assuming one part of the equipment reflects the condition of the entire unit is a dangerous strategy. Preventive maintenance plays a huge role in preserving equipment as well as saving lives. A simple task of applying grease to threads and similarly exposed surface reduces the risk of corrosion by prevention. The practice of Denso-Tape application on larger thread and hammer unions has been in slow decline for several years.

Denso Tape™ is a proprietary form of a cold applied anti-corrosion tape using a petroleum compound to prevent corrosion when stored. It also has the benefit of reducing physical damage to the areas wrapped.

7.2.5 Monitoring

The process of monitoring equipment corrosion is similar to that used in the erosion process as described and relies on wall thickness measurements to detect deterioration. Corrosion develops slower than erosion, and due to this nature, the detection process is generally carried out using manual probes at a lower time interval – every three to six hours as standard – and a graph of measured thicknesses plotted to identify any areas of concern.

Despite the ability to use technology here, it does not compensate for visual inspection of equipment. The exposed areas where ruts and corrosion may occur with the equipment should be thoroughly checked by personnel.

Employing a certified third-party inspection company with ultrasonic thickness testers and expertise in the field would prove beneficial as part of asset integrity management, which helps prolong the equipment lifecycle and saves lives with early detection of failure. It is becoming even more common for operating companies to request such regular certifications.

7.3   EROSION

Where any fluid flows then there is the potential of erosion. This can be from the flow of liquids alone but increases with any solids or particles in the flow and is accelerated as the flow velocity increases.

High-pressure and high-volume gas flows are particularly susceptible to this type of pipework damage.

Pipework should always be rigged up in straight lines, where possible, to minimise flow direction changes, not only to reduce potential back pressure issues but also to minimise potential erosion at the point of direction change.

Normally this is an elbow but it can occur at the choke manifold. Lead-lined "T Pieces" can be used to reduce erosion effects but straight lines are always the better option.

7.3.1   THERMOWELLS

The use of thermowells where high velocity erosion is expected should be discouraged. The thermowell protrudes into the flow stream and becomes exposed to the erosive effect of the flow. As the thermowell is usually softer steel, stainless, more prone to erosion, it can fail easily.

Often thermowells do not have a certificate of compliance or pressure rating which should exclude them from well test operations.

Most electronic temperature sensor probes are spring loaded with the temperature sensor, a thermistor, mounted on the end so a simple clamp can be made to mount a temperature sensor on the outside of the pipework as shown in Figure 7.2 and Figure 7.3.

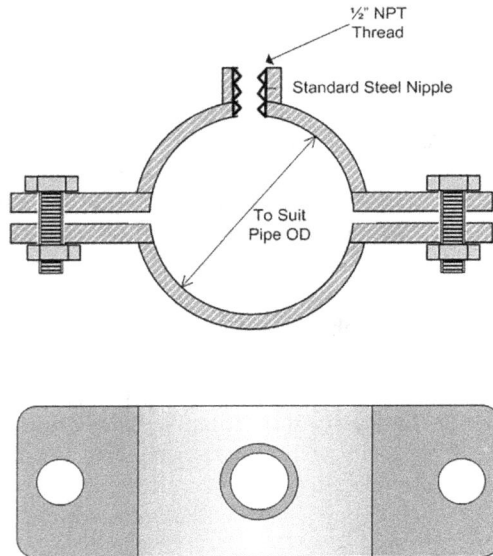

FIGURE 7.2   Temperature Clamp Details.

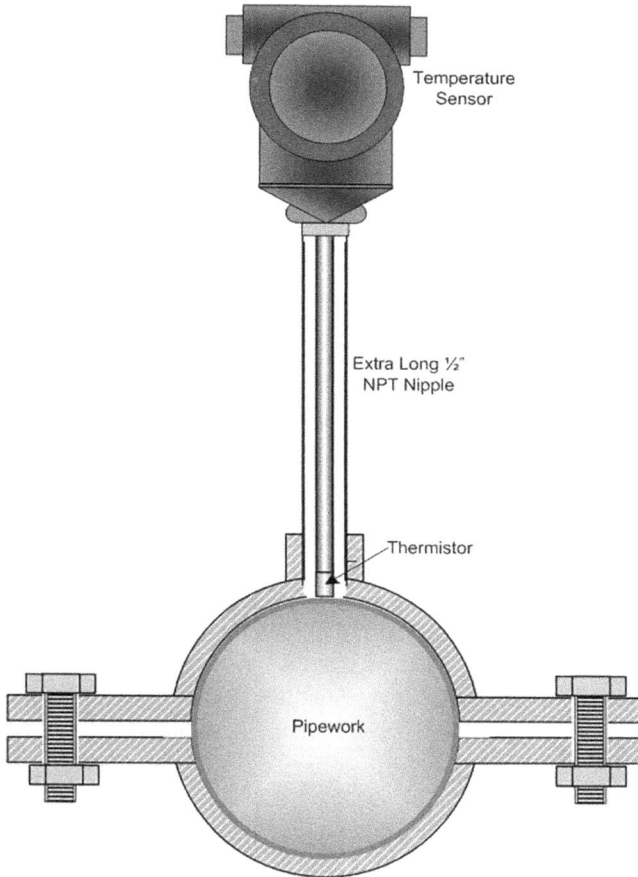

FIGURE 7.3   Temperature Sensor Mounted on Pipe.

The extra long nipple has to be selected – or made – to suit the length of the temperature probe. The nipple here is not pressure bearing so a 1/2" steel electrical conduit can be used.

The spring mounted inside the temperature sensor body keeps the thermistor in contact with the pipework. It is recommended than any paint or coating is removed where the thermistor makes contact with the pipework.

7.3.2 PARTICLE DETECTION

Because it is not easy to see particles or sand in the flow using standard BS&W sampling, detection probes are used.

There are two different types of sand detection probes standardly used in well testing although there are others available in general:

i. Intrusive detection
ii. Non-intrusive detection

Sand detectors rely on actual impacts of the particles with the detector probe or pipe wall in order to measure sand/particle volume in the flow.

What has to be taken into account is that, because the sand detector does not effectively cover the full flow area, only a percentage of the solids will hit the sensor and the rest of the particles will bypass the sensor. These bypassed particles are therefore not detected.

Recognising the cross-sectional area of the pipe versus the detector it becomes apparent that the majority of potential particles will bypass the actual sensor. Consequently, any results from any sensor must take into account the "missing" impacts.

Increasing the area of the detector will improve the issue but then restrict the potential flow rates as it blocks the flowlines causing increased back pressures.

7.3.2.1 Passive Intrusive Sand Detectors

Intrusive sand detectors, as the name implies, physically intrude into the well flow as shown in Figures 7.4 and Figure 7.5.

This type of sand detector, often referred to as a coupon, is a softer metal sealed tube that isolates the flowline pressure from the dial gauge or indicator. The absence of line pressure on the gauge indicates that the tube is intact.

Different tubes are used according to the line pressure and expected erosive particles. All and any fittings must be capable of withstanding the full working pressure and pressure testing.

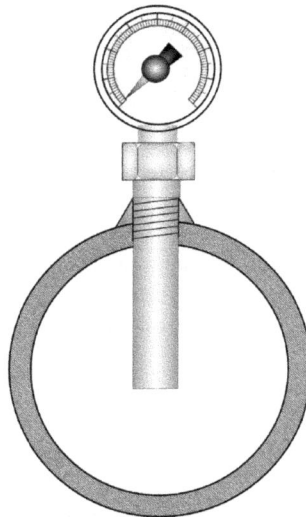

FIGURE 7.4   Cross-Section across Pipe with an Intrusive Detector.

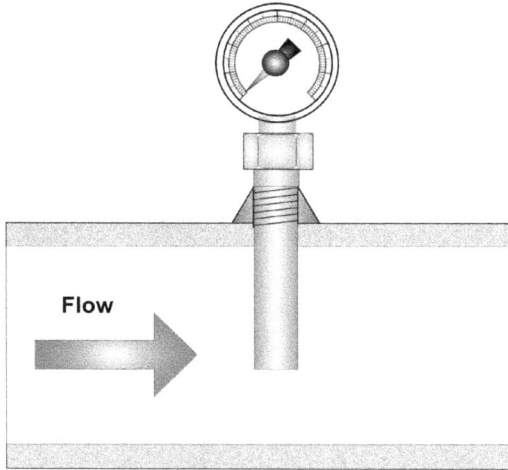

FIGURE 7.5   Basic Intrusive Sand Detector.

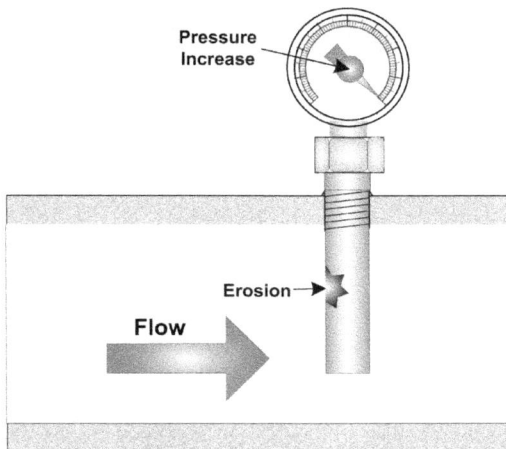

FIGURE 7.6   Basic Intrusive Sand Detector with Erosion.

If the erosion penetrates the wall of the coupon, as shown in Figure 7.6, this allows pressure ingress and the pressure gauge, or indicator, shows the full line pressure. This can be expanded to include the data acquisition system with the inclusion of a pressure sensor, shown in Figure 7.7, capable of reading the full line pressure.

If the electronic sensor is connected to a data acquisition system with an alarm function the system will automatically alert any sudden applied pressure.

Because of the limitations of national pipe thread (NPT), with respect to pressure, the threaded option can be replaced with a flanged option that is capable of handling the line pressure, as depicted in Figure 7.8. The capability of coping with the high line pressure extends to all of its components.

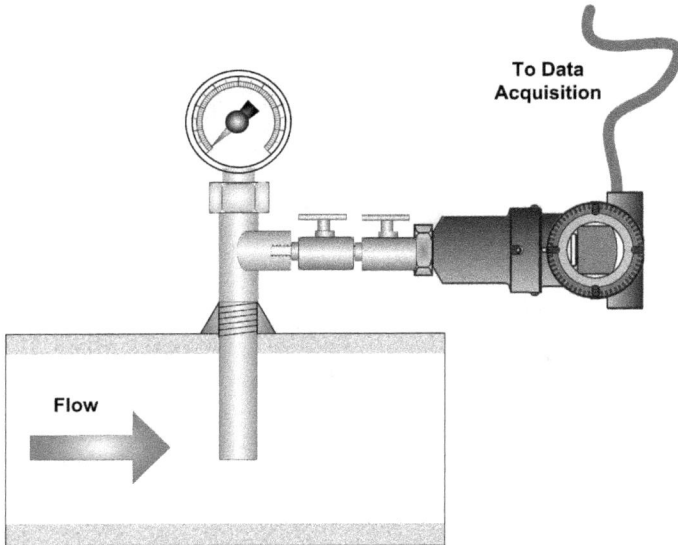

FIGURE 7.7   Basic Intrusive Sand Detector with Active Pressure Sensor.

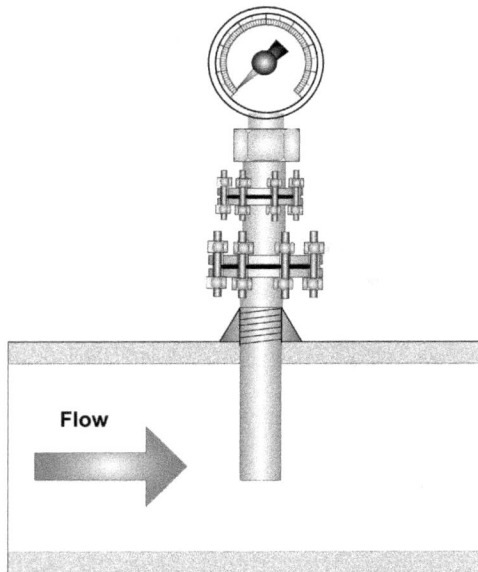

FIGURE 7.8   Basic Intrusive Sand Detector Using Flanges.

Again, this option can be expanded to include an integral pressure sensor.

The main problem with these types of sand detectors is that there is only an indication of erosion after the damage is done. In order to actually monitor erosion in progress, an active detector is necessary.

7.3.2.2 Active Intrusive Sand Detectors

Active sand detectors also intrude into the flowline but have built in sensors to calculate the amount of erosion taking place in real time, illustrated in Figure 7.9.

As these are electrical devices, all the electrical safety certifications and restrictions have to be in place and the sensor marked accordingly. This is usually an intrinsically safe (IS) system.

This type of sensor also suffers from the fact that a high percentage of the particles actually do not strike the sensor so again they are not true readings.

The active sand detector uses sensor strips, illustrated in Figure 7.9, embedded into the probe which are exposed to the flow and the impact from particles. These probes tend to have flat or squared faces to allow greater exposure to the particles, as shown previously.

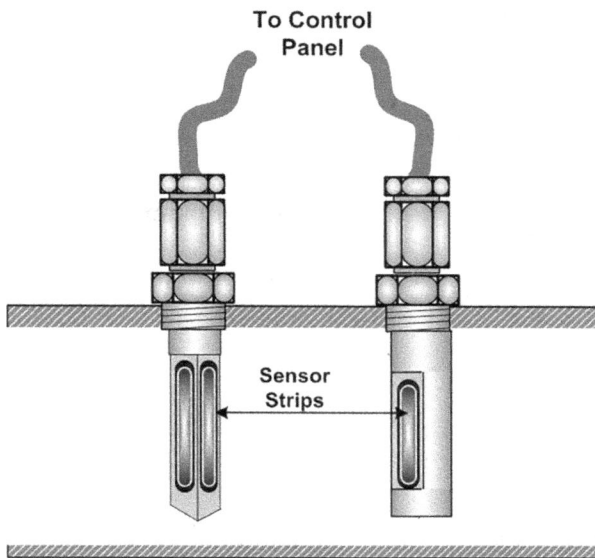

FIGURE 7.9 Basic Types of Active Intrusive Sand Detectors.

The detector strips are usually resistive in nature and function on the abrasive action of the flow. The flow removes material from the strip altering the electrical properties.

These readings are then transmitted to the control panel for calculation of erosion and display. These detectors can then usually be connected to a data acquisition system for storage and alarms.

This type of sensor is proactive and gives the users better warning if erosion is taking place and allows the operators to correlate sand production against flow rates and choke sizes. Using this technique, the well testers can reduce the choke size until no sand is produced and look to increase the choke as sand production gradually diminishes.

7.3.2.3 Non-Intrusive Sand Detection

This is one of the more common types of sand production monitoring, where a sensor is clamped onto the pipe using removable mounts and uses acoustics to count the number of particles that impact – or hit – the side of the pipework's inner walls.

The main advantage of this is that it is easy to install, remove, and use.

Figure 7.10 shows two particles of sand flowing in a pipe; the sand particle "Impact Particle" hits the side of the wall and causes impact noise and therefore is counted. The "Non-Impact Particle" sand particle does not impact on the wall causing no impact noise and therefore is not counted.

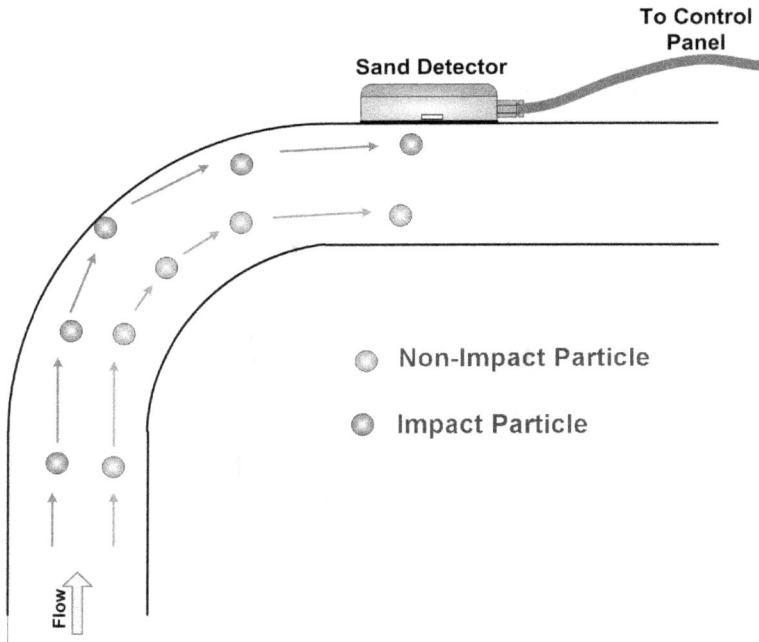

FIGURE 7.10 Non-Intrusive Sand Detection System.

This type of sensor has to be mounted on a bend or directional flow change and requires there to be sufficient flow velocity to generate impact while considering that the property of the carrying liquids, viscosity, etc., does not dampen or reduce impact noise.

This technology has evolved from its original development by the Rolls Royce Aero-Engine group who used the impact technology to test the stability of the turbine blades in its newly developed plane engines. What was happening was that the turbine blades were disintegrating under heat and stress and the method was developed to hear the blades impact against the housing.

This was acquired by British Petroleum (BP) who, in conjunction with an instrumentation company, patented the concepts and developed it into the sand detector.

Many thanks to the engineers at Pulsar for confirming the history of the acoustic sand detector. https://pulsarmeasurement.com/oil-gas/sand-detection-sand-holding-pools.html

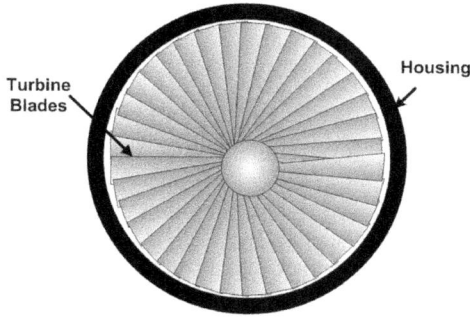

FIGURE 7.11 Turbine Blades.

As illustrated in Figure 7.12, the noise from the impact of a sand particle hits the pipe wall near the change of flow direction and produces impact noise that is transmitted by the pipe wall. Clamped to the pipe wall is a piezo-electric type crystal microphone *tuned* to the frequency of the impacts. The control electronics then filters out all background noise and counts the impacts from the sand particles.

Obviously, the sand that does not impact on the pipe wall does not contribute to the sand measurement, which makes actual measurements of sand mass not truly feasible.

It is for this reason that the sand detector cannot be used to determine quantitative volumes of sand, without the ability to compensate for or extrapolate the sand bypassing detection, which by the ratio of areas, probe area/pipe area, is probably greater than those detected.

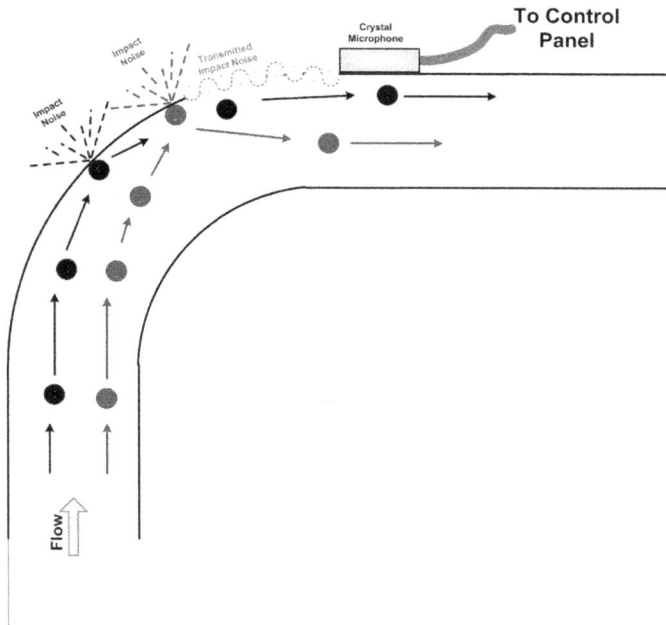

FIGURE 7.12 Sand Impact Noise Transmission.

These types of sand detectors give the following type of results:

- No sand detected
- Some sand detected
- More sand detected
- Lots of sand detected

The acoustic sand detector is best used to measure the increase or decrease in sand after a mechanical operation such as a choke change or the efficiency of a sand filter.

The unprocessed output from the sand detector is impacts per second (IPS). The output from the control panel can be connected to a data acquisition system using a serial or digital interface allowing the data to be recorded. Thus, analytical plots of IPS vs time or choke size can be produced to optimise the maximum flow rate for minimum sand.

Sand Detector Option

In order to "get around" the original patent by other companies the crystal microphone was placed on a tuned rod system, shown in Figure 7.13. Both versions of the sand detectors work well if installed properly.

These techniques do not work efficiently on Target Tee's and pipework that use a lead fill, because the lead dampens the acoustic impact signals.

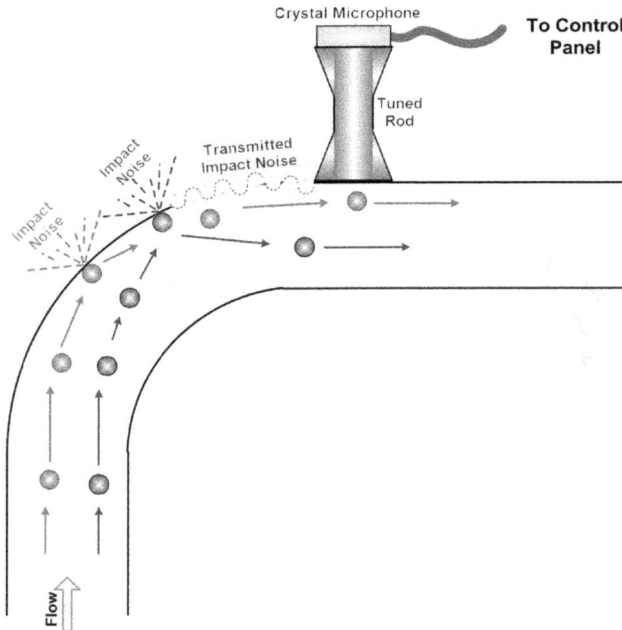

FIGURE 7.13   Impact Noise Transmission Using a Tuned Rod.

7.3.3 WALL THICKNESS MEASUREMENT

The wall thickness is the ultimate safety factor in pipes and measuring sand production only warns that pipe thickness erosion is potentially happening, therefore it makes sense to physically measure the wall thickness.

This involves one of the methods utilised in non-destructive testing (NDT) for determining the wall thickness, involving a transducer that contains two sensors:

  i. Transmitter
  ii. Receiver

The transmitter emits a pulse that travels through the wall of the pipe from the outer wall until it encounters the inner wall where it is reflected back through the wall to the receiver.

The time taken from transmitting to receiving is a function of the wall thickness dependant on the speed of sound in the pipe material, normally steel. The material types and respective acoustic conductivity are programmed into the meter.

If one of the well test crew is responsible for taking the thickness measurements, then it is important that they undergo some training in the use of the equipment by the manufacturer or a qualified engineer. Any relevant sections of API 570 and NACE 3410 etc. should be covered.

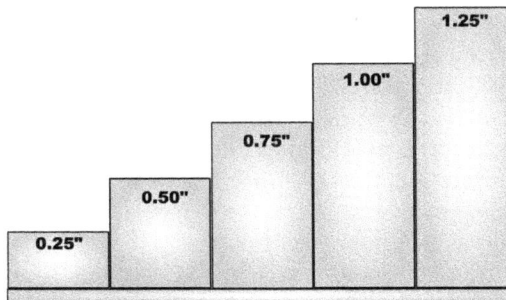

FIGURE 7.14 Typical Thickness Calibration Block.

Calibration blocks, illustrated in Figure 7.14, with certified thicknesses, are supplied with the NDT system. The sensor and display should be checked and calibrated against these before carrying out any measurements. Occasionally these calibration blocks are supplied separately from the NDT system. Regardless, they should be procured and available on location.

After calibration against the certified blocks, the system's accuracy would be in the region of ±0.002 inches (0.05 mm). The true accuracy for the systems should be verified with the manufacturer.

Dependent upon the intended thickness to be measured, it may be necessary to change the probe for different frequencies; these details should be checked with the sensor manufacturer.

TABLE 7.1
Example of Probe Specifications

Metal Thickness	Probe
0.06 to 10 inches 1.5 to 250 mm	Standard
0.03 to 1 inch 0.7 to 25 mm	Miniature
0.2 to 15.7 inches 5 to 400 mm	Low frequency

Any paint or coating must be removed so the measurement is made directly with the metal face.

This type of measurement uses an acoustic couplant or gel between the probe and the measured sample; this ensures that a good acoustic contact is made. The probe should not be used without the correct couplant in place.

Circular wall thickness testers exist that effectively wrap around the pipe and take multiple thickness readings at the same time, but these are not usually used in well testing. It is more common that an operator will repeatedly walk through the installation taking readings at the same specific points.

The equipment used in this must be an intrinsically safe device, probe, and sensor, with appropriate documentation.

The points to be measured must be decided on, prepared, and ready before the test starts. It should be remembered that this is a single point measurement and significant erosion could take place anywhere. Therefore, it is important that the selection of the measurement point(s) is in an area with the highest probability of erosion. Downstream of choke manifold or any flow direction changes, etc. are likely points.

The sample points must have the paint completely stripped to the bare metal – illustrated in Figure 7.15 – and a paint circle drawn round the sample area. This is used to mark the position of the thickness measurement and should be numbered so that samples are taken from the same location.

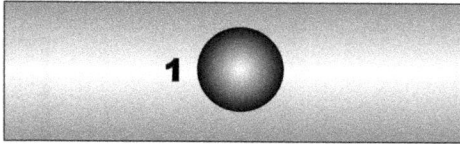

FIGURE 7.15   Thickness Pipe Preparation.

The thickness MUST be taken before the job starts and should be repeated to verify the readings. These then are the base readings. All subsequent readings are referenced to the base readings to give any material erosion or loss over time.

Multiple readings are taken at each individual spot and the average reading recorded. Any obviously erroneous readings should be discarded. When the readings are taken during the flow period all safety procedures must be in place including a "buddy system" where an operator is designated to watch the process in case of issues or accidents.

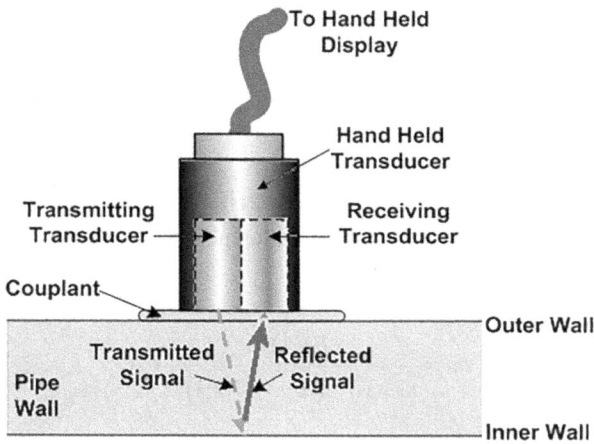

FIGURE 7.16   Thickness Measurement.

All electronics that are used or taken into a hazardous area must have the relevant safety certification approved by the country or region, normally an intrinsically safe device.

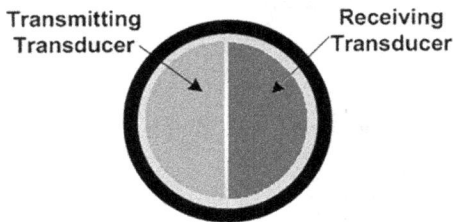

FIGURE 7.18   Example Handheld Display.

FIGURE 7.17   Example Thickness Transducer Face.

Some of the display (Figure 7.18) units have the capability to log the data and time in memory otherwise the readings will have to be recorded on a data sheet. A time period of an hour between readings is normal.

Any low readings should be brought to the attention of the senior well tester on shift immediately.

7.3.3.1   Wall Thickness Equation for a Specific Pressure

Wall thickness is given by the equation from the pressure rating formula, ASME B31.3 as per Section 304.1.2, Equation 3a.

$$tm = \frac{(P \times D)}{\left(2 \times \left(S \times (E + P) \times Y\right)\right)} + CA$$

Where
tm = minimum allowable wall thickness (inch)
D = outside diameter (inch)
P = pressure (psig)
S = allowable stress (FROM ASME B31.3)
Y = Temperature Coefficient = 0.4
E = joint efficiency factor
CA = corrosion allowance = 0.0625"

The thickness equation is fairly simple but it contains three variables that should be known before attempting any calculations:

- The allowable stress of the material taken from ASME B31.3
- The temperature coefficient
- The corrosion allowance

Results will be erroneous if these factors are not properly known.

The better approach is to have the thickness data included in the actual pipework documentation. The thickness of all pipework along with the pressure test data are routinely checked in a yearly inspection.

The included data should be

- The minimum permissible wall thickness for the maximum operating pressure, worst case
- The actual wall thickness at the time of testing

The inclusion of the actual wall thickness will give the service company an indication of the wear of the pipework during use.

7.3.3.2   360 Degree Pipe Thickness Monitoring

This type of system, shown in Figure 7.19, does not work using the ultrasonic sensors described earlier; it relies on acoustic lamb waves between the two sensor blocks.

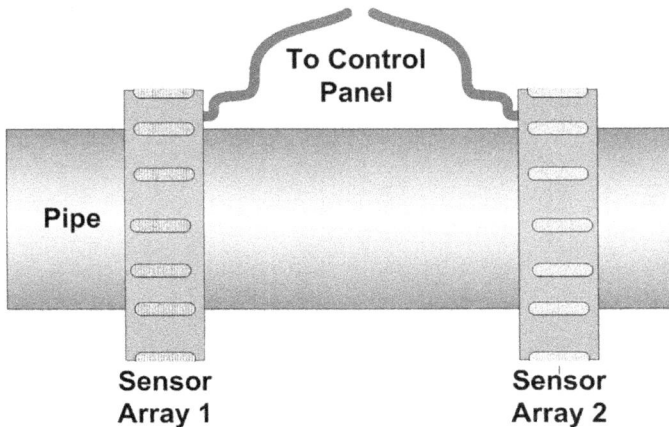

FIGURE 7.19   Lamb Wave Sensor Array.

Generally, this methodology is installed and operated by a specialist service company, not normally by well testers. The output, however, should be interfaced with the well testing data acquisition system. This will involve liaison between both companies before the job starts. The sensitivity and accuracy of this system is dependent on the manufacturer who should be approached for full specification.

7.3.3.3 Wireless Thickness Monitoring

More recent developments have resulted in a small battery-powered wall thickness monitor that can be strapped to the pipework targeting areas to be monitored, Figure 7.20. They use the WirelessHART communication protocol. This is an intrinsically safe system.

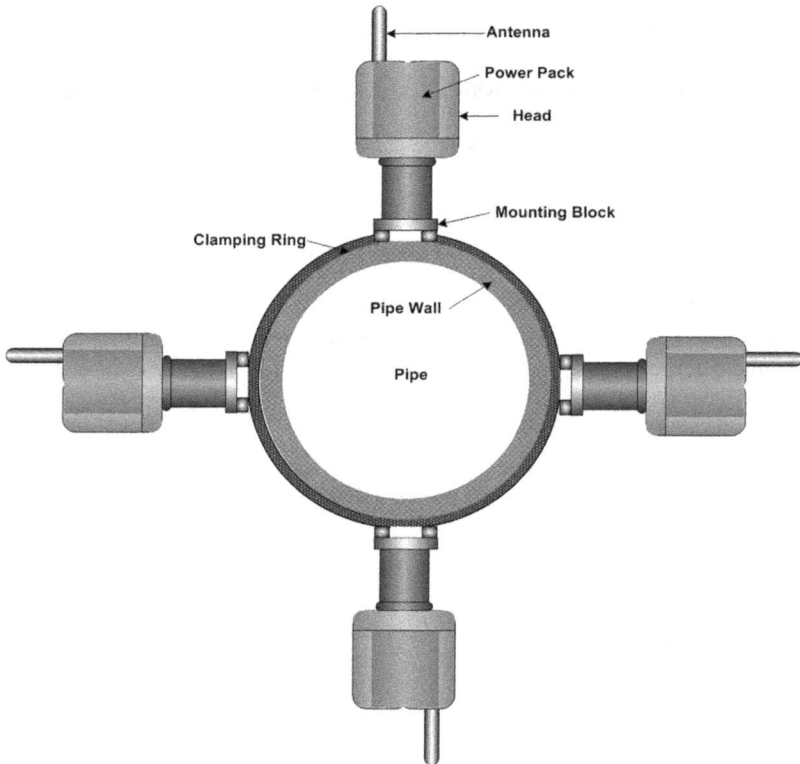

FIGURE 7.20 WirelessHART Thickness Monitor.

The WirelessHART thickness monitors can be installed using a steel clamping mechanism as shown in Figure 7.20 or by using a more temporary nylon type belt, shown in Figure 7.21, which can be adjusted to the pipe diameter. The use of a nylon type strap allows the sensors to be mounted on bends and odd-shaped structures.

This type of technology is not usually available/compatible with all data acquisition systems and may require the use of a secondary service company.

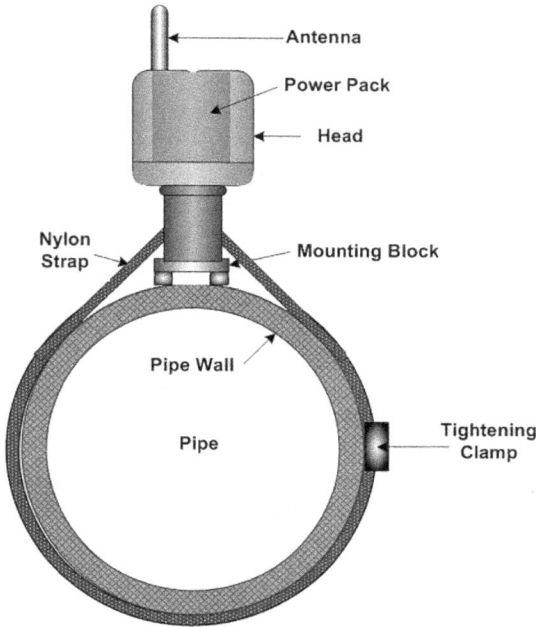

FIGURE 7.21 WirelessHART Thickness Monitor Using Nylon Strap.

7.3.4 SAND FILTERING

If the well is producing sand in significant amounts, a separate piece of equipment, a sand filter or sand separator, can be used to remove the sand from the well flow.

Different technologies are used:

- Cyclonic de-sander
- Large body sand separator
- Sand filter
- Also, generic sand catchers

7.3.4.1 Sand Filters

A dual pot sand filter assembly option is the most common used in well testing because of the ease of use and configurability.

The first versions of the sand filters were introduced by Baker Production Services in 1983 (now merged with Expro) and tested on the Shell gas platforms. The original models proved the concept of the screening but lacked the necessary capacity using small 15,000 psi vessels, so larger versions were later developed.

The original screens were constructed with reverse wound wedge wire which allows for higher differential pressures than the conventional wire commonly in use today. The use of conventionally wound wire effectively puts the effect of the differential pressure on the outside of the screen, effectively in collapse mode, so heavy duty mandrels or cages have to be used to support the mesh.

Many thanks as always to Mr Robert Gooch for his help and feedback regarding the history of the sand filters with Trianglia Oilfield Services. An experienced 25-year-old engineer with 50 years' experience – so he tells us.

These sand filters have evolved into three pressure options of 5,000, 10,000, and 15,000 psi. The higher pressure ratings will have heavier wall pipework, all associated fittings, and unions. Instead of normal elbows sand filters usually use solid blocks which can better withstand high erosion from change of flow direction.

The venting of sand filters will involve gas with particles of sand/proppant included in the vented stream; this has to be considered in the venting system design as opposed to the simple venting to atmosphere. Not only is there a potential for ignition of the gas from the sand/proppant particles hitting the steel, but there is a static electricity issue. The venting hose can be a simple flexible hose that has to be tied down in multiple points to avoid any possibility of whipping. Venting is potentially hazardous and any personnel involved should be aware of the hazards involved. Buddy system and full PPE are essential.

It is essential that the sand filters are earthed correctly with the correct cable at all times.

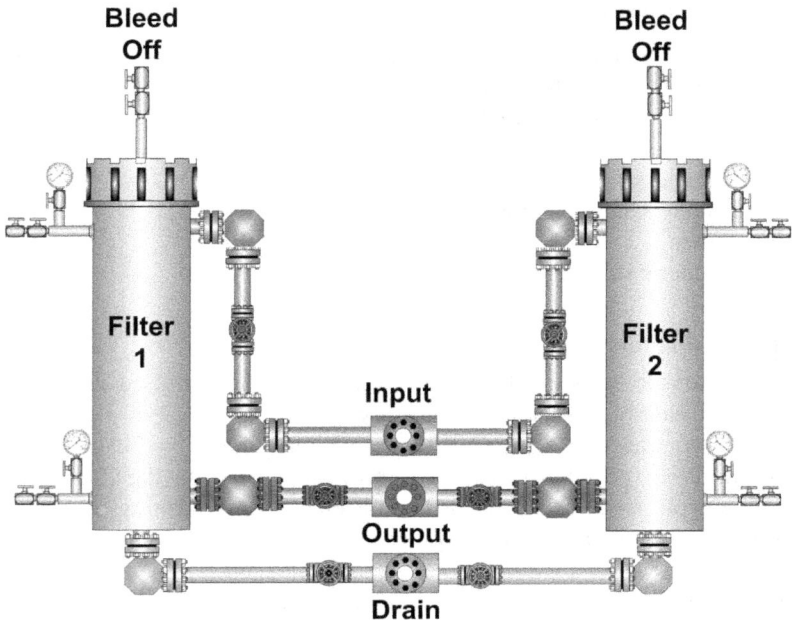

FIGURE 7.22   High-Pressure Dual Pot Sand Filter.

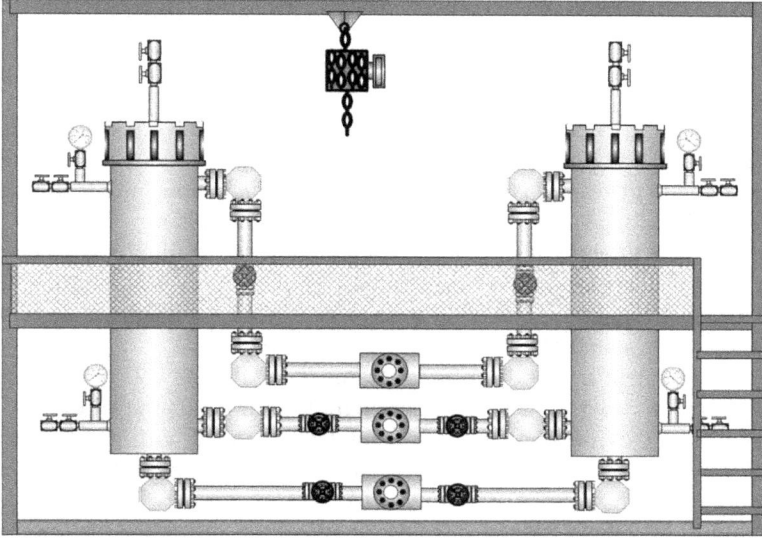

FIGURE 7.23 Dual Pot Sand Filter with Operational Framework.

The sand filters work by using a stainless-steel mesh that allows the fluids to pass through but traps the sand particles inside the mesh filter. Because the mesh tube is not strong enough to withstand the mechanical handling and damage, it is mounted inside a stainless-steel tube with large holes as shown in Figure 7.24.

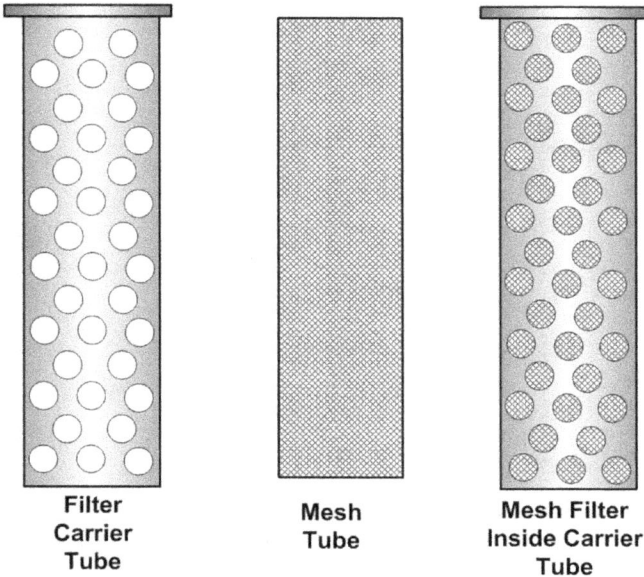

Filter Carrier Tube

Mesh Tube

Mesh Filter Inside Carrier Tube

FIGURE 7.24 Sand Filter Carrier and Mesh.

Based on various manufacturers' specifications the sand filters will work with an approximate maximum 10 lb/min of sand suspended in 50% solids slurry. Designs may vary so functionality must be verified with the specific manufacturer.

When using the sand filters the produced fluids should be controlled to an estimated maximum 1 bbl/minute whiles the sand/proppant volumes should be continually monitored to give a solids volume rate. The maximum flow rates that the system is capable of handling is again manufacturer and size dependant.

Some manufacturers will supply the sand filters with the filter pots piped in parallel in order to increase the flow capabilities. The filter screens are available from 50–800 μm, dependent upon particle sizes to be filtered, although a 200 μm screen size is supplied as standard.

The filter system comes in a metal framework, Figure 7.23, with an upper-level platform in order to use a hoist or chain blocks to lift the filters once they need changing. Obviously, the filters cannot be changed with pressure trapped inside the vessel so pressure must be vented, generally gas, from the top of the vessel.

There must be a double bleed valve on the top of the vessel and the lower needle valve opened first. Any bleed off control should be carried out using the upper needle valve. The reason for this strategy is that the bleed off is likely to be a very abrasive

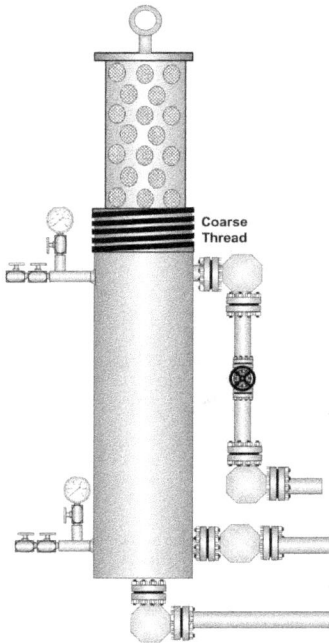

FIGURE 7.25 Inserting a Filter Tube.

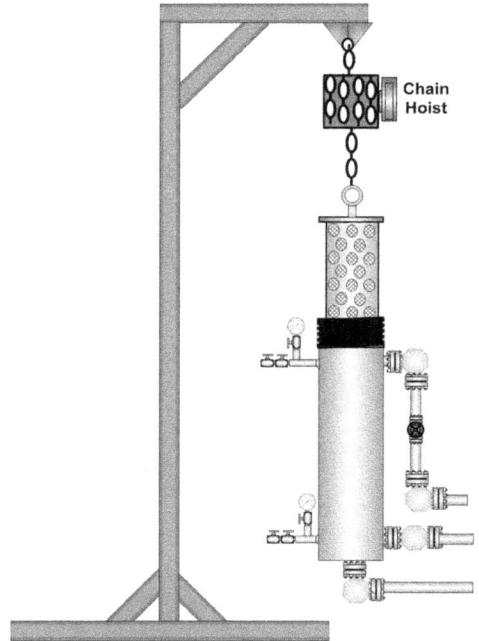

FIGURE 7.26 Lifting a Filter Tube with a Chain Hoist.

flow and will erode the needle valves very quickly. When the upper valve is used it erodes, leaving the unused lower needle valve undamaged to be closed and to shut off pressure, whereas with the high erosion potential of needle valves, it is recommended that a large number of spares are carried.

One of the methods of detecting when a tube is full is to measure the differential pressure (dP) across the filter tube, Figure 7.27.

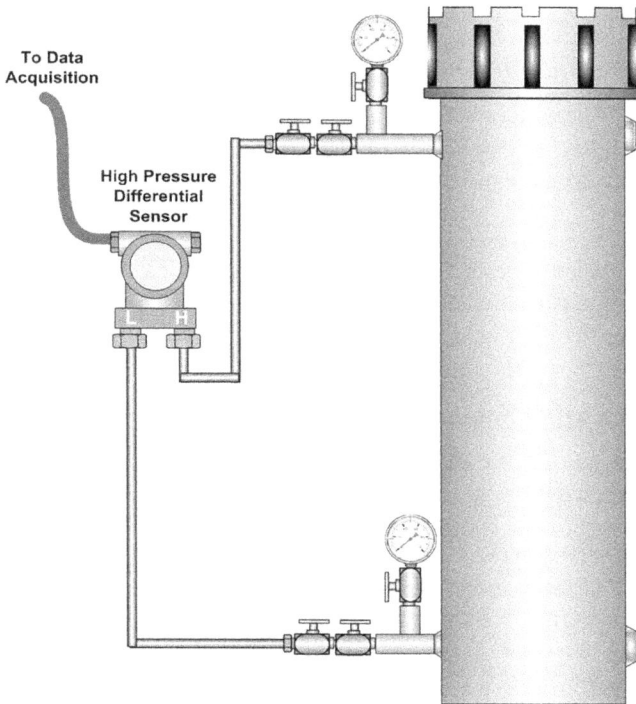

FIGURE 7.27   Differential Pressure across a Filter Tube.

When the tube is empty the differential pressure will be at its lowest. The value will depend upon flow rates, fluid types, and viscosity etc. so a benchmark value of an empty tube is required. When emptying the tube, the differential and corresponding level of sand in the tube should be entered into a table to get a better tailored relationship between sand in the tube and differential pressure.

The sensor used must be a high-pressure model; a differential similar to the one used on a separator may not have sufficient operating pressure. This should be supplied with the filter system to ensure compatibility and design.

Depending on the flow rates, high volume of sand/proppant and fluid will result in a high delta pressure (dP) across the filter which can damage the filter. If the dP is increasing towards a high reading it is better to switch filters immediately rather than waiting.

Sand Filter Positioning

The sand filters are normally placed before the choke manifold where the fluid velocity is lower, eliminating as much sand before it can damage the equipment involved. Some operational factors may influence the positioning of the sand filters, operational pressure, the type of operation, operational requirements/objectives, etc. All this should have been agreed during the TWOP meetings before the equipment are mobilised.

Checking Efficiency of Sand Filters

To verify the efficiency of the system sand detectors should be placed before and after the sand filter, to monitor the output sand compared to the input sand. They would normally be mounted on the flowlines of the sand filters, immediately after elbow blocks.

Monitoring the sand detectors is another way of determining whether the filters need changing or not. The higher the downstream sand detection, making it closer to the upstream sand detection, means less sand is being trapped in the filter and most probably means the filter is nearing full capacity.

Swapping/Changing Sand Filters

Changing filters is one of the most dangerous operations during a well test because the venting of the sand filter pots introduces potential for ignition from sand in a high-volume gas stream. The sand filters' configuration must have the correct earthing system connected to reduce the potential of a spark.

All these sand filter operations and potential hazards must be discussed in the pre-job safety meeting, tool box talk, before shift start or flowing the well. These risks are heightened if H_2S – or any other poisonous gasses– are present. Then a vent line from the bleed off valves to a safe area must be used to remove any vented gas away from the working area. Obviously, all safety precautions including breathing apparatus and gas detectors must be included in the standard PPE with these operations.

The use of any electronic devices during this operation is banned. This especially applies to mobile phones and non-certified radio sets. Also, the coveralls – or other clothing – worn by the operators should be anti-static, unable to build up static electricity and/or sparks to avoid risk of combustion. In some operations, personal earthing straps are used.

The switching operation is similar to the choke manifold and is a two operator or more operation. Changing filters is not a single man operation; a minimum of two trained well test operators are needed, who are trained in the operation and the location of all valves and safety features. Untrained personnel are not allowed on or to assist in these operations. All untrained personnel should also keep their distance during sand filter operations.

All sequence of events and operations must be recorded on the manual readings sheet.

General Requirements:

- There should be an emergency shut down (ESD) located on the safe working platform which should be easily accessed from one or more permanently installed access points
- With regard to extra safety while the filter is in place, in the sand filter body the act of leaning over the opening should be avoided. Especially when the cap is first taken off
- Ensure all PPE, especially safety glasses or masks, are in place in case of sand ejection
  It is worth mentioning again that mobile phones and non-certified radios must not be used; this is mandatory
- Before starting any mechanical operations, take a note of the differential pressure reading across the sand. This should be noted on the manual readings sheet

Sequence of Operations:

- The first step is to switch between filter pots so the valves and settings can be verified, especially the drain valves

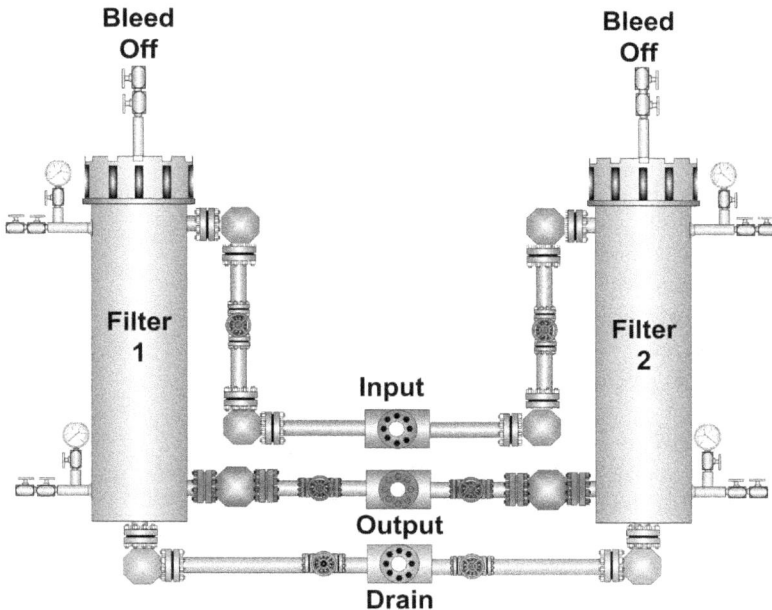

FIGURE 7.28   Sand Filter Valve Location.

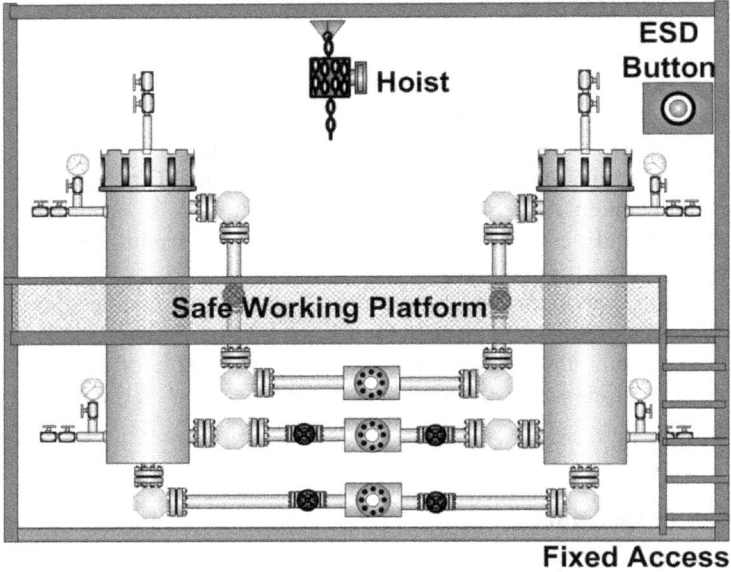

FIGURE 7.29   Sand Filter Safe Working.

- Open and verify that the outlet valve of the empty sand filter is fully open
- Slowly open the inlet valve of the "empty" sand filter until it is fully open. This is slightly different to the choke manifold where both sand filters should be open and in line

> **NOTE:** *This may cause the back pressure to reduce.*

- When fully open, close the inlet of the "used or full" filter and verify it is fully closed
- When the inlet valve is fully closed, begin closing the output valve of the "used or full" filter
- Safety must be the overriding concern as operators will be working at height immediately next to equipment under high pressure
- Using the "buddy" system where one operator watches the other, the lower needle valve of the bleed off should be opened completely. Refer to Figure 7.30
- Only when the lower needle valve is fully open should the upper needle valve be slowly opened and a percentage of the pressure slowly vented off to a safe area. The actual amount bled off is not critical as long as the pressure on the gauges drops significantly

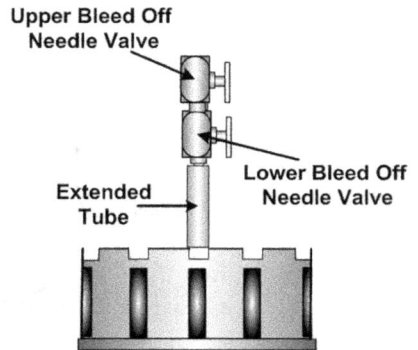

FIGURE 7.30 Sand Filter Cap Illustration.

- Close the upper and then the lower needle valve while monitoring the pressure to ensure it is not increasing. If the pressure increases there is a leak and it is too dangerous to continue until the leak is rectified
- Only when the dropped pressure is holding (not increasing), open the lower needle valve to fully open; then the upper needle valve can be slowly opened so that all the pressure is slowly vented off to a safe area

NOTE: *Due to an accident in the first versions of the sand filters, an extended-length nipple tube/pipe is installed on the top of the bleed valves. The accident occurred because the operator leaned over to check the venting and the sand in the stream abraded his forehead; the long nipple tube prevents leaning over.*

- Only when the vessel has no pressure, check ALL the connected dial gauges and any electronic gauges; an attempt can be made to remove the top cap
- The caps use a coarse thread similar to that used on slickline lubricators and should undo easily if there is no trapped pressure
- As the cap is heavy, the needle valves are normally taken off and a lifting eye inserted in their place to allow the weight of the cap to be lifted by the hoisting equipment, normally a block and tackle
- After removal the cap is lowered onto the working platform, normally on wood to ensure no sparks are created
- Before attempting to lift the filter from its housing the drain valve should be opened and any liquid drained to a suitable receptacle at atmospheric pressure. This may take some time
- By draining any fluid and with the bottom now open to the atmosphere, hydraulic effects are reduced/eliminated, easing the removal of the filter assembly
- It may be necessary to wash round the filter using a water hose to ensure no sand bridges or blockages are in place; check that the water flows out of the drain system
- When all clear, connect the hoist system to remove the filter, and if there are no issues slowly begin lifting the filter from the pot. Just in case of any trapped pressure, pay attention to when the filter begins to lift or increase in weight

   A weight scale is often used on the hoist to check for hang ups and ultimately to record the weight of the filter
- Continue to slowly lift the filter from the pot, taking all safety precautions
- Check the filter for any samples required
- Measure the level of sand in the filter and record the value on the manual reading sheet next to the differential pressure reached across the filter prior to emptying; weight also if taken

   By recording differential pressure and length/volume of sand, a look up table or graph can be created to give an estimation of sand in the filter without any disassembly

- Lower the filter to the floor and wash out the sand in a safe area where fluid and sand can be disposed of. Safety precautions should still be in place, ensuring glasses and gloves are being worn

NOTE: *If the weight of the dried sand is noted and a cumulative record kept then a reasonable estimation of the sand produced can be deduced. This can be important in sand/proppant flow-back after a hydraulic fracturing operation.*

- Clean the filter; refer to Figure 7.31, making sure no mud or similar material is trapped in the mesh
  - A high-pressure washer or steam cleaner may have to be used
  - Ensure the high-pressure washers have a safety certificate to enable them to work in a hazardous area
- After cleaning the filter check for any signs of damage or erosion on the filter. Report to the well test supervisor to verify if the filter needs repairing or replacing

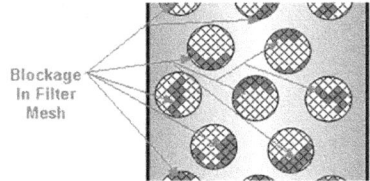

Blockage In Filter Mesh

FIGURE 7.31   Sand Filter Blockage.

All filter mesh should be completely clear before returning. If in doubt the filter should be changed if possible.

- Lift the now clean filter using the hoist
- The filter's internal body should be cleaned with water (washed down) so there are no remaining solids inside. This water should be flowed out the drain line, again to a safe area
- With the drain valve still fully open, lower the filter assembly into its housing
  - The drain line is left open to minimise compression effects
- Remove any lifting fittings
- Check and clean the threads on the actual housing to remove any potential solids; this should be with light oil, such as diesel
- Lift the cap using the hoist and similarly clean the threads before screwing it on top of the housing
- Clean and replace any Teflon tape or sealing compound and refit the needle valves and close both of them
- Close the drain valves
- The assembly should now be ready
  - Inlet valve status – closed
  - Outlet valve status – closed
  - Drain valve status – closed
  - Bleed off valves status – closed

7.3.4.2 Debris/Junk/Proppant Catchers

One of the purposes of the debris catcher is to offer a quicker measurement, within an hour on average as a smaller scale equipment, of any produced debris, sand

and/or proppants, instead of the traditional large-sized sand catchers/filters. The quicker measurements of debris production can assist in operations when quick judgement calls on most efficient choke sizes are crucial. The larger-sized sand catchers would require many hours until proper sand production measurements could be deduced.

The series of debris catchers now in use have designs that vary from manufacturer to manufacturer and are a derivation of the original sand filter designs. The debris catcher is basically a choke manifold with two miniature filter pods in place of the flow pipe, designed to catch small quantities of debris, sand, and/or proppant.

The manifolds similar to the current designs of choke manifolds have independent upstream and downstream manual isolation gate valves as well as an independent bypass valve that enables the filters to be isolated if not in use.

There are two fundamental configurations of the debris catcher, of the same operation principal:

i. As a modified choke manifold shown in Figure 7.32, where the filter pods are mounted horizontal to the choke manifold. This configuration usually has all its mechanisms on the same plane and as such is safer and easier to operate

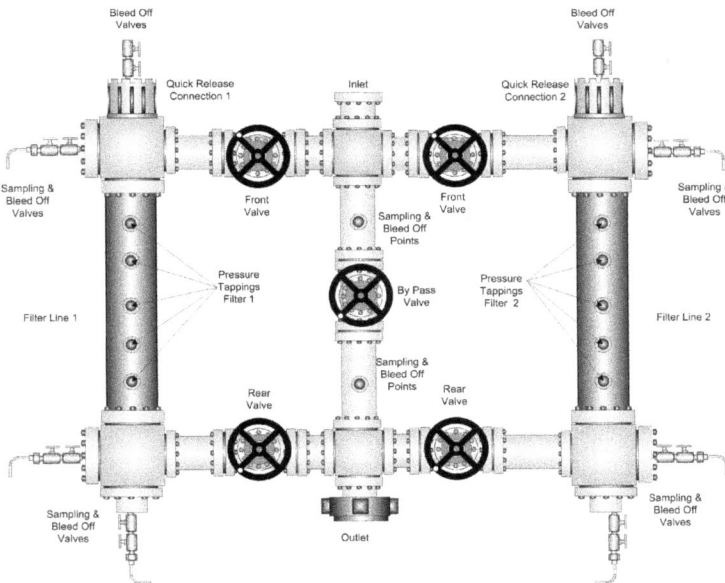

FIGURE 7.32 High-Pressure Debris Catcher.

ii. Where the debris catcher is designed as a miniature or smaller sand catcher, shown in Figure 7.33, as described earlier. These are operated in an identical way to the sand filters/catchers. This configuration can have considerably more substructure and frames to safely operate and to empty or change the filters due to the height from the ground of the filter pots

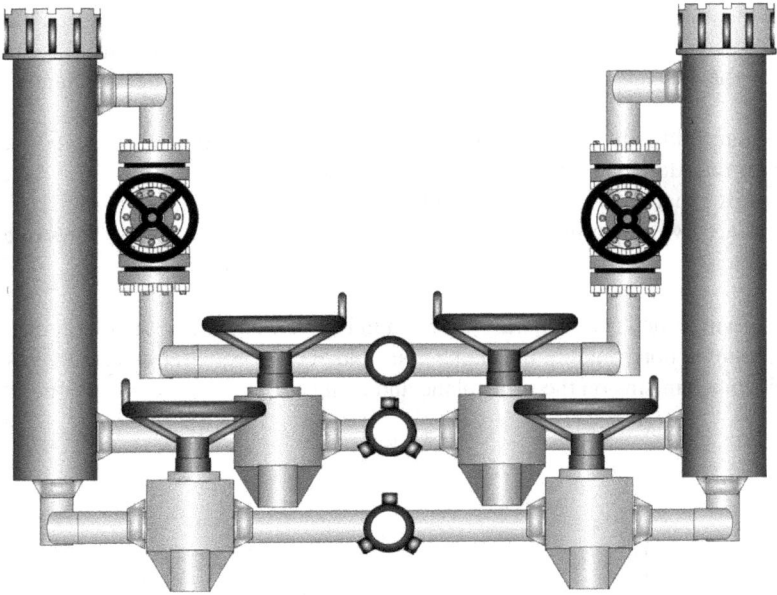

FIGURE 7.33   Vertical Debris Catcher.

The design of the Debris Catcher mimics the sand traps, but in a smaller scale, where the flow is into the centre of the filter and out through the sides, the bottom of the filter is normally a solid plate to aid in rigidity, illustrated in Figure 7.34. The use of a dual filter system allows for the removal of the filters for periodic debris production measuring, cleaning, maintenance or replacement.

FIGURE 7.34   Filter Tube.

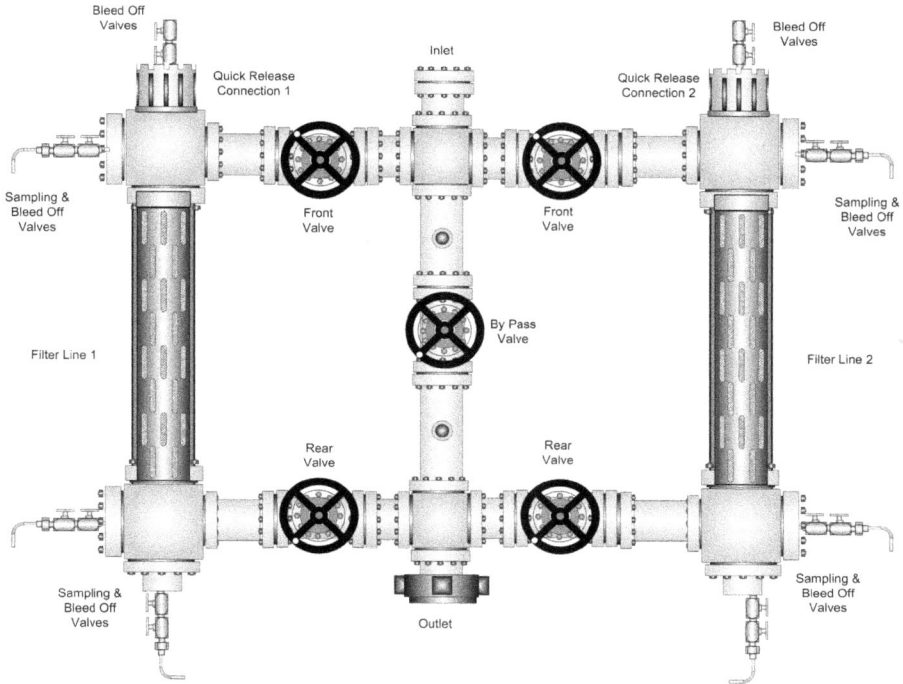

FIGURE 7.35   Debris Catcher Showing Filters in Place.

There are multiple ports on the debris catcher assembly to facilitate

* Pressure measurement
* Pressure bleed off
* Sampling
* Draining
* Flushing

On the filter lines, there are usually a number of ports to ensure efficient flushing and cleaning can take place. The flushing and cleaning are usually with a water-based liquid using a small-bore pump although high-pressure jet washers can also be utilised. All pumps used must have a safety certificate in order to be used in a hazardous area with particular attention to water ingress into any cabling or switching systems.

Operating the debris catcher with flow is an identical process to that of the choke manifold in the manner in which the valves are operated, closed/opened, and switched. Refer to Section 5.8.

If using a Debris Catcher to trap sand or proppant it must be understood that because of the smaller cross-sectional arc of the debris catcher it is possible for the debris catcher to experience blockages more frequently than would occur with a sand trap. Therefore, the differential pressure across the filter and inlet pressure should be monitored for increased pressure, indicating either a blockage or a filled filter.

This type of differential pressure monitoring, shown in Figure 7.36, would not function if the debris catcher is used for its original purpose, trapping debris. Debris in this case is scale, small stones, etc. and would not effectively block the filter to extents of increasing the differential pressure.

Cleaning and flushing of the debris catcher can be hazardous due to sand and other particulates that can become airborne during washing. Therefore, full PPE must be worn by all personnel that are in the area with particular emphasis on:

Safety glasses
Face mask
Appropriate gloves
Wet suit
Rubber safety boots

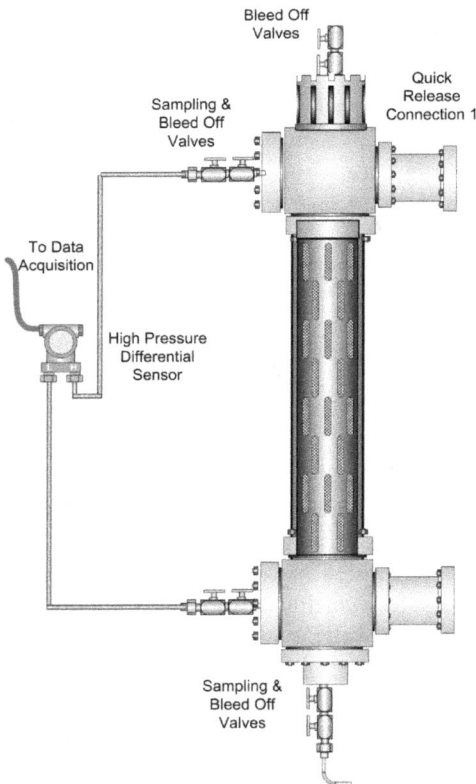

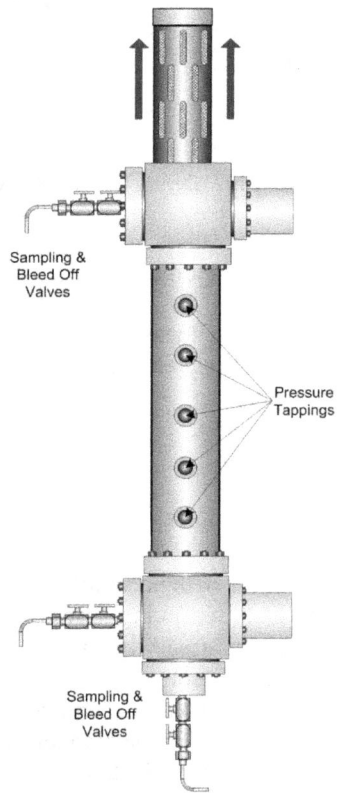

FIGURE 7.36   Differential Pressure Monitoring.   **FIGURE 7.37**   Removal of Filter Tube.

Removal of the filter tubes may involve injection of liquid into the pressure tappings to assist in freeing the filter tubes, especially if being used as a miniature sand filter.

Once the filter tubes have been removed, they should be cleaned and processed as laid out in the sand filters section.

Low-Pressure Version

If used as a debris/junk catcher at low pressures, the materials and valves will usually be de-rated to suit the application. This pressure is nominally 1,440 psig to match the separator's maximum operating pressure.

This version (Figure 7.38) has applications in operations such as filtering burner coolant to ensure the burner does not get blocked by debris from the rig pumps.

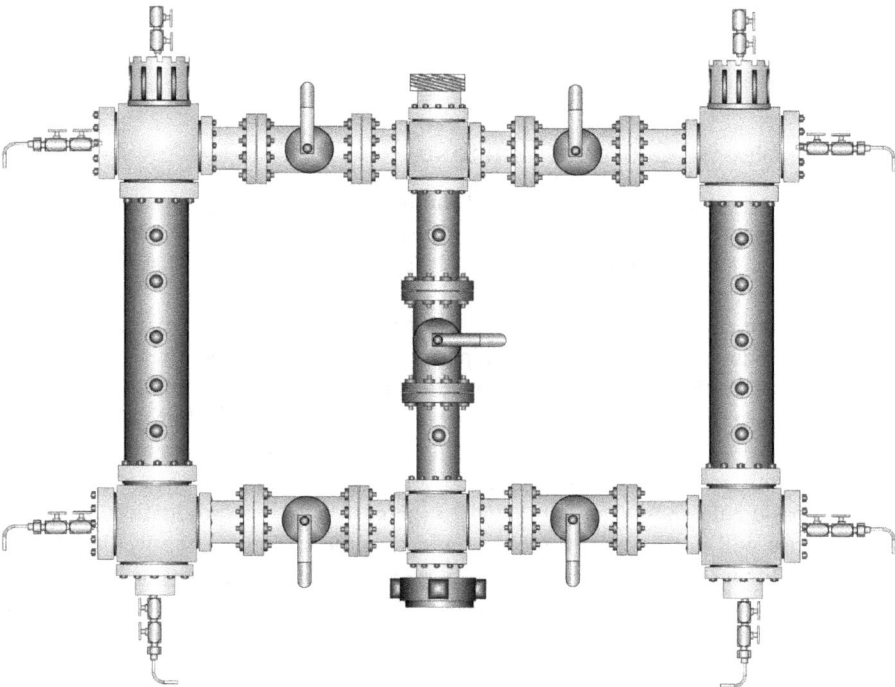

FIGURE 7.38   Low-Pressure Debris/Junk Manifold.

Example Specifications

All figures quoted are general and are manufacturer specific; consult with the service company.

Description	Specification
Working pressure	0–5000 psi 0–10,000 psi 0–15,000 psi
Filter details	Dependent upon manufacturer
Flow rates	0–50 MMSCF/D (Gas) 0–5,000 Bbl/D (Liquid)
Inlet/outlet process connection	Hammer union Flange Grey loc (Dependent on process pressure)
Filter mesh	316 stainless steel
Filter mesh size	400, 800 or 1,200 microns (standard) Check with service provider

7.3.5 Increasing Choke Size with Sand

When flowing sand/proppant, the use of the fixed choke is recommended due to its higher erosion resistance and lower susceptibility to flow damage. Proppant/sand production is usually limited to 400–500 lbs/hour for this type of sand filter. Again, this should be clarified at the TWOP.

Only when the wellhead pressure is relatively stable and the monitored sand/proppant production over a 3–4-hour period is reduced to less than 50 lbs/hour can the fixed choke be increased in increments 1/64ths (if possible). Continue to slowly increase the choke sizes in this way until an increased sand/proppant is measured at the surface. It must be stressed that only small increases of choke should be attempted, otherwise the solids production could plug the tubing.

This basic rule should be continued to increase proppant production, although the increase will not be instant but dependent on the volume of the wellbore, flow rates, etc. During the flow stages, if surface sand/proppant production is in excess of approximately 400–500 lbs/hour (or the sand filter's hourly capacity), then the choke size should be slowly decreased to return sand/proppant production within limits.

This flow control to gradually increase sand/proppant production within the limits of the equipment should continue until the natural decline in the production of sand/proppant reaches a defined target.

8 Basic Sediment and Water (BS&W)

Bad practice in taking BS&W readings result in the most frequent and largest errors in well testing as both actual readings and flow calculations depend upon its accuracy. Most personnel have not been instructed in the correct methods of taking a BS&W reading and rely on inherited practices, which are normally the easiest and quickest ways – yet incorrect. Taking a single BS&W sample instead of two is the most common fault that is not declared or remedied.

All BS&W sampling must be carried out in accordance with the safety practices and setup on the location. Operating company representatives should also randomly witness sampling from start to finish.

Example:

A well testing company was commissioned to monitor the clean-up stages of a production well that had recently been worked over using a trailer mount package.

The service company carried out the work as normal however only one (1) BS&W reading was taken at the beginning of the flow and applied to all consecutive flow rates.

The initial BS&W reading was 40% water, but the water component was reduced to less than 5% at the end of the allocated flow periods. The well test service company had applied the 40% reading to the entire test, whereas the operating company had measured and recorded the BS&W at the gathering station every hour, a large discrepancy.

This is because the well test crew involved were used to carrying out flow verification tests on the wells where the BS&W tended to stay constant. So, they did not bother to check and reported all the same readings, including gas and oil gravity, throughout the flow period.

The service company were not paid and were fined. The supervisor was suspended and the crew financially penalised.

This shows the importance of the supervisor physically checking the readings. A manual data sheet was also not used.

DOI: 10.1201/9781032623689-8

8.1 BS&W DEFINITIONS AND STANDARDS

Basic sediment and water (BS&W) has many names and descriptions within the oilfield:

- BS&W
- Water cut
- Percentage water
- Etc.

Whichever name is used, there should be a recognised process and structure for how it is taken and interpreted. Errors generated because of the haphazard, uncontrolled, and unchecked methods employed are the prime cause for errors in oil and water flow rate calculations. Therefore, it is essential for all personnel to understand the importance of BS&W and its effects on flow rates.

There are set customer approved and API standards that apply to the process of determining a BS&W sample in the field, such as API MPMS Chapter 10.4 There are also other standards that utilise different approaches such as API MPMS Chapter 10.3, which specifies BS&W determination in the laboratory.

Some of these procedures are normally specified for fiscal purposes and it is often impractical to utilise these methods on a well test. However, the well test supervisor should be aware of these procedures and if they are specified by the client, then they must be used in the place of any other service company procedure.

As there does not appear to be a set standard for the taking of BS&W samples by hand in atmospheric conditions, such as the case in surface well testing, a standard convention has been utilised by the well test crews from at least the late 1960s onwards.

The well test method described in this chapter should be common throughout the service companies involved. Unfortunately, the lack of training, knowledge, and wrongly inherited procedures contribute to significant errors within the reported results and the subsequent calculation of liquid rates.

8.2 USE OF BS&W

Within well testing BS&W monitoring is used for two main purposes:

i. The initial phase of the test to see if the well is "cleaning up" so the flow can be diverted through the process equipment without causing any damage

   This is sampling at the choke manifold, only. This BS&W reading is just an indicator of cleaning up and must not be used for calculation of water rates in three-phase separation

ii. BS&W taken at the choke manifold can be used to correct liquid flow rates when operating the separator in two-phase flow, although a better sampling position would be just before the separator, especially if heaters are used

   In the event of any chemical injection in the surface flow lines that affects or contaminates the BS&W readings used for two-phase flow correction, it is preferred that the BS&W be taken upstream of the choke manifold and before the chemical injection point

8.3   BS&W SAMPLING POINT

During clean up conditions and when the separator is operating in two-phase flow, the BS&W readings should be taken at the choke manifold. However, the choke manifold has two sampling positions.

- Upstream of choke manifold
- Downstream of choke manifold

Therefore, the question is often asked which sampling position should be used. Often, well testers use the upstream sample points, but there is no true standard and there is no correct answer.

Well tests with higher pressures (HPHT) are becoming more common and we must take into account the increased risks with higher pressures.

8.3.1   BASIC OPERATIONAL ISSUES OF BS&W SAMPLING

Where PVT sampling is to be carried out, especially with the presence of condensate or light oils, then sampling should be taken upstream of the choke manifold. This comes with its own safety issues and should only be carried out by trained or specialist staff.

For the argument of whether upstream or downstream of the choke sampling is a better option, the following points weigh heavily. They are based on fluid sampling for BS&W and fluid gravity measurements and not for analysis sampling:

i. High-Pressure Valves

ALL safety procedures specify that it is safer to open and use a low-pressure valve rather than a high-pressure valve and that the lower pressure option should be followed.

ii. Inlet to Separator

The fluid flow from downstream of the choke manifold actually goes into the separator and therefore is more representative, not the upstream fluid flow.

Also, in the case of a heater (or heat exchanger) where high viscosities are present and the fluid is heated, the BS&W sample must be taken *downstream of the heater*, so the heated fluid is sampled. This becomes a mandatory procedure and is not optional. It is apparent that the fluids would dramatically change after being heated and any upstream sampling would not at all be representative.

iii. Shrinkage

With the lower pressures involved downstream of the choke manifold the BS&W is less likely to have high shrinkage errors than a sample taken at a higher pressure upstream of the choke manifold.

iv. Multiple Samples

Often well testers will take BS&W from both the choke manifold and separator oil line.

a. If the separator is in three-phase mode, the BS&W taken at the choke manifold is generally taken for determination of solids in the well flow only

If this BS&W is taken and reported, then it should have a separate column clearly labelled that it is from the choke manifold and not used in any calculations

NOTE: *With the separator in three-phase mode the BS&W from the choke manifold MUST NOT be used in the separator fluid calculations and corrections. Again, this is important*

b. If the separator is in three-phase mode, BS&W is *only* taken from the oil flowline for calculation purposes

Any water in the oil flowline is used to compensate the water flow rates which are deducted from the calculated oil flow rates

c. If the separator is in two-phase mode the BS&W must only be taken upstream of the separator

NO BS&W sampling should be taken from the separator, as by running the separator in two-phase operation it means exactly what it says: gas and a liquid phase, where the liquids (water and oil) cannot be separated by the separator. This is important, no liquid separation

Therefore, the output liquid is treated as single phase from the oil flowline, and the upstream separator BS&W is used to calculate individual liquids' flow rates

If there is measurable water then the separator **MUST** be run in three-phase mode

d. All BS&W samples from multiple points are doing is confusing the junior well testers and showing that the supervisor is unsure of the procedures involved

If the well test separator is doing its job correctly and separates the fluid phases, the BS&W at the separator oil flow will have no relation to the BS&W at the choke manifold. With perfect separation, the BS&W at the oil flowline should be 0%, while BS&W at the choke manifold could be any percentage.

If water carryover is suspected or observed then the water level should be checked on the separator water sight glass, and if necessary the water level should be lowered using the water level control valve. The level should be revised after allowing time for the water level to be lowered.

NOTE: *This is a reason why the separator sight glasses must be kept clean and operable.*

From the previous statements the correct place for intended BS&W sampling would be downstream of the choke manifold. There are potential issues with chemical injection here but these are easily solved if the test crew are trained and competent.

In Summary BS&W Sampling Procedure:

a. When the separator is in two-phase operation, BS&W is taken from down-stream of the choke manifold. However, the procedures and methods are identical for upstream sampling if specified or required, with safety procedures for the use and operation of valves under high pressure that must be followed

b. When the separator is in three-phase operation, BS&W is taken from the separator oil line only. This BS&W is then used for calculation and correction of the fluids exported from the separator. BS&W taken from anywhere else is only considered as information and MUST be clearly indicated on the reports so no mistakes can be made

8.4 BS&W PROCESS CONSIDERATIONS

If there are issues in separation of the oil and liquid phases or with a lot of carry-over from the oil phase to the water phase that cannot be corrected mechanically within the separator, then run the separator in two-phase mode. This is important.

By running in two-phases mode all the liquid is flowed through the oil line only and there is no flow through the water line. The water is calculated from the BS&W taken from the choke manifold or before the separator.

The manual measurement of BS&W affects the associated flow rates and cumulative calculations:

   i. Water rate
  ii. Water cumulative
 iii. Oil rate
 iv. Oil cumulative
   v. GOR
 vi. GOR2

For these calculated results the correct and precise methods for determining the BS&W must be employed under all conditions.

The responsibility of taking these samples is usually given to the most junior and inexperienced member of the well test crew without proper training or, more importantly, understanding as to the effect their actions have on the rate calculations.

What is more worrying is that, often, the experienced personnel on the crew do not appreciate the magnitude of these errors. This is a product of poor training in many cases.

To explain:

The flow rate through a meter is compensated for the amount of BS&W in the flow stream, either automatically – as is the case with the new generation of flow meters (MPFM, Vx, and Coriolis) – or manually. Whichever is the case, an accurate manual sample of the BS&W must be taken at regular intervals, if only for cross checking.

Table 8.1 shows the degree of flow rate correction that takes place according to the BS&W reading. More often than not this takes place automatically within a data acquisition program or within the company's Excel-based manual well testing calculations, neither of which are properly monitored by the senior or supervisory staff.

TABLE 8.1
BS&W Flow Rate Error Table

Flow Rate	BS&W Reading (%) and Flow Rate Correction					
Bbls/D	1	2	3	4	5	10
1,000	990	980	970	960	950	900
2,000	1,980	1,960	1,940	1,920	1,900	1,800
3,000	2,970	2,940	2,910	2,880	2,850	2,700
4,000	3,960	3,920	3,880	3,840	3,800	3,600
5,000	4,950	4,900	4,850	4,800	4,750	4,500
10,000	9,900	9,800	9,700	9,600	9,500	9,000

In most cases the BS&W reading taken is lower than the true reading because of poor and non-standard practices. This is the worst-case scenario for new wells or new fields as decisions are made on field development and perforation intervals based on these results.

Using Table 8.1, here is an example with a 3,000 Bbls/D flow rate:

If the reported BS&W is reported as 2% this gives a corrected rate of 2,940 Bbls/D.
If the true BS&W is 5% this gives a corrected rate of 2,850 Bbls/D.

This is a difference of 90 Bbls/D in the flow rate. These errors increase proportionally and dramatically with higher flow rates.

8.5 LARGE QUANTITIES OF WATER IN OIL LINE

Occasionally when monitoring the BS&W in the oil line we find amounts of water; this can be caused by the configuration of the separator level controllers.

Figure 8.1 shows the fluid levels normally associated with the separation of oil and water within a separator. The interface of the two fluids is below the top of the weir plate, the height of which depends upon the separator and configuration. Thus in three-phase mode only the oil flows over the weir plate and into the oil compartment and the water here is flowed out of the water outlet. This is a standard three-phase separation.

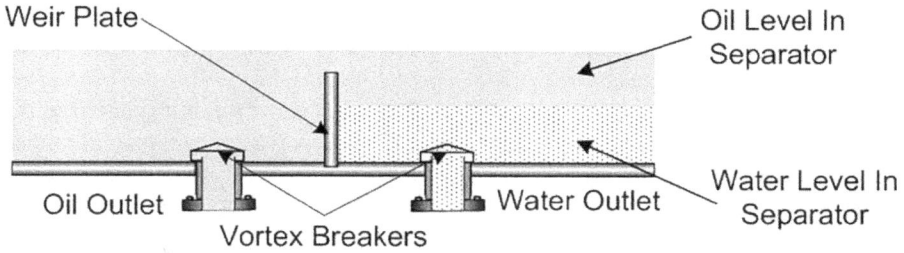

FIGURE 8.1 Depicting Oil and Water with Three-Phase Separation.

However, if the oil water interface is set too high to the top of the weir plate it is possible for the water to flow over the weir plate and into the oil compartment as shown in Figure 8.2. This water overflow can be simply corrected by lowering the interface level on the water controller.

Again, this shows the importance of using the separator sight glasses with a weir plate height indication to ensure correct levels.

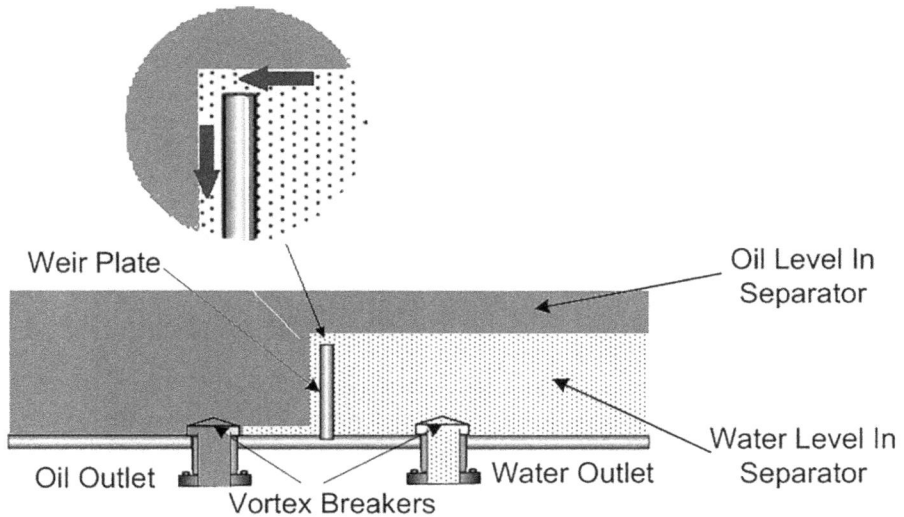

FIGURE 8.2 Oil and Water Interface above Weir Plate.

NOTE: *This like all operations has to be noted on the manual readings sheet as it may affect the flow rates when the water level is lowered. Water rate increases, oil rate decreases.*

8.6 MAJOR FACTORS AFFECTING BS&W

8.6.1 TIMING

It is generally accepted that a BS&W sample, along with gravity samples, should be taken every time there is a choke change. However, if the fluid at the surface is sampled immediately after the choke change, then the sample is not representative of the produced fluid at the perforation or sand face but rather the fluid column within the well from the previous flow.

What needs to happen is that time is allowed to pass before sampling, based on the flow rate and internal tubing size, allowing for the inflow from the reservoir to reach the surface, a "bottoms up" approach.

Another point that is often neglected is the fluid left in the separator vessel and flowlines from the previous flow. It is necessary to know the volume of the separator body, which indicates the residual volume of fluids left in the vessel. Sufficient flow time should be allowed for the residual volume to be displaced by the fluids that are to be sampled so there is no cross contamination of fluids within the separator, leading to errors in fluid properties.

Occasionally residual fluids are left in the well testing equipment from pressure testing, any flushing operation, or gaseous blankets, and attention should be paid to the existence of such fluids. Their existence contaminates any produced fluids and leads to fluid property changes that would be difficult to explain.

Only then, will the fluid that is sampled represent the fluid produced by the choke change.

Tubing Volume Equation

The volume of the tubing is given by the equation

$$\text{Tubing Volume} = \left[\frac{d^{\wedge}2}{1029.4} \right] * D$$

Where
  Tubing volume in barrels
  d = tubing internal diameter in inches
  D = depth of tubing in feet.

Knowing the flow rate, the time for the produced fluid to get to the surface is

Time to Surface Equation

$$\text{Time to Surface} = \left| \frac{\text{Tubing Volume}}{\text{Flow Rate}} \right| * 1440$$

Where
  Time to surface in minutes
  Tubing volume in barrels
  Flow rate = barrels/day

With today's computing power these equations are easily solved.

8.6.2 BS&W LOCATION

Depending on the well flow, type of test, and the operating company's requirements, the flow through the separator will be measured in either:

Three-phase This is where all three-phases, oil, gas, and water are metered and measured independently.
Two-phase This is where only the gas and combined liquids are measured.

Combined liquids are defined as a mixture of the oil and water which are not individually measured but rather the sum of the parts. There is a simple rule that is very easily – and often – ignored.

If there is measurable water then the separator should be run in three phases.

Some well test operators always take the easy option of running the separator in two phases even though there is measurable water, especially in GOR/PGOR tests. This always needs to be verified, as three-phase operation measures the water flow and two-phase derives it from the BS&W readings.

The phase the operating mode runs through the separator determines where the BS&W sample is taken.

Only in three-phase flow should BS&W be taken from the separator oil flowline.

NOTE: *Running the separator in three phase can significantly reduce the water content in the oil line, which can assist in improving burner performance.*

8.6.2.1   BS&W and Two-Phase Flow

With two-phase flow, the BS&W sample must be taken *before* the separator. Normally it is taken from downstream of the choke manifold. However, if a heater or steam exchanger is in the process stream, the better place to take the BS&W sample is after the heater as the fluid temperature has been artificially raised and in turn the fluid properties changed. The fluid property changes also assist the water to segregate from the fluid during centrifuging.

The illustration of the choke manifold, Figure 8.3, shows the sampling points being downstream of the chokes and using dual needle or ball valves. The upstream points are for bleeding off pressure only.

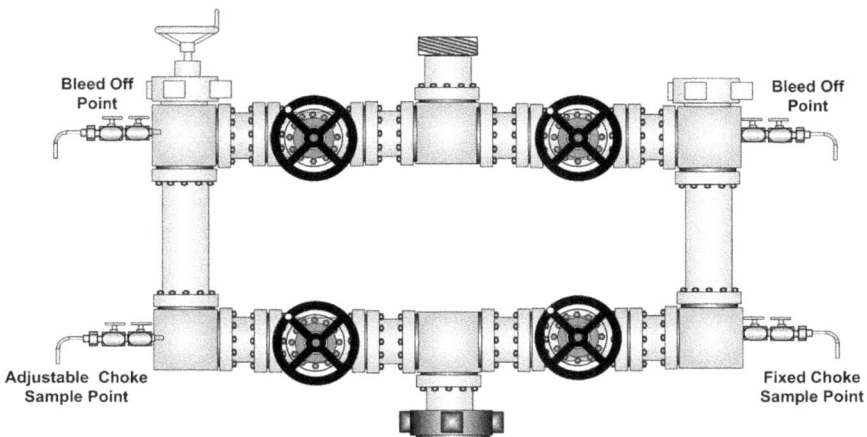

FIGURE 8.3   Choke Manifold with Sampling Points.

This setup is based on four factors.

i. It is safer to sample from a valve that has lower pressure than one that has a higher pressure
ii. The physical act of fluid passing through a choke will churn and agitate the produced fluid making it a more representative sample of the whole flow and easier to separate in centrifuging
iii. The use of dual valves is an essential safety measure

Figure 8.4. shows the two valves in series. Valve 1, connected to the pipework, should always remain open and only Valve 2 is opened and closed to take the sample or bleed off

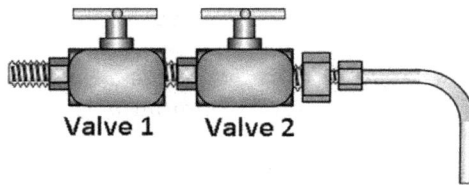

Valve 1   Valve 2

FIGURE 8.4   Sampling Valves.

The precautionary two-valve setup is because the well may produce sand and debris which may cause any valve to fail. In this configuration, though accessibility is an advantage, if Valve 2 fails, then Valve 1 can be closed and Valve 2 replaced without the need to depressurise the flowline to replace the valve. Valve 2 in this configuration is considered a sacrificial valve

iii. If chemical injection is introduced upstream of the choke manifold, then sampling at the upstream point can have a lower concentration of chemicals in the sample when compared with the downstream point

The chemical injection point on the upstream choke data header should be located after (downstream) the sampling point. If this is not the case, then injection should be ceased while a sample is taken. The process should be recorded on the manual data sheets.

If there is a sampling point nearer to the separator or process plant this can be used – again with dual sampling valves – but not from any of the separator outlet lines.

Two-phase separator mode's BS&W sample must not be taken from the separator.

8.6.2.2   BS&W and Three-Phase Flow

With three-phase flow, the BS&W samples are ONLY taken from the separator oil flowline; any other BS&W samples are not required.

This must not be confused with the BS&W readings taken at the choke manifold, which in three phase can be used to assess the separator efficiency and operation. For example, if the BS&W reading at the choke manifold is the same as the BS&W taken at the separator oil flowline it is obvious that the separator is not actually separating the liquid phases.

8.6.2.3 Position of BS&W Sampling Point

After determining which part of the process equipment the sample should be taken from, there is another consideration which can result in erroneous readings.

Crude oil will tend to naturally segregate in a pipe, especially over a distance, with the lighter component at the top and heavier at the bottom, illustrated in Figure 8.5.

FIGURE 8.5   Fluid Segregation in a Pipe.

Therefore, the position of the sampling point becomes important. If the sampling point is at the top of the pipe, then a gaseous phase will dominate. If at the bottom, then the heavier water will dominate.

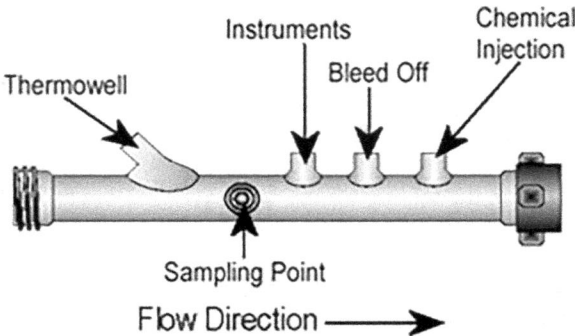

FIGURE 8.6   Example Data Header Configuration.

The action of the fluid passing through a choke will cause the crude oil to mix so the sample becomes more representative of the flowing fluid, therefore taking the sample downstream of the choke is important.

It is recommended that a data header, Figure 8.6, be installed immediately before and immediately after the choke manifold to facilitate all types of sampling; this is particularly important if a specialised sampling company is used.

Supplying two data headers is not normally standard practice when quoting a well test package and *should* be requested by the operating company.

8.7   CORRECTION OF FLOW USING BS&W

This has been touched upon earlier, but reiterated here under a more comprehensive understanding of operations and the variables in play.

8.7.1   TWO-PHASE WATER FLOW CALCULATION USING BS&W

For two-phase flow, the BS&W reading at the choke manifold is used to calculate the amount of water flowing in the separator oil line. The water and oil phases will flow out of the separator as a combined flow from its oil outlet.

The water constituent of the flow from the separator is therefore

$$\textbf{Water Flow} = \textbf{V}_s * (\textbf{BS\&W as a fraction})$$

Where

V_s = the total two-phase liquid flow rate through the oil meter.

The BS&W from downstream choke manifold as a fraction is meant that a BS&W of 12% is equal to 0.12.

Water flow rate can often be expressed as Q_w.

Therefore, if the total fluid flowing through the meter is 2,000 Bbls/D and BS&W is 12%. We would find that:

$$\text{Water Flow } (Q_w) = 2000 * 0.12 = 240 \text{ Bbls/D}$$

This of course is totally dependent on the accuracy and reliability of the BS&W reading used.

When the flow is put through the separator, the water separation is ignored and the total flow is output through the oil line. The water component of the flow is then derived using the total BS&W, taken downstream of the choke manifold. There still can be water flowing in the oil line which *must not* be used in any calculations when in two-phase flow.

Although the separator may actually internally separate some water from the flow, most of this separated volume is retained in the separator vessel. Therefore, the BS&W in the oil line can be different to the total BS&W as water has been extracted.

The total BS&W actually accounts for the water separated and the water retained in the separator body. Any separated water has been accounted for and should not be included in calculations or cumulative volumes; the BS&W from the oil line must therefore not be used.

8.7.2   THREE-PHASE WATER CALCULATION FROM BS&W

With the separator operating in three-phase there can be two separate components for the water calculations. The total water flow is measured by a water dedicated meter.

- Water flowing in the water line
- Water mixed with oil in the oil line

As no separator is usually totally efficient there may be water exported in the oil line when operating in three phases. This must be corrected for both the oil flow rates acquired as well as the water flow rates.

The formulae for compensating water flow rate entrained in the oil flow is:

Corrected Oil Flow (Q_o) =Oil Vs *(1-BS&W as a fraction)

Where
   Oil Vs = the separated oil flow rate through the oil meters. (Oil flow rate can
      often be expressed as Q_o.)
   BS&W from the oil line as a fraction – a BS&W of 12% is equal to 0.12.

Example

$$\text{Corrected Oil Flow } (Q_o) = 2000 \times (1 - 0.12) = 1760 \text{ Bbls/D.}$$

From this figure the difference in the flow of 240 Bbls/D is due to the water in the oil line that has not been separated out properly by the separator.

The water entrained in the oil line constituting part of the total can be simply calculated by:

Water in Oil Line = Vs * (BS&W as a fraction)

Where
   Oil Vs = the separated oil flow rate through the oil meters. (oil flow rate can
      often be expressed as Q_o.)
   BS&W = BS&W from the oil line as a fraction
   A BS&W of 12% is equal to 0.12.

$$\text{Water in Oil Line} = 2000 * 0.12 = 240 \text{ Bbls/D}$$

Therefore, the total water flow from the separator is:

Total Water Flow = Water Flow in Water Line + Calculated Water Flow in Oil Line

This gives, with a measured water flow rate from the separator water line of 480 Bbls/D,

$$\text{Total Water Flow} = 480 + 240 = 720 \text{ Bbls/D}$$

This is why accurate, correct, and repeatable BS&W measurements are critical during the entire well test. To re-iterate all this data including time, BS&W reading, location of sampling points, and liquid meters readings should be written on the manual readings sheets.

Several company representatives will insist on the BS&W being taken from the choke manifold while the flow is through a three-phase separator. This can and does lead to misunderstandings and confusion. To clarify:

- The BS&W from the choke manifold is raw and unprocessed
- The BS&W from the separator has been processed and separated so the two can never be compared

There is nothing wrong in carrying out BS&W measurements from two different sampling points as long as they are clearly labelled in the report and no inference is given on dissimilar results.

Refer to the sampling sheet detailed in the reporting section.

A high BS&W reading from the separator oil line can indicate the separator is not separating the phases correctly and may require mechanical intervention to the separator settings. As described earlier, if water carryover is suspected or observed then the water level should be checked on the separator water sight glass and, if necessary, the water level should be lowered using the water level control valve. This should be revised after allowing time for the water level to be lowered and stabilised.

8.8   BS&W SAMPLING

The following section gives an indication of the sequence and actions that should be undertaken in taking a BS&W sample either from the choke manifold or the separator. Several service companies supposedly have their own procedures but in practice, in the field, the operators are either unaware of the proper method or are too complacent to adopt them.

8.8.1   BS&W EQUIPMENT REQUIRED

Depending on the area or type of test, the equipment used may vary. The list given here is with ideal conditions.

  i. Exd certified electric centrifuge with heating capability
 ii. 4 x Centrifuge tubes and spares
iii. 2 x 1 litre measuring cylinders (***NOTE:*** *1 litre and not smaller*)
 iv. Demulsifier and associated solvents
  v. Cleaning materials
 vi. Drip trays and buckets
vii. Disposal container (45-gallon drum)

It is important that the vessel in which the fluid sample is taken in has a wide mouth with no restrictions so that any entrained gas can easily escape. The practice of using a disposable water bottle that has a neck restriction will restrict escape of gasses. An even worse effect with closed bottles or cylinders is with the shrinkage effect on BS&W samples; the pressure is not completely to atmosphere. A measuring cylinder is specified and must be used.

8.8.2   BS&W SAFETY

- All BS&W sampling must be carried out in accordance with the safety practices and arrangements on the location

- Before attempting to obtain a sample, from whichever designated sampling point, ensure that the operator is wearing the full and proper PPE with special attention to H_2S, safety glasses, and gloves
- Until the well has been tested and no dangerous contaminants are confirmed, H_2S, acid etc. BS&W should be conducted with extreme caution and all sampling should be considered to contain hazardous materials and gasses, so all monitors and precautions must be in place
- Drip trays or suitable buckets must always be placed beneath all the sampling points, to contain spillages
- The correct PPE and adherence to safety procedures must be in place and followed at all times by all personnel

8.8.3 BS&W Sampling Procedures

There are five major ways of determining a BS&W; these methods are combined in a number of ways to improve results:

i. **Cutting or dilution** – This is where the raw sample is diluted or cut using a solvent – or diesel – to reduce the viscosity so that the water component separates under centrifugal force

ii. **Demulsifier** – A demulsifier is a chemical that assists in breaking any emulsions or mixtures of oil and water that will not separate under normal conditions

iii. **Heating** – Heating the oil in a bath reduces its viscosity

iv. **Untreated** – The sample is centrifuged without any additives, although heating may be applied to enhance the process

v. **Settling** – This is where a larger sample is left to settle over a generally long period of time, as a reference

These methods are revisited later in this chapter with further details and explanations.

8.8.4 BS&W Different Types of Water Content

There are two distinct types of BS&W samples.

- Conventional with low water cut – water in oil
- Water content greater than oil – oil in water

These require different analysis methods and will be detailed separately in the liquid gravities section.

8.8.5 Initial BS&W Samples

Throughout the oilfield environment there are many different types of oil from very low gravities to very high gravities; consequently some of the BS&W methods work where others may not.

It is important that when the well first starts flowing hydrocarbons, an initial BS&W sample is taken, measured, and recorded. The next BS&W sample taken from the same location should use a different BS&W method to verify the first.

Example

BS&W Sample 1 – Depending on the oil type the method of determining a BS&W is as follows:

   i. If the sample is with light oil, then use the sample untreated
  ii. If the sample is with heavy oil, then the sample should be determined using the dilution or cutting method
          If at all unsure use the cutting or dilution method

BS&W Sample 2 – If sample 1 was determined using the cut method with a solvent like toluene the next sample should not be diluted at all but used untreated or with a demulsifier to verify the first sample result.

All results with the BS&W method used are to be recorded on the manual readings sheets.

Now we have two different results from two different methods but they should both have the same BS&W results. Thereafter, the supervisor should decide which method should be used for the duration of the test.

The operating company may have a record of wells in the field. If it is an existing or known field it could be possible to compare the current readings with findings and results from other wells in the same field and reservoir.

Where routine reservoir performance testing (GOR/PGOR) is conducted it is common practice to utilise BS&W, gravity (oil and gas), etc. from the results from previous well tests. This defeats the objective of the test; the well could have been subject to work-over or stimulation so the previous data will be invalid. This practice must be discontinued. There is no reason why a comparison between a previous test and the current test should not be noted on the report, but each test is conducted on its own merits.

It is important that company personnel, whatever the type of test, witness at least one BS&W sample taken and processed to verify the procedure and reported results.

If there is any doubt with the result, the sample and testing method can be repeated, or another method can be used to confirm results. If doubt still pursues all of the results, the settling method can be employed; taking a one-litre sample of the fluid, have it labelled and kept for a period of time to see the segregation or have it sent to a PVT lab for analysis.

In all instances, the method of obtaining the BS&W sample in a one-litre measuring cylinder should *not* be changed.

8.9   BS&W SAMPLE RECORDS

The practice of keeping a record of all the samples is often not followed. This has two main functions:

   i. A record and summary of all the samples taken during the test in one place
  ii. To monitor any fluid changes that occurs during the test period

It is common assumption that the % of BS&W can decrease with the time from when the well was opened, especially with new wells. Similarly, the gas, oil, and water gravities can change, and these fluid properties' changes are equally important to record.

It is strongly recommended that a sample sheet is used, similar to Table 8.2. All the results including date, time, sampling point, and any other pertinent information (comments) should be included in the main results sheet and the sample sheet.

TABLE 8.2
Example of BS&W Record Sheet

Date and Time	Sample Number	Sampling Point	BS&W		Spin Time	Heat Time	Cutting		OIL SG	Temp DEG ()	Salinity	pH	Gas SG	H₂S ppm	CO₂ %
			S	W			C	D							

From Table 8.2, both BS&W samples are recorded for accuracy and clarity as well as the other readings associated with well testing. Detailing these variables allows any changes or potential wrong readings to be easily identifiable as the well cleans up. This data filled sheet should also be included in the data acquisition package report.

8.9.1   COMMON PRACTICE OF BS&W SAMPLING

This procedure is an example of the common approach to atmospheric fluid sampling and incorporates steps and methodology that are often omitted or left out. The methodology applies to all liquid sampling whether for liquid gravity, BS&W, salinity, or other measurements.

The first step is taking the sample from the flow.

All BS&W samples should be taken in a *clean and dry one-litre measuring cylinder; this is mandatory.*

Why Use a Measuring Cylinder?

The reason for the one-litre measuring cylinder, Figure 8.7, being specified is that the well testers will use the same cylinder to acquire liquid gravity measurements using instruments, and smaller cylinders do not have enough room or clearance for valid hydrometer gravity readings to be taken.

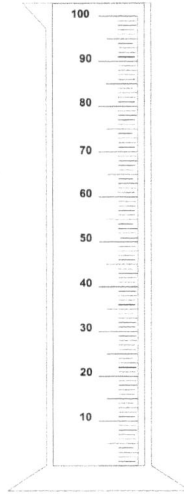

FIGURE 8.7   Measuring Cylinder.

The measuring cylinder should be manufactured in a clear or translucent form of plastic, most commonly nylon, as glass sampling apparatus are not safe to use in the field, because:

i. They are slippery with fluids
ii. If the sample surges or spurts out of the valves, then there is a good chance that the glassware will be forced out of the holding hand, causing the glass to break with a high risk of injury

NOTE: *If the measuring cylinder has become opaque or damaged in any form it should be replaced.*

The measuring cylinder is also a calibrated graduated vessel that gives the operator the chance to actually measure the quantity of liquid being drawn. Also, when all else fails and as a quality check, one litre of oil left in the measuring cylinder will separate if given enough time. Since the cylinder has a graduated scale, the level of separation between oil and water can be measured easily.

It is often necessary to get a rough idea of shrinkage using a measuring cylinder; with a combined meter shrinkage factor the actual shrinkage cannot be extrapolated from the reading. The graduated scale on the measuring cylinder assists with this.

If the sample is used for shrinkage using the measuring cylinder, then the oil sample MUST be taken from the separator oil line. After settling, the final level must be compensated for temperature to bring it to 60 degrees F conditions.

In some cases, the shrinkage is not measured but determined from a computerised equation – or from tables – for use in the GOR2 estimation; then using a sample left in a cylinder will act as a validating/comparative purpose.

FIGURE 8.8   Sample Left in Measuring Cylinder.

Heavy Gravity Oil

In some parts of the world the oil is too viscous to pour into the BS&W centrifuge tubes. In order to reduce the viscosity, normally achieved using heat, the measuring cylinder should be placed in a container containing near boiling water and allowed to heat up.

Timing and care should be taken because while the crude oil becomes less viscous, gas and water can be released from the sample. If possible, the sample should be removed from the heat before any separation occurs.

Instead of simple agitation, the measuring cylinder should have the top covered and the whole sample shaken in a circular motion to ensure the sample is fully mixed while being aware of potential gas release.

NOTE: *Full PPE must be used when taking BS&W samples.*

Dead Oil

A main issue with BS&W sampling is the "dead oil" as depicted in Figure 8.9.

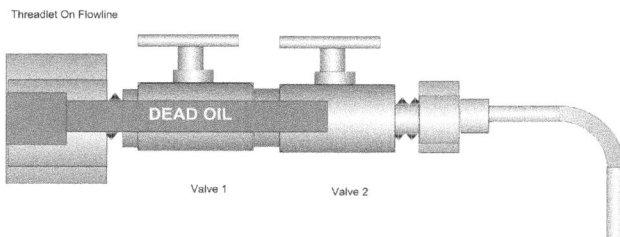

FIGURE 8.9   Dead Oil.

After a sample has been taken and the front valve closed there is a section of oil that is trapped within the fittings, shown in Figure 8.9. The trapped oil remains in place until another sample is taken which could be up to an hour.

This means that the fluid from the sampling point has to be flowed to a tray or similar container to ensure that the "live oil" is sampled and the dead oil does not affect the reading.

Taking a sample directly into a small volume centrifuge bottle causes errors as the oil used is not a true sample, being mostly dead oil.

BS&W Sampling Technique

- The correct PPE and adherence to safety procedures must be in place and followed at all times
- This procedure is based around a standard one-litre measuring cylinder that should be used for all fluid sampling procedures

Place the measuring cylinder under the gooseneck at the sampling point as shown in Figure 8.10. Position the gooseneck against the side of the measuring cylinder wall so the sample flows round the internal wall of the measuring cylinder and not directly into the bottom of the cylinder, as this can cause splashback. The operator should not bend over the cylinder as the potential for gas escape is high and if there are any dangerous contaminants then breathing apparatus should be worn.

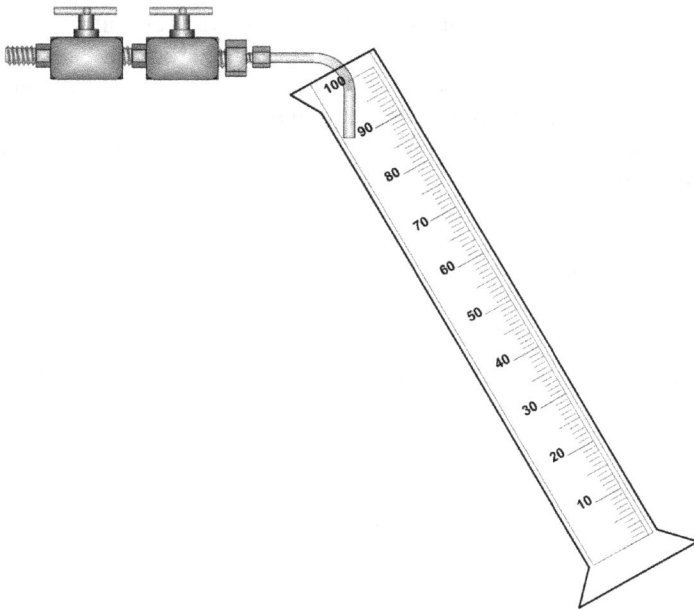

FIGURE 8.10   Measuring Cylinder in Position for Fluid Sampling.

The valves used on BS&W sampling ports should be a needle valve or similar construction. Ball valves are not recommended, since with needle valves a turbulent flow is induced which assists in the breaking of emulsions and entrained gas in the fluid mix, similar to the effects of a choke. Also, from a safety point of view there is more control when dealing with a pressurised fluid using a needle valve than a full flow or ball valve.

The measuring cylinder should be held at an angle, as shown, with the gooseneck held firmly on the measuring cylinder wall so the fluid is forced to flow in a circular motion around the inside of the measuring cylinder wall. The needle valve should be opened slowly so that the flowing pressure does not cause the cylinder to be forced out of the operator's hand.

The causing of fluid to flow into the cylinder with a circular motion is assisting the dissolved or entrained gas to be released. Entrained gas present in the centrifuging process is a significant cause of error.

The operator should take the minimum volume of liquid, approximately 500 mL (0.5 litre), which is considered to be a representative sample of the liquid flow, more if the sample is to be used for other tests.

NOTE: *We should not acquire the dead oil in the sampling lines, this cannot be used.*

If possible, use the remaining sample for the oil gravity measurement. This is another purpose of a large sample, so, we can get all the fluid results from the same sample.

It is good practice to take enough sample to cover the requirements of all the applied tests. A second or top up sample can be taken if necessary but this has the potential to change the measured results. If a second sample is taken and used then like all events it should be highlighted on the manual reading sheets and DAS sequence of operations.

The practice of taking a sample directly into a centrifuge tube MUST NOT be used as it is both very dangerous and does not represent a true sample of the produced fluid.

Once a sample has been taken the sample in the measuring cylinder must be kept agitated and not allowed to settle under any circumstances. This should be done by a continual hand rotary action on the cylinder so the fluid swirls round inside allowing any dissolved or entrained gas to escape the fluid while keeping the liquids mixed. Without this is being performed BS&W will have measurement errors.

The sample must be put into the centrifuge tubes as soon as possible so no segregation takes place that can affect the BS&W reading. As much entrained gas as possible should be allowed to escape from the sample without liquids segregating before any BS&W operations; this is achieved by continuing agitation. Entrained gas in the fluid sample WILL cause shrinkage in the BS&W tubes.

The remaining fluid in the measuring cylinder should be left to sit before any gravity measurements that are taken after segregation of oil and water and de-gassing of the liquid to avoid their effects on gravity readings.

No demulsifiers or cutting liquids may be put in the measuring cylinder.

Figure 8.11 shows pictures taken for a gassed sample acquired; the dissolved gasses can be seen still exiting the sample and should not be used for gravity measurements. The degassed sample as photographed is the one that should be used for gravity and any other measurements.

Gassed Sample   De-Gassed Sample

FIGURE 8.11   Crude Oil Samples.

As the gasses escape from the sample the volume will shrink, as pictured in Figure 8.12; temperature effects and separation will also account for shrinkage.

FIGURE 8.12   Sample Showing Shrinkage If Left to Settle.

NOTE: *If the shrinkage in a measuring cylinder is to be used for comparative purposes to the shrinkage equation, then the oil sample MUST be taken from the separator oil line.*

If the centrifuge tube is filled with oil that has gas content, as shown in Figure 8.13, then when it is spun in the centrifuge the volume will shrink. This is often a missed influencing factor resulting in BS&W reading errors.

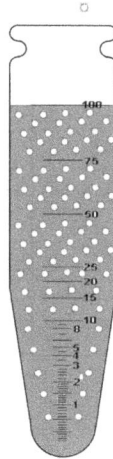

FIGURE 8.13   Centrifuge Tube with Entrained Gas in Oil.

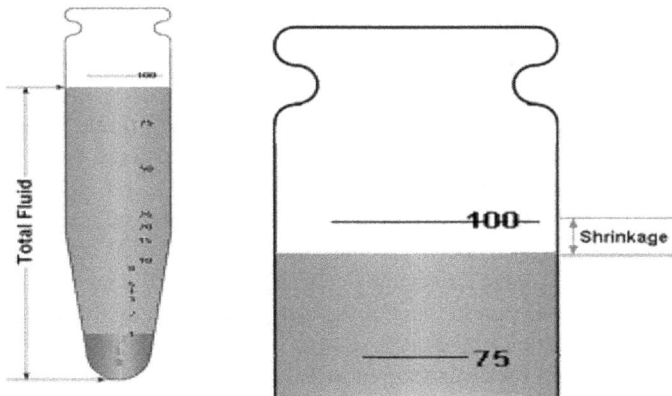

FIGURE 8.14   Centrifuge Tube Showing Gas Shrinkage after Centrifuging.

After centrifuging, the water will separate from the oil giving a percentage of BS&W. However, the entrained or dissolved gas will escape the oil volume, resulting in the oil volume shrinking. As an example the oil has shrunk to 90% of its original volume so a 1% BS&W reading originally acquired on the tube is not a true value any longer. The BS&W should be a fixed percentage from a total volume that does not change.

If there is a constant shrinkage effect on the BS&W tubes then an Excel type sheet should be utilised to calculate the correct amount of BS&W as shown in Table 8.3. This is often missed by operators and an uncorrected BS&W is reported to the DAS operator.

TABLE 8.3

Example BS&W Correction Sheet

	Tube 1					Tube 2				
	Reading from Tube			Corrected		Reading from Tube			Corrected	
Time	Total Fluid	H_2O	Solids	Cor H_2O	Cor. Solids	Total Fluid	H_2O	Solids	Cor H_2O	Cor. Solids
12:15	97	6	1	**6.19**	**1.03**	97	5.5	1	**5.67**	**1.03**

Table 8.3 will keep a record of the BS&W readings and correct them for shrinkage. This correction sheet is not normally included in the report but should be kept in the well folder for future reference.

8.10   BS&W CENTRIFUGING

A centrifuge is a laboratory device used for the separation of fluids. The fluids while in a vessel are spun at high speed, generating a centrifugal force on the liquids and separating them based on density.

There are myriad different types of centrifuges available and it is generally up to the service company to provide a suitable model. Any electric centrifuges used in hazardous areas MUST have a relevant electrical safety certificate.

Regardless of the make and model of the centrifuge used, the tubes must be clean and dry with no residue left in the tubes. The settings, methods, or techniques applied to the samples intended for centrifuge must be consistent and constant. The settings on a centrifuge mainly fall under:

    i. Heating time or temperature
   ii. Speed of rotation
  iii. Duration of rotation

All these factors affect the result and must remain constant. If any changes occur, then they should be noted, explained, and continued from then on, especially across shifts.

Initially on a new well all aspects of centrifuging preparation, demulsifier, cut, etc. should be tried until the one that suits the composition of the produced oil is concluded. This method should then be continued throughout the test as the default method.

The centrifuge holder normally has four positions, as shown in Figure 8.15; there can never be an odd number of tubes as this will throw the centrifuge out of balance. Two tubes should be used, and these can only be in the positions of:

A and B or C and D

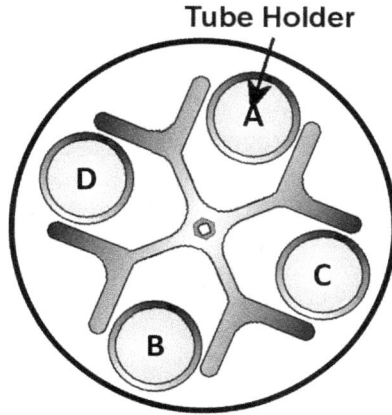

Tube Holder

FIGURE 8.15   Centrifuge Tube Holder.

Only if the positions are marked can four tubes be used with two different preparations.

There is a certain amount of "loose practice" here with only *one* sample being taken and a placebo or dummy put in the centrifuge equipment to act as balance. Two samples *MUST* be taken at all times as this acts as a quality control measure by being able to compare the BS&W in both samples for similar results.

This malpractice should be discontinued as soon as possible; it increases the chances of error in the BS&W sample and comparisons cannot be performed. Failure to do so should be considered a disciplinary offence.

8.11   BS&W IMPROVEMENT METHODS

Often the gravity or the viscosity of the oil is too high for the components to be separated during the centrifuging stages. The oil gravity or viscosity of the sample are reduced by mixing with a known volume of a lighter solvent to aid separation before the samples are placed in the centrifuge. This also applies to emulsions that are difficult to break.

8.11.1   Fluids Used for BS&W Cutting

NOTE: *Protective rubber gloves should be used here in addition to standard PPE.*

There are several fluids used for cutting the BS&W samples; these can be from:

  i. Xylene
  ii. Toluene
  iii. Benzene
  iv. Varsol

If none of these are available, then diesel is an option. Gasoline or petrol should not be used due to safety issues (flammability).

All chemicals should have their safety data sheets available on location and all recommended handling procedures should be followed. If there is a specified and standard process at the location regarding cutting procedure and regarding the fluid used to cut the sample, then this must be continued so that comparisons of measurements make sense.

The centrifuge tubes should be clean and dry before use. If this is impractical the tubes should be cleaned or rinsed with the chemical used for cutting to minimise introduced errors.

Both of the two centrifuges tubes should have the cutting fluid poured into the tubes to the required level which is normally a standard of 50% by volume as shown in Figure 8.16.

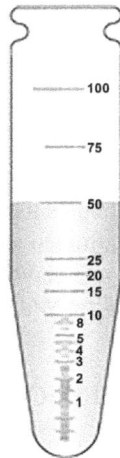

FIGURE 8.16   Centrifuge Tube with Cutting Fluid.

The cutting fluid should be prepared *prior* to actually taking the BS&W sample from the sampling point, so no time is given for the sample to sit and settle in the measuring cylinder until it is combined with the cutting fluid; this can lead to erroneous results. Also, it is important that the cutting fluid is poured into the centrifuge tubes first, to allow the operator to record a precise measurement of the cutting fluid used.

Fifty percent cutting fluid dilution is the standard for cutting the crude oil in a centrifuge tube that is to be centrifuged for ease of calculation as detailed in Figure 8.17.

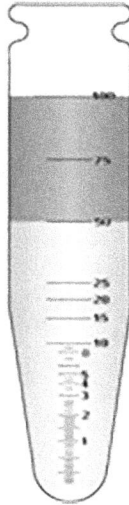

FIGURE 8.17   Cutting Fluid and Fluid Sample.

All samples and measurements must be treated in the exact same manner and follow the same process; this is critical for comparable results. It is occasional practice to cut one sample and not the other, using two centrifuge tubes. This is an area where confusion can occur as, unless care is taken, it is often difficult to determine a benchmark and obtain a comparative result between samples.

The sample of fluid in the measuring cylinder should have been continuously agitated prior to pouring the liquids into the centrifuge tubes. If the sample in the measuring cylinder has been allowed to sit then it should be discarded for a fresh sample.

Using the agitated sample in the measuring cylinder, fill the two centrifuge tubes to the 100% graduation mark, ensuring that the sample in the measuring cylinder is kept agitated by spinning it, especially between tubes.

After pouring the sample from the measuring cylinder into the centrifuge tube, the mixture of cutting fluid and sample must be mixed. The normal method to do this is to put a gloved thumb on the top of the centrifuge cylinder and invert and shake a few times until the fluids are visually mixed. There is nothing wrong in this method if the operator is wearing suitable protective gloves and eye protection to perform it.

After mixing, the operator should be aware that the mixing of the two liquids may cause a significant change in viscosity and enable gas to be released. Consequently, the fluid may spray out of the tube, therefore PPE is essential, especially eye protection.

The release of gas is indicative that the sample has undergone shrinkage; once again the sample has shrunk and the total volume has changed. In this case, if the sample in the measuring cylinder has been kept agitated then the centrifuge tube can be "topped up" to the 100% volume mark, with the sampled fluids and mixing repeated. If the sample in the measuring cylinder has been left to settle, then the entire process should be repeated from the start at the sampling phase. Topping up with the cutting agent at this stage is incorrect and will give false readings.

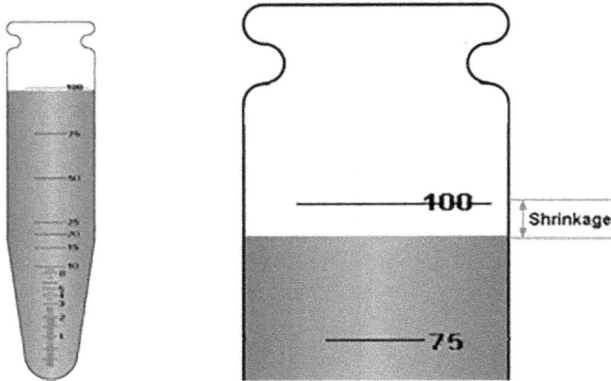

FIGURE 8.18   Shrinkage Effect of Gas Released during Mixing with Cutting Fluid.

When the cutting method has been adopted it is generally sufficient; heating the BS&W samples would no longer be required, placing the two samples in the centrifuge as a final step before results. Occasionally heating and demulsifying the BS&W samples to a cut sample may be an adopted technique; then this strategy should continue for the complete well test to ensure consistency and proper comparison.

Again, here we emphasise that both BS&W centrifuge tube samples must be treated the same way; one should not be prepared with cutting agent while the other one without. With dissimilar samples the comparisons are invalidated.

8.11.2   BS&W Demulsifier Method

A demulsifier is a chemical that is introduced into the crude oil emulsion to aid and accelerate the separation of its oil and water phases. They can range from very simple to complex chemical additives. Demulsifiers are not just used in BS&W sampling but are often used by chemical injection pumps prior to passing flow through the separator for the same purpose.

As per standards, fill a clean and dry centrifuge tube to 100% from the measuring cylinder sample, as shown in Figure 8.19 and agitate by rotating to help remove any entrained gas. Again, this is done while being aware of the possibility that entrained gas can cause a spray or splash out. If the volume is reduced due to shrinkage, refill from the sampling cylinder provided the sample has not been left to settle.

Normally only a single drop of demulsifier is added into the centrifuge sample and then shaken again with caution to allow gas escape.

The use of demulsifier is usually uncontrolled and it is common to see operators just tipping or shaking demulsifier from a bottle rather than by a controlled method.

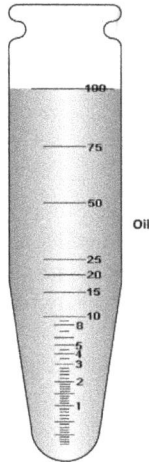

FIGURE 8.19   Centrifuge Tube Full of Raw Crude Oil.

The act of adding demulsifier adds to the total volume of the sample and if it is not properly recorded and monitored can lead to erroneous measurements.

This is usually a very minor error, but the error is magnified when operators use different amounts and a variety of demulsifiers for different samples, dramatically affecting the results. It is important that the demulsifier additions are controlled and consistent.

8.11.2.1   Types of BS&W Demulsifier

There are several types of demulsifiers on the market. These range from Tretolite, first introduced in 1914 by Baker Hughes Ltd. to the newer viscosity reducing chemicals such as NuFloTM 187 from NuGeneration Technologies LLC.

However, there are occasions where a commercial emulsion breaker is not on site; then the more common and easily available substances that can be used are:

i. Washing Up Liquid or Liquid Soap

The addition of soap helps break the chemical bonds that enable the oil to bond to the water molecules, aiding in their separation.

ii. Salt

Salt is absorbed by the water in the sample, increasing its specific gravity, making it easier to separate from oil when using a centrifuge.

iii. Raw Oil

This is the preferred method as long as the correct PPE and adherence to safety procedures are in place and followed at all times

When using the raw oil method, fill a clean and dry *centrifuge tube* to 100% level from the measuring cylinder with the oil sample (which has been agitated constantly to help remove any entrained gas) and leave it aside before carrying out the centrifuging. This makes it possible to top up any shrinkage that has occurred in the BS&W tubes.

This can be done as long as the sampling cylinder has not been left to settle or any additives used.

8.11.3 BS&W HEATING

Some centrifuges have heated centrifuge bowls which keeps the sample temperature constant during the centrifuge cycle. If available on the centrifuge it is good practice to heat the sample for a fixed period or to a fixed temperature before the centrifuge cycle is started. Heating reduces the viscosity of the oil making the centrifuging process more effective.

If heating is used it should be at a fixed temperature and for a fixed duration so all samples are processed the same way, otherwise different results would be induced with different samples.

8.11.3.1 BS&W Temperature

Heating the sample before centrifuging helps the process of oil/water separation and accelerates the process. Increasing the temperature of the BS&W sample to be centrifuged can be combined with other methods of improving the process and has the following benefits:

i. Reduced viscosity of liquids, primarily oil
ii. Heating the sample will increase the mobility of the water either in an emulsion or as drops, so the water will coalesce and drop out of solution faster
iii. Heating will increase or accentuate the difference in density of the component parts, which improves separation of fluids

If the samples have NOT been previously heated, the same non-heating strategy and practice should be continued. If heating has been introduced, then it must be continued on all samples in order for the results to be consistent and comparable. The heat treatment of the sample must be repeated consistently throughout, mainly the duration of heating and temperature.

NOTE: *A heating time of five minutes before centrifuging may be considered as a starting point.*

8.12 BS&W CORRECTION EQUATION

This is a simple calculation where the corrected H_2O and solids can be added together as a percentage for the BS&W measurement:

$$\textbf{Corrected } H_2O = (H_2O/\textbf{Total Fluid}) \times 100$$
$$\textbf{Corrected Solids} = (\textbf{Solids/Total Fluid}) \times 100$$

NOTE: *Only the corrected H_2O is used for correcting flow rates.*

It is good practice to leave the remaining fluid in the measuring cylinder for use in gravity measurement. In this case the fluid should be allowed to sit and settle for several minutes to allow the oil and water to separate. The separated oil can then be utilised for the measurement of the specific gravity. This ensures that both readings are contiguous.

When using the sample for gravity readings, it is critical that no demulsifier, cutting agent, or any other chemical or substance is introduced into the sample as it will affect the results.

It is good practice to have multiple one-litre measuring cylinders within the well test lab, allowing the use of the remaining BS&W containing measuring cylinder to be used simultaneously with further sampling cylinders, leaving the older samples in their measuring cylinder to separate liquids.

Before carrying out the physical centrifuging of the liquid a simple check is to place your thumb, using rubber gloves, over the end of the centrifuge sample tube and invert it or shake it. Be aware of potential trapped gas or escaping of gasses and fluids, so PPE, especially glasses, is mandatory.

When the thumb is removed if there is a "pop" of gas, similar to a soda bottle, then there is entrained gas in the sample and it will shrink during centrifuge operations giving errors as described earlier.

8.13   COMPARISON OF BS&W SAMPLES

After the centrifuge cycle has been completed then both the samples should be visually compared. They should be similar or very close, as shown in Figure 8.20.

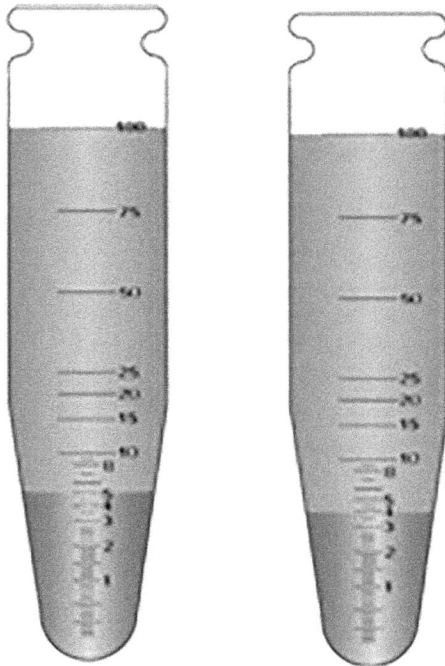

FIGURE 8.20   BS&W Samples from Centrifuge.

The BS&W samples from the centrifuge should now be compared and upon closer examination it can be seen that there is a slight difference, as illustrated in Figure 8.21.

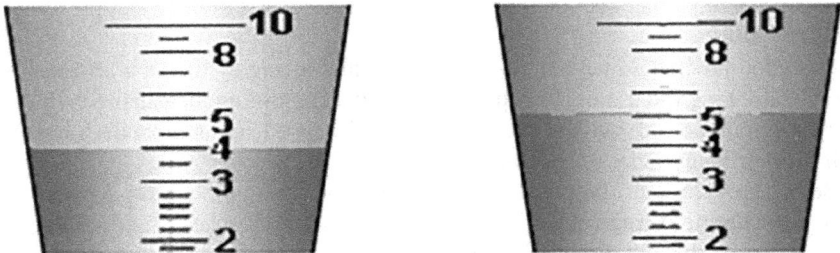

FIGURE 8.21   Detailed Examination of Samples from Centrifuge.

When the two samples are examined, it is common to have a minor difference between the levels of water and oil. However, if there is a significant difference between the two samples, nominally 3%, then the sample should be discarded and a new sample taken.

Figure 8.21 shows the results of two samples from the same centrifuge process – or spin out – where there is a small difference between the BS&W.

The first sample shows 4% water.
The second sample shows 5% water.

8.13.1   BS&W Samples Averaging

With two different results the average reading from the two samples is calculated by:

$$BS\&W = [(\text{High BS\&W Reading} - \text{Low BS\&W Reading})/2] + \text{Low BS\&W Reading}$$

$$BS\&W = [(5-4)/2)] + 4$$

$$BS\&W = 4.5\%$$

Both individual readings and the average should be recorded on the well test manual readings sheets.

8.13.2   BS&W Readings Using Cutting

The results from using 50% cutting or dilution mean that the readings from the centrifuge should be doubled.

The first sample shows 4% water with cutting in place; the true reading is then $4 \times 2 = 8\%$

The second sample shows 5% water with cutting in place; the true reading is $5 \times 2 = 10\%$

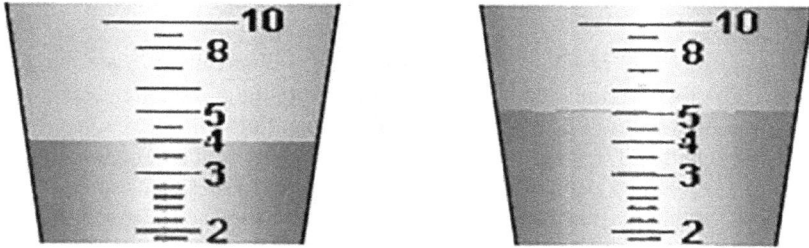

FIGURE 8.22   Cut or Diluted BS&W Results.

The average of the BS&W reading is now:

$$BS\&W = [(\text{High BS\&W Reading} - \text{Low BS\&W Reading})/2]$$
$$+ \text{Low BS\&W Reading}$$

$$BS\&W = [(10 - 8)/2)] + 8$$

$$BS\&W = 9\%$$

There is often a discrepancy in reported BS&W readings as the effect of cutting was never – or consistently – applied. The calculations demonstrate why 50% cutting is employed as standard; it makes the calculations simpler.

8.13.3   BS&W HIGH PERCENTAGE WATER CONTENT

Throughout the oilfield there are wells that have water content greater than the oil content; this now becomes an oil in water measurement but the term BS&W is still employed.

The normal method of centrifuging the sample often does not yield the correct results; it is here that the technique of settling by employing a measuring cylinder on its own works best. The measuring cylinder as per the standard practices is filled to the one-litre mark. This sample is then left to segregate for a specific time of no less than 30 minutes.

It is recommended that the level of water separation be taken every 30 minutes until there is little or no change in the volume of water. The length of time it takes for no further changes in the water volume gives a standard time for settlement for the entire well test, which should remain constant.

It is fairly common to see use un-calibrated bottles or sealed bottles being used for this purpose, but these will not give an accurate reading so only graduated measuring cylinders should be in use. The measuring cylinder has a wide mouth that allows for any gas releases and is graduated so an accurate measurement can be obtained.

As mentioned earlier, this is an example of why two measuring cylinders are specified, so one BS&W sample can be allowed to sit and settle/segregate while the other cylinder can be used in standard operations.

8.13.4 INJECTED CHEMICAL EFFECTS ON BS&W

If there is a high level of injected chemicals in the well stream, hydrate prevention, defoamer, etc. that are injected upstream of the sampling points, then this can have an effect on the measured BS&W. This can also affect fluid gravity readings as well.

Normally chemical injection rates will have no or little effect. What is important is that any increase or decrease of chemical injection must be documented in the sequence of events or well test diary and included in the final test reports so any anomalies can be explained.

When and if the injected chemical is found to be influencing the BS&W or gravity then:

   i. The injection should be stopped during the sampling phase
   ii. The sample should be taken upstream of the injection point

Depending upon the flowing pressures and rates this can be a hazardous operation and full safety precautions must be taken.

From the Data Header schematic in Figure 8.23, the chemical injection port is the last port before the flow enters the choke manifold. This is mandatory, otherwise the act of injecting chemicals will cause a pressure increase/surge that will be picked up by the data acquisition system, causing erroneous readings. It will also be part of the sample taken for measurements. The correct sampling point is in the centre of the flowline and away from the injection port.

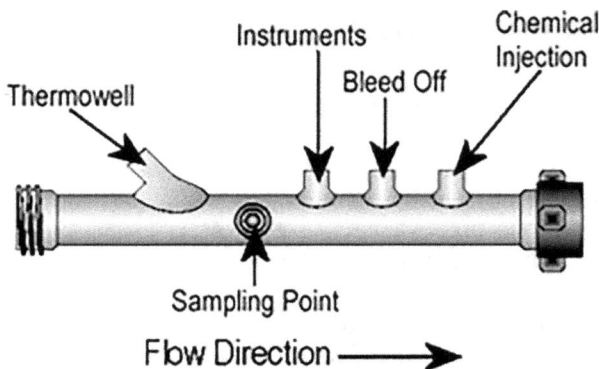

FIGURE 8.23   Typical Data Header.

8.13.5 CLEANING

After the BS&W measurement has been completed, the BS&W tubes should be emptied and cleaned with a suitable safe solvent, usually cutting fluid, until all traces of liquids and sediments are removed. The same is done with the measuring cylinders.

The BS&W tubes and cylinders are normally left in an inverted position, supported on a wooden peg, nail, or slickline shear pin, to dry ready for use.

8.14   ELECTRONIC MEASUREMENT OF BS&W

There have been major advances in the technology of measuring the amount of water in an oil line over recent years. These are sometimes used to replace manual BS&W readings, especially with in-line real-time instruments used in DAS. These water cut reading meter technologies range from:

*   Optical
*   Capacitance
*   Conductance
*   Multi phase flow meters (MPFM)
*   Ultrasonic
*   Radioactive

It is important to note that some of these real-time meters require a manual intervention for correction, also known as seeding, of the measured values.

This "seeding" is taken from the manual BS&W readings which are used to correct the calibration of the in-line instrument. The use of the manual BS&W readings as a reference point for these meters' calculations re-enforces the need to adhere to a standard method of BS&W measurement.

There are issues with a lot of these instruments that are not mentioned, especially when water content is greater than 50%. This is when it becomes an oil in water meter instead of a water in oil meter so the measurement techniques used do not always work across the full 0–100% range.

8.14.1   ISSUES OF ELECTRONIC MEASUREMENTS

These meters have been developed for use in a stable controlled process environment, and they face a number of issues when used in the environment of surface well testing.

i. Entrained Gas in BS&W

Most of the meters are manufactured for the purpose of measuring liquids that do not have any gas dissolved in their flow. Entrained gas has a significant effect of the electrical properties of the fluid. The presence of gas induces new fluid properties as sensed by the meter and changes the accuracy of the meter, bringing these meters' viability for the intended purpose into question.

This is why most water in fluid meters are used in export scenarios where the liquids have been stripped of most phases and contents that impede the meters' measurements.

ii. Sediments of BS&W

The meters are usually sold and described as water cut meters that measure the actual water in the flowline. It is obvious here they do not measure BS&W as a whole; only the water component is measured and the solids are not.

Therefore, relying on these meters can prove costly if the solids increase.

iii. Water in Oil

Most of the meters measure water in oil. However, some cease to function correctly when the water content exceeds the percentage of the oil. This is known as an inversion, an oil in water scenario.

iv. Accuracy

The accuracy response of several of the meters is not a true linear function but varies by the percentage of water in the fluid, the density, etc. This makes taking a digital reading without correction invalid.

v. BS&W Manual Readings in Digital Meters

It is also common practice to take a manual BS&W reading and feed it back to the electronic meter, referred to often as calibration. However, this is not calibration but effectively seeding the meter to a single point benchmark and the meter is now only as good as the manual samples.

The so-called calibration entails a manual reading of the fluid, often taken the day before in routine PGOR/GOR tests, and this reading is entered into the computer program to correct the actual reading of the meter.

In such a case it is not a true reading but the water cut/BS&W meter is being used as a comparator, where the true reading is compared to a manual reading and corrected to suit.

This applies to all the previous technologies.

vi. BS&W Online Correction into Digital Meters

Another comparison technique involves, once more, taking a manual reading during the actual test and using a manufacturer-supplied computer program to feed the manual reading back into the water cut meter via a serial link. This has the additional effect of correcting the digital displays.

The manual readings are used to update the water cut meter, as the readings are taken, giving the impression that the water cut meter and manual readings both agree.

On a more technical note, this communication often has the additional hazard of lacking proper electrical protection in the line. The excuse is it is just a signal line and very low voltage, etc. This is a breach of electrical safety regulations as ALL electrical lines that enter a hazardous area have to comply with the required electrical standard protection methods with both the cabling and fittings.

9 Liquid Gravity Measurements

Similar to BS&W measurements the gravities measured from the produced liquids have a significant effect on the calculated results.

In order to calculate the oil rates and calibrate, BS&W meters, MPFMs, etc. correctly, it is essential that accurate oil gravities are obtained. To achieve this, we must use the correct procedures, apparatus, and methods, while always assuming that this liquid property is changing. These variables are rarely constant; values at the beginning of the well test differ from the last results, and neither should be used constantly or for the whole well testing period.

If the well is being retested, referring to the previous test results is a good quality check and indicator for this new well test, but the habit of using these old results as new readings must be discouraged and discontinued.

Like BS&W measurements, these readings are often taken in error attributable to mistaken practices from a lack of proper training or knowledge. For example, like the BS&W timing error, a sample for gravity measurement is taken after a choke change without allowing time for the inflow from the reservoir under the new choke to reach the surface.

In addition to the operational variables such as timing, there are associated standards behind the procedures used to measure and report the fluid properties of and associated components for liquids and gases. While strict observance of the standard is not always possible, the personnel should adhere to the recommended procedures as close as possible.

API gravity measurement	ASTMD1298
Gas gravity using a Ranarex	ASTM D1070
H_2S by chemical tube	ASTM D4810
CO_2 by chemical tube	ASTM D4984

This again is an area of training not fully covered by many of the service companies or part of competence schemes.

9.1 MAJOR FACTORS AFFECTING GRAVITY MEASUREMENTS

Normally in a well test, the liquid samples taken and prepared for BS&W determination are taken from the same sample that is used to measure the gravity of the liquids.

Therefore, by association, much of the methodology involved in the sampling of BS&W which can adversely affect the results also applies to the measurements of the liquid properties discussed here. Some of these variables are revisited in this chapter to emphasise their importance and any adaptation suited to liquid property measurements.

9.1.1 TIMING

The well test operators will often take the required gravity reading almost exactly after changing the choke. This they perceive to be standard procedure.

The fluid being sampled instantly after choke change is not the fluid that is being produced under the new choke but rather a remnant from the previous choke still in the production string/tubing.

As described in the BS&W section, the operators should wait until the fluid from the new choke flow reaches the surface before any sampling procedures take place. This should be described in the report as new fluid to surface after the choke change.

9.1.2 SAMPLE COOLING

ASTM recommends that when taking the specific gravity, the more volatile of crude oils should be cooled to at least 18°C (65°F) before the sample is measured. While this is not a standard practice amongst well testing it should be encouraged when testing light crudes in a hot climate.

9.1.3 SURFACE SAMPLING

The surface sampling process is as described in the BS&W section. It is important that the BS&W determination is carried out first and the results obtained before attempting to measure the fluid gravity.

The agitation, specified in the BS&W procedure, is used to help remove any dissolved gas in the sample which is extremely important for measuring the liquid gravity. Unlike the BS&W case it is recommended practice to let the sample sit and settle before carrying out any gravity measurements in order for any water in the sample to segregate and entrained gas to leave the liquid.

If the sample in a measuring cylinder is left to settle for a long period of time, the ratio of the separated liquids in the measuring cylinder will give an "approximate" BS&W, which can be compared with the measured BS&W result for consistency. If the liquid sample in a one-litre measuring cylinder is also to be used for gravity measurements, then the quantity of liquid should be sufficient for both tasks.

The standard volume of a BS&W centrifuge tube is 100 ml. If the standard two BS&W centrifuge samples are taken then the total volume left in the measuring cylinder should be in the region of 800 ml, or slightly less, satisfying the total volume needed for all measurements.

Depending upon the liquid gravity, percentage of water, and hydrometers available, this volume may have to be adjusted – or new samples taken – to permit correct hydrometer operations.

9.1.4 MEASURING CYLINDERS

Re-emphasising here, samples should be taken in a 1-litre measuring cylinder rather than a 250- or 500-ml cylinder, to avoid limitations with separation and gas escape by the diameter of the measuring cylinders. The dimensions of the one-litre measuring cylinder also comply with the regulations of physical space between the cylinder wall and the hydrometer, whereas the smaller measuring cylinders do not.

Measuring cylinders should be constructed of a plastic material – resistant to discoloration or attack by oil samples – and shall not affect the fluid being tested. They shall not become opaque under prolonged exposure to sunlight. New or good condition measuring cylinders should be employed; their costs are insignificant in the scheme of operational costs but crucial for proper results.

The measuring cylinder is the very basis of *all* fluid and BS&W measurements, and given the low cost of plastic measuring cylinders these should be used exclusively with at least two spares on the location. One can be used for the shrinkage test comparison as described earlier.

9.1.4.1 Hydrometer Gap

To avoid obstructive friction with the measuring cylinder wall, there MUST be a distance of at least one cm between the walls of the measuring cylinder and the hydrometer, as depicted in Figure 9.1. This alone makes the practice of using a cylinder less than one litre impractical and invalidates the hydrometer reading.

The one-cm gap is to ensure that the hydrometer can spin freely without undue drag exerted by the walls of the measuring cylinder. Smaller volume cylinders cannot accommodate this specified gap of one cm effectively, ruling them out of use.

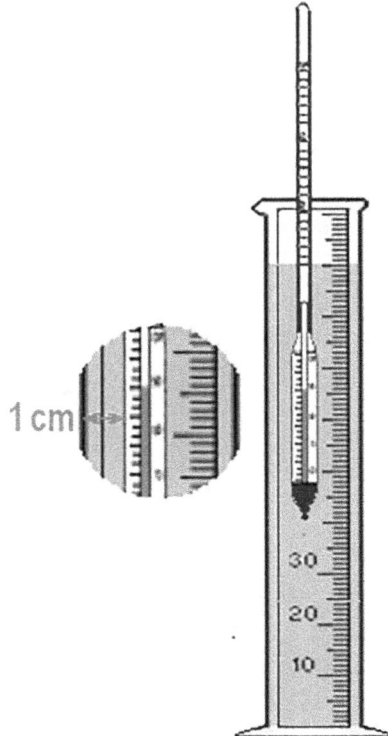

FIGURE 9.1 Measuring Cylinder with Hydrometer – One-cm Gap.

9.2 OIL WATER SEGREGATION

To acquire an adequate volume of oil to measure its gravity, we would use the measuring cylinder sample we took for the BS&W measurements taking into account the volume removed for the BS&W centrifuge tubes; 100 ml per tube is standard. There are a number of points that should be ensured at this stage:

i. Correct BS&W measurements were acquired and confirmed before continuing with gravity measurements
ii. The oil volume is segregated from any water sampled for proper use and immersion of the hydrometer
iii. There should be no trace of any dissolved or entrained gas
iv. The hydrometer floats freely
v. Adequate oil volume is available to measure oil gravity

After the BS&W readings have been determined then this result should be noted on the manual readings sheets and compared to the segregation of oil and water in the measuring cylinder after it has been left for a longer period of time to segregate under gravity, assuming all dissolved gasses have been removed by agitation.

NOTE: *Leaving the oil and water to segregate in the measuring cylinder is not a standard practice and is only usually adopted if there are any doubts with the centrifuged BS&W validity. Small quantities of demulsifiers can be used if required but only for BS&W, not for hydrometer gravity measurement.*

After the sample has been allowed to settle and the water and oil content has segregated, as shown in Figure 9.2, then the water volume can be measured and converted to a percentage of the total volume of the measuring cylinder to get an approximate BS&W.

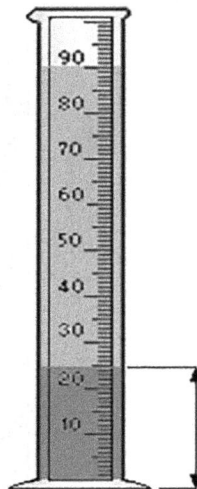

FIGURE 9.2 Segregated Water Volume.

9.2.1   CONFIRMING BS&W FROM MEASURING CYLINDER

This procedure gives an "estimate" of the BS&W and must *not* be considered a replacement of the standard BS&W centrifuged readings. If the BS&W readings had been carried out correctly then this estimation would not be necessary, only a verification. Centrifuged BS&W should be carried out every one hour as a standard or on a choke change where flow rates change, after allowing for the tubing volume displacement for the inflow.

In the case of the previous example (Figure 9.2) the water volume is 26 ml and the total volume is 90 ml therefore the percentage water is

$$\frac{Water\,Volume\,In\,Measuring\,Cylinder}{Total\,Volume\,In\,Measuring\,Cylinder} \times 100 = \textbf{Corrected BS\&W Reading}$$

$$\frac{26}{90} \times 100 = \textbf{28.8}\%$$

It is this percentage that should be used as a comparison and not the volumes from the measuring cylinder.

Because there is no agitation or centrifugal force on the measuring cylinder, the separation under gravity will take a lot longer than for a final BS&W spin out. This comparison may take several hours to complete and should be done early in the flow period especially with high viscosity oils.

If the measured percentage is not similar to that from the centrifuged BS&W readings then the gravity measurement can be in error. The comparison may not be an exact match, but it is a close verification that the BS&W results are correct.

The sample in the measuring cylinder was taken at a specific time and sampling point then left aside for some time, which has to be factored in. When compared with the centrifuged BS&W results, records care must be taken to ensure that the correct corresponding records to that sample and its time are used; all these details should be recorded on the manual readings sheet for reference.

If demulsifiers or similar chemicals are used in the centrifuged sample then it should also be added to the measuring cylinder sample allowed to settle for BS&W results; a like-for-like approach for accurate comparisons can be made.

If the percentage in the measuring cylinder is greater than the percentage from the centrifuged BS&W then the centrifuged BS&W result is amiss. This could be for a number of reasons most likely from the handling process; the sample had been allowed to settle or had not been agitated whilst transferred to the BS&W tubes. In this case the entire BS&W and gravity samples must be discarded and re-taken following correct procedures. This is the main reason for a static comparison.

Other reasons can be insufficient heating and insufficient time or speed of rotation in the centrifuge. Again, a new sample must be taken and all times and speeds increased if possible.

9.2.2   ADEQUATE OIL VOLUME

If the water percentage is significantly lower than the BS&W reading there are two options:

i. Leave the sample to settle longer. Additives will alter the specific gravity and should only be used in BS&W determination

ii. Transfer some of the oil from the measuring cylinder to the **TWO** BS&W centrifuge tubes then perform centrifuge while heating. No demulsifier or additive must be used in this BS&W process

After the centrifuge is completed, transfer the segregated oil only to another clean measuring cylinder.

Empty the BS&W tubes back in the original measuring cylinder. Repeat the process until there is sufficient centrifuged oil in the second measuring cylinder to take a gravity reading.

9.3   OIL GRAVITY VALIDITY

Originally when conducting a well test the oil gravity could only be obtained from a glass hydrometer. Today's technology allows the gravity to be determined by dedicated instruments or as a by-product of several metering systems. These measurements are then cross checked against a standard glass manual hydrometer reading.

9.3.1   HYDROMETER FUNCTIONALITY

A hydrometer is an instrument used to determine the specific gravity of a liquid, based on the concept of buoyancy.

Hydrometers function using the Archimedes principle which basically states that a solid body will displace its own weight within a liquid in as long as it floats. Therefore hydrometers are normally split basic types: liquids heavier than water and liquids lighter than water.

The construction is usually of glass with a weighted bulb at the bottom with the specific gravity scale on the stem, as illustrated in Figure 9.3.

The specific gravity is measured from the scale as it leaves the liquid. The scale can generally be in specific gravity (s.g.) or API terms but can also be in specialist units as for the ones used in wine making

Specific gravity (s.g.) in the oilfield is defined as the ratio of the density of a crude oil (or other liquid) to the density of pure water. Another term for specific gravity is relative density.

Hydrometers come in different ranges and scales for fluid densities; there does not exist a one-for-all type of hydrometer. The proper hydrometer scale/range fitting to the liquid being measured must be used.

FIGURE 9.3   Illustration of Hydrometer.

Where the fluid densities are previously known a smaller range of glass hydrometers can be utilised to suit. However, with a well test package that encounters a wide range of oil wells, a full set of hydrometers, with spares, should be available to cover all possible liquid densities.

Table 9.1 is an example list of the ranges of glass hydrometers ranges that should be available with the well testing package:

TABLE 9.1
Hydrometer Ranges
Normally Required

SG Range
0.60 – 0.65
0.65 – 0.70
0.75 – 0.80
0.80 – 0.85
0.85 – 0.90
0.90 – 0.95
0.95 – 1.00
1.00 – 1.05
1.05 – 1.10

The well test unit should have a single hydrometer for each range, but spares should also be included especially in remote locations.

Although specific gravity (s.g.) is the standard unit of gravity measurement of a liquid in the industry, the American Petroleum Institute (API) refer to a separate system to measure density of crude oil described as Degree API; there are no units attached to API. Again, this is by comparison to water. A crude oil can be described as 40 degrees API.

API density has the following relationship to the standard s.g. measurement:

$$\text{API} = \frac{141.5}{SG} - 131.5$$

The ASTMD1298 standard specifies that the hydrometers should comply with the E 100 or ISO 649–1. The standards and compliance certificates should be available with the well testing unit paperwork.

What is often overlooked is that unless the hydrometers have a built-in thermometer, then at least two thermometers should be included with the supplied hydrometer set for respective temperature readings.

9.3.2 INSERTING THE HYDROMETER

The hydrometer should be gently lowered into the measuring cylinder and spun as it is released, as depicted in Figure 9.4. Spinning the hydrometer assists in removing any bubbles of air or gas that might be clinging to the surface of the hydrometer.

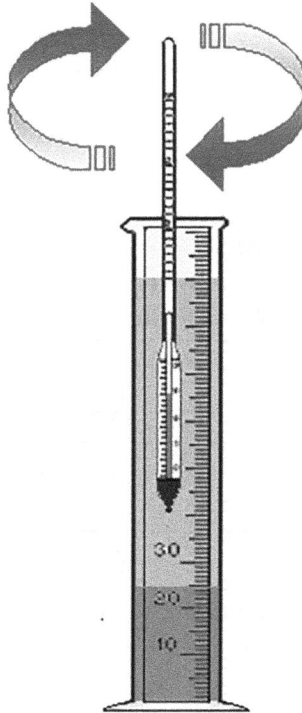

FIGURE 9.4   Spinning the Hydrometer.

The gas bubbles would affect the hydrometer's resultant buoyancy, effectively lightening the specific gravity of the fluid – a potentially false reading.

When measuring the specific gravity of the oil it is critical that no cutting agents and no demulsifiers are used as these affect the measured gravity. (This also applies to digital hydrometers.)

9.3.3   HYDROMETER TEMPERATURE ERRORS

Hydrometers have a calibrated working temperature range and in the case of high temperature crudes it is possible for the hydrometer to be in error.

If in doubt with the temperature operating range of the hydrometer, a standard practice is to check higher range hydrometers, which can also measure the gravity of water, against a measuring cylinder full of water at the same – or close to the same – temperature of the intended oil sample. The hydrometer in the water should read close to one (1).

Any slight difference between the reading and the true gravity of water should be noted, and this difference should be used to correct the future measurements of oil gravity. If the hydrometer in water reads:

0.990 then the error is $1 - 0.990 = + 0.010$
1.010 then the error is $1 - 1.010 = - 0.010$

These errors should be added to the reading of future oil samples measured with a hydrometer, at the referenced temperature.

The corrected gravity to 60 degrees is used as another technique to compensate for temperature errors, measuring the gravity at the taken temperature and then correcting it to 60 degrees. Note this initial result down and repeat the process for confirmation after letting the oil sample cool down.

If there is no error with the hydrometers, the readings at 60 degrees F should be similar.

9.3.3.1 Gravity Correction to 60 Degrees F

A property of most oils is that its volume expands and contracts relative to the applied temperature, therefore all measurements of oil are converted to a standard temperature. With the API system the standard temperature is 60 degrees F; in the metric system the standard temperature is 20 degrees C.

What this effectively means is that if a tank has 60 Bbls stored in it at 90 degrees F the measured volume will reduce as the temperature decreases. To combat this effect, the oilfield uses a correction formula to correct the volumes to 60 degrees Fahrenheit. This corrected volume is referred to as a standard barrel. The correction for volume is referred to as the volume correction factor

The specific gravity also changes with temperature as well. The correction for specific gravity, in well testing is referred to as the k factor.

By using these two corrections in the measurements and calculations, they correct the results for temperature differences. Before the advent of computers these correction factors had to be looked up in a reference book, often using manual interpolation, however with the advancements of computers, volume corrections and k factors are easily calculated.

In this section we will concentrate on the correction of the oil's SG, k factor, by using the formulae:

$$SG_{60} = SG_T + (0.00069 - 0.000372\ SG_T)\ (T - 60)$$

Where
SG_{60} = corrected gravity to 60 degrees F
SG_T = specific gravity at observed temperature T
T = observed temperature

Limits of Validity
T = 0 – 110 degrees F
SG_T = 0.68 – 0.92

Example of the Effect of Temperature:

On the Gulf of Mexico (GOM) there was a discrepancy between the volume of crude oil flowed from the well testing separator when compared to the volume of oil stored on the tanker. The operating company had a separate calibrated turbine meter on the export line to verify the total volume flowed to the tanker. The discrepancy was between the measured flow volume from the meter and the dipped volume on the tanker.

Shrinkage was initially thought to be the reason but carrying out a shrinkage test on the separator proved that shrinkage would not be sufficiently responsible.

Eventually the temperature of the oil in the tanker and the temperature flowing through the turbine were taken, and then the volume correction factors were applied based on the individual temperatures. When compared as standard barrels at 60 degrees F this brought the differences to acceptable levels.

9.3.4 GASEOUS SAMPLE

When there is trapped, entrained, or dissolved gas in the intended oil sample it can cause a significant difference in the gravity reading, as shown in Figure 9.5.

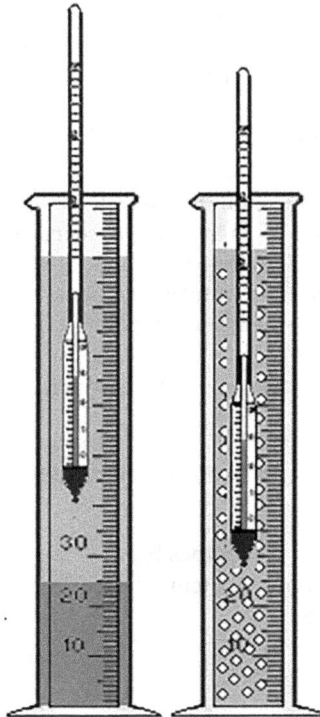

FIGURE 9.5   Entrained Gas.

Figure 9.6 shows two measuring cylinders with identical oil in each one. The measuring cylinder on the left has a significant amount of gas still dissolved within the sample.

This effectively lowers the gravity reading on the hydrometer by a reduction in the buoyancy of the liquid. A main reason why the sample has to be agitated after sampling to assist in the reduction of the gas, then left to allow any liquid to segregate.

Gassed Sample   De-Gassed Sample

FIGURE 9.6   Fluid Samples in Measuring Cylinder.

9.3.5   HYDROMETER IN WATER

The most common error when using a hydrometer is to have the hydrometer partially in the water or other liquid. More often this is because the measuring cylinders are old, discoloured, and stained, making it difficult for the operator to recognise the liquid interface or different liquid. This is an easy fix by using a new measuring cylinder.

The illustration, Figure 9.7, shows the hydrometer tip entering the separated or segregated water; this has the effect of increasing the buoyancy and giving a higher fluid gravity.

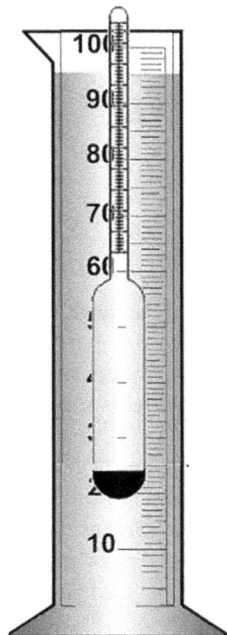

FIGURE 9.7   Hydrometer Partially in Water.

9.3.6 READING THE HYDROMETER

People rush at this step when patience is a virtue. The reading should be taken when the hydrometer is stationary or rotating very slowly. The operator's eye level should be at the top of the oil (liquid) level, as depicted in Figure 9.8.

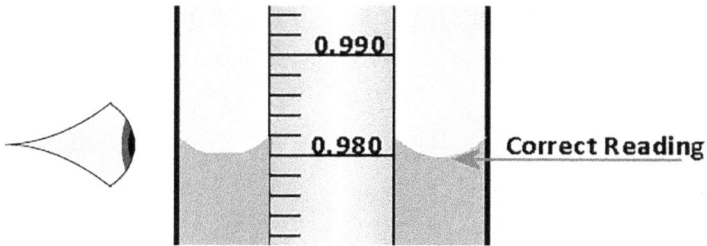

FIGURE 9.8   Correct Reading of Liquid Gravity.

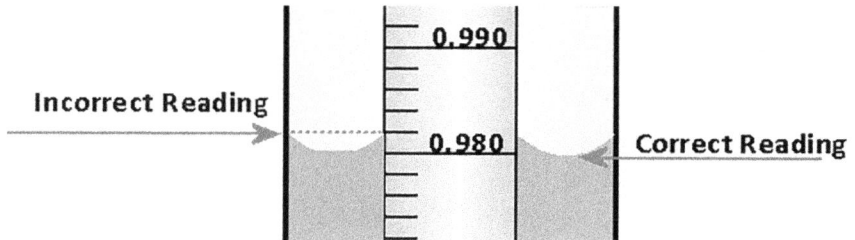

FIGURE 9.9   Dual Meniscus.

NOTE: *A meniscus is a curved surface of the liquid.*

The reading across the bottom of the meniscus is the correct gravity of the oil, as shown in Figure 9.9. In the case of a double meniscus then the lower reading should always be used.

The reading taken from the hydrometer is the uncorrected oil gravity which has to be converted to a gravity at standard conditions of 60 degrees F. Therefore, the temperature of the oil is taken at the same time as the gravity is measured.

The thermometer should be suspended in the measuring cylinder – as shown in Figure 9.10, before the oil gravity is taken – and given time to read the liquid temperature.

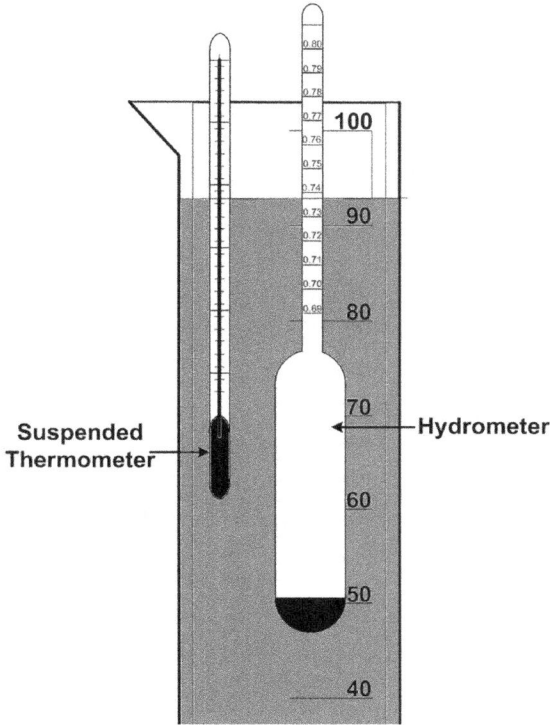

FIGURE 9.10   Simultaneous Gravity and Temperature Reading.

With both instruments submerged into the oil it allows both readings to be taken for the gravity and temperature simultaneously. The use of a paperclip to hold the thermometer in place is common practice.

If the thermometer is put in the sample later, then it should be given time to stabilise and the oil gravity reading checked.

With some thermometers there is an "Immersion Line" marked on the thermometer, shown in Figure 9.11. This is the depth limit the thermometer has to be immersed into the liquid in order to achieve an accurate reading.

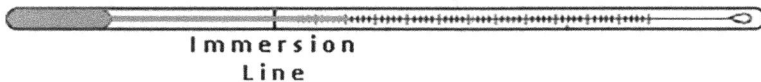

FIGURE 9.11   Thermometer Immersion Line.

It can take a while for the temperature to stabilise and it is burdening to hold the thermometer
in the liquid in the absence of a holder, so often a
rubber/elastic band is wound round the top of the
thermometer and an expanded paper clip used to
hang the thermometer in the liquid.

Some of the hydrometers available, particularly
those with an API scaling, can have a built-in thermometer (Figure 9.12).

These can be used as normal hydrometers, but
time must be allowed for the thermometer to stabilise before taking its reading.

9.3.7   HIGH WATER CONTENT SAMPLES

Where there is a high water content in the well flow
it is often the case that there is not enough oil in the
sample to determine the oil gravity. In this case the
oil must be separated from the water before trying
to measure its gravity.

FIGURE 9.12   Hydrometer with
Internal Thermometer.

In circumstances like these, surface sampling should be done quickly to try to
keep the same oil constituency.

Often it is enough to allow the sample to separate under gravity on its own in the
measuring cylinder. When separation has taken place the oil component is poured
into a new measuring cylinder; more than one measuring cylinder filled samples
should have been taken at the same time and this process repeated for each, until sufficient oil volume is available to measure the gravity.

9.3.7.1   Using a Separation Funnel

From the separation funnel and after titration the separated oil can be drawn to build
up the required volume needed. The final oil sample volume should be sufficient to
measure using a hydrometer.

The water in the funnel, once separated, should be drained and the level in
the funnel refilled from the original sample. This has to be repeated until a measurable quantity of oil is left in the funnel that will enable an oil gravity to be
measured.

If a separation funnel is not available then the oil can be separated using the
centrifuge. This involves using multiple flow samples and pouring the separated oil
into a measuring cylinder until a sufficient quantity of oil is accumulated. This accumulated oil may not always be just tipped alone; thus, this accumulated oil should be
allowed to separate in the measurement cylinder.

Although Figure 9.13 shows a measuring cylinder, any clean vessel will suffice.
The measuring cylinder is suggested to allow for the separated water to be saved for
later use, measuring its salinity and specific gravity.

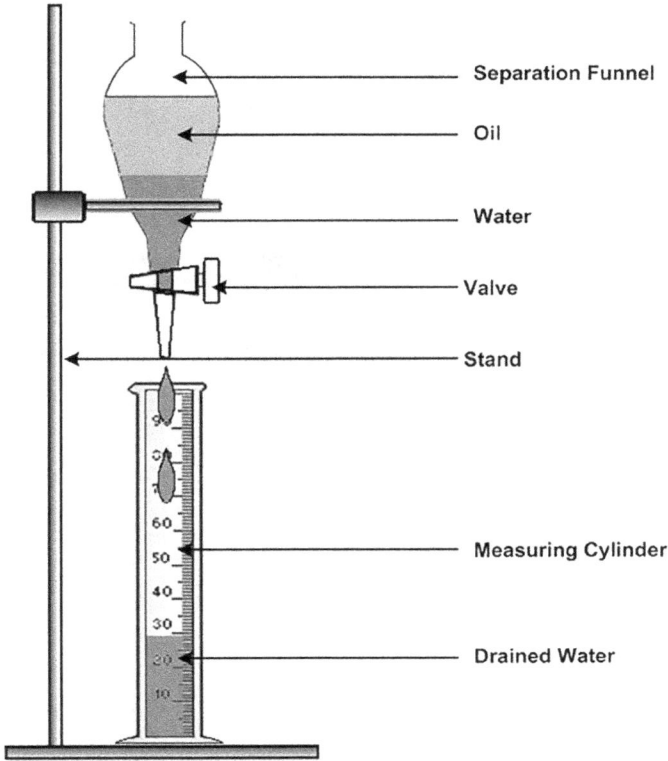

FIGURE 9.13 Separation Funnel in Use.

9.4 DIGITAL SPECIFIC GRAVITY INSTRUMENTS

There now exist several instruments that can digitally measure the s.g. of the oil. These are generally very good and an aid to well testing.

Like all instruments they are only as good as the way they are used and the operating instructions should be followed. It is important that the full details of the instrument are available on location: manual, calibration, etc.

Similarly to the hydrometer, if the oil sample used is not pure oil, then the reported specific gravity would be incorrect.

These instruments utilise a small sample, a few ml, therefore the sample can be drawn directly from the BS&W tube using the built-in sampling syringe. This saves a lot of time and effort when dealing with high percentage water cut flows. Again, no cutting agents must be present in any sample used for measuring gravity with these devices.

In some cases of high specific gravities, the sampling syringe is not capable of drawing a representative sample. In this case, the sample should be injected into the meter using an external syringe. These external syringes normally come with the equipment but can easily be replaced with other types.

It is important that no demulsifier or cutting of the sample is introduced; this will have an effect on the gravity. If needed, heating and spin time in the centrifuge are better options to get a true sample.

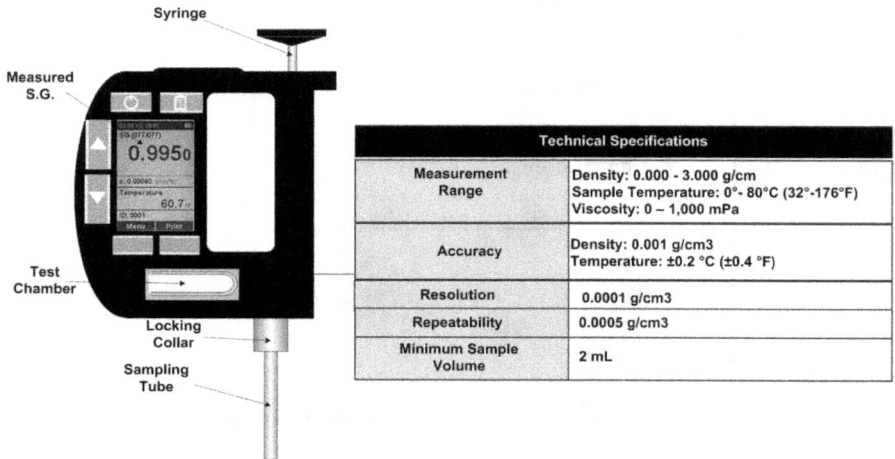

	Technical Specifications	
Measurement Range	Density: 0.000 - 3.000 g/cm Sample Temperature: 0°- 80°C (32°-176°F) Viscosity: 0 – 1,000 mPa	
Accuracy	Density: 0.001 g/cm3 Temperature: ±0.2 °C (±0.4 °F)	
Resolution	0.0001 g/cm3	
Repeatability	0.0005 g/cm3	
Minimum Sample Volume	2 mL	

FIGURE 9.14   Digital Hydrometer and Basic Specifications.

From the specification panel there is no declaration or certificate of electrical safety, normally intrinsic safety (IS), so unless otherwise specified these devices must be used in a safe area.

The meter must be thoroughly cleaned immediately after use. The measured sample is small and any residue from previous samples – or cleaning fluids – will affect any future sample reading.

It is this chamber, shown in Figure 9.15, that must be kept clean with no residue or solvent remaining before use.

FIGURE 9.15   Digital Hydrometer Test Chamber Example.

The sampling syringe, once cleaned, is used to draw and purge cleaning fluids through the instrument. In the case of difficulty purging the instrument, a normal syringe can be used which generates a more positive flushing action. The internal sampling syringe should be cleaned after each use.

If digital hydrometers are used to measure the gravity of oil and water, then a separate/individual digital hydrometer should be used for each purpose, avoiding any cross contamination of fluids.

Even with the availability of a digital meter it is recommended at least once per shift that the digital meter reading is compared to a manual reading.

9.4.1 TECHNOLOGICAL SG ADVANCEMENTS

Some of the more advanced metering systems, MPFM, require that the oil and water gravity are inputted into them, in order for them to calculate the flow rates and associated functions.

Therefore, the importance of correct oil gravities has not diminished from the humble hand calculations but has expanded to include the newer digital technology, and it is often the case where any measured gravities have to be verified always using the "old fashioned" technique of glass hydrometers. The correct procedures have to be documented and complied with.

9.5 WATER MEASUREMENTS

Similar to oil properties, the produced water has to be measured for its own properties that reflect important and required information for well testing and reporting.

9.5.1 WATER SPECIFIC GRAVITY

The specific gravity of the water is necessary although most well testers do not automatically take water data. Produced water would be part of the flow sample taken and separated using the separation funnel or from centrifuge tubes. The produced water specific gravity measured is especially important during clean up phases where it can be used, with other factors, to determine when true reservoir fluids are being produced.

9.5.2 WATER SALINITY

Water salinity is normally measured as well to reflect further details of the water properties. This can be carried out by two methods.

i. Refractometer
ii. Chemical titration

9.5.2.1 Refractometer

A refractometer, illustrated in Figure 9.16, is an optical device that works by reading the angle of light refraction from its internal prism and a sample of fluid placed on an optical glass which is the only window of light into the instrument. The different concentrations of solutions result in different angles of refraction – referred to as the refractive index of the solution. The refractive index is proportional to the salinity of the liquid.

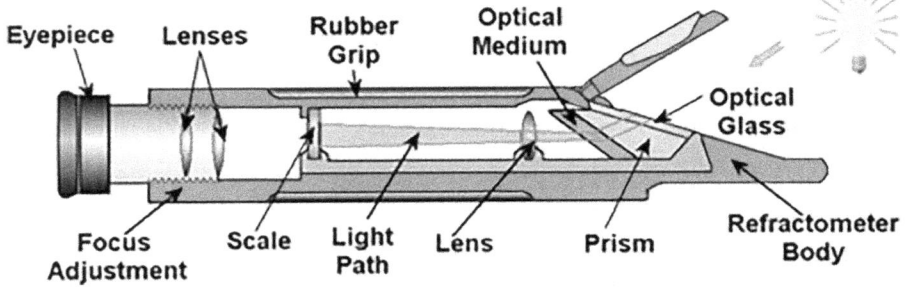

FIGURE 9.16   Basic Internals of a Refractometer.

Refractometers are generally used in preference to the chemical titration method as there are no potentially hazardous chemicals involved.

The refractometer should be clean with all optical parts clear, with no discolouration or scratches that could impede the light's travel and in turn the readings.

Before applying the sample to the refractometer ensure that the acid or alkali concentration cannot damage the apparatus by using a litmus (pH) test. Normally liquids of pH 6 and below can damage the refractometer seals and cause liquid ingress. The applied liquid should be free from particulates. If necessary, filter the liquid first. Kitchen roll or tissue can be used if no proper filter paper is available.

Plastic materials should be used to place the droplet of water onto the optical glass of the refractometer. Use of plastic or paper drinking straws as a pipette have proved a good substitute for a pipette, to prevent damage to the glass face. As the surface of the refractometer is a soft optical-grade glass, the use of hard materials such as metal, glass, etc. should be avoided.

Figure 9.17 shows the internal scaling of a refractometer; both the salinity and the specific gravity can be determined. The refractometer should be calibrated with

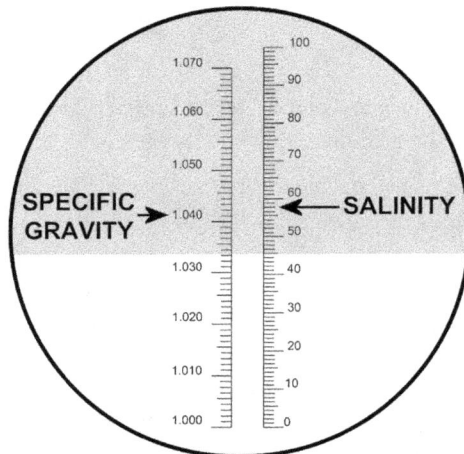

FIGURE 9.17   Typical Refractometer Internal Scale.

distilled water at regular intervals. It is important that the refractometer be cleaned with distilled water or a suitable solvent and dried after use, as prolonged exposure to some well fluids can cause damage to the optical surfaces.

Where possible it is recommended that the chemical titration method is used on the test for one reading at least as a quality assurance reference.

9.5.2.2 Chemical Titration

This is the original method used for calculating salinity either used by the mud person, the mud loggers, or the well testers. The process of titration is quite simple, even for a non-chemist, and it requires the following chemicals:

- Potassium chromate
- Silver nitrate
- Distilled water (not drinking water)

The appropriate PPE and safety sheets should also be available and implemented. The equipment required are:

Measuring Pipette
Stand
Mixing Bowl
Glass Stirring Rod

Basic Procedure – This should be detailed in the service company procedures:

i. Add silver nitrate to the measuring pipette and record level on manual readings sheet
ii. Add the sampled liquid to a clean and dry bowl
iii. Add distilled water if a small volume
iv. Add four to five drops of potassium chromate to bowl
v. Mixture will turn milky
vi. Slowly drop a small amount of silver nitrate into bowl and mix
vii. Repeat adding silver nitrate (mixing) until the fluid in the bowl turns red
viii. Measure amount of silver nitrate used and record on manual reading sheet
   Chlorides (ppm) = silver nitrate (ml) x 10,000

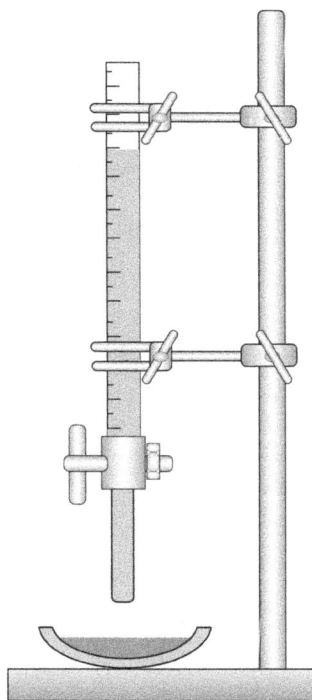

FIGURE 9.18 Illustration of a Titration Apparatus.

Both the chemical titration result and refractometer readings should be similar. If not the refractometer should be inspected, cleaned, and re-calibrated before any new readings are taken.

9.5.3 pH Measurement

In order to make sense of pH readings it helps to understand what pH is. The term stands for *potential hydrogen* and is a measurement of how many hydrogen ions (H+) are in the liquid.

- The greater number of H+ ions, the more acidic the liquid
- The greater number of hydroxide ions (OH-), the more alkaline the liquid

Where there is an equal amount of H+ and OH- then the liquid is said to be neutral. Water is the prime example of a neutral liquid.

Standardly in well testing litmus paper is used to indicate the pH of a liquid, although the litmus indicator is also available as a liquid.

Litmus paper is a strip of chemically impregnated paper that is dipped into the liquid sample. Depending on the acidity or alkalinity of the liquid, the litmus paper turns a different colour. Litmus paper measurement only provides a relatively coarse measurement of pH which is only suitable for making simple determinations.

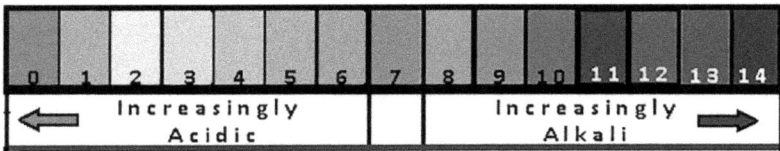

FIGURE 9.19 pH Paper Scaling.

What is important, though, is the quality of the pH paper. It should be sourced from a reputable supplier of laboratory instruments and not the cheapest available.

pH paper has a use-by date and should be kept in a clean dry environment without prolonged exposure to sunlight. It should also be kept in a watertight container when in use in the field.

10 Gas Measurements

The gasses produced by a well are composed of hydrocarbon components and can contain non-hydrocarbon gasses such as, H_2S, CO_2, etc., which are referred to as gas contaminants. The properties of the produced gas are normally measured in two ways within well testing:

The gas specific gravity – with a gas a gravitometer
The gas contaminants – with individual gas reaction tubes.

These measurements are then used in the calculation of the gas production rates. Consequently, any error in these measurements will have a direct effect on the gas flow rate, gas cumulative, and gas oil ratio calculations.

Random gas gravity samples and associated contaminant measurements should be witnessed by the operating company representative to ensure:

i. Correct procedures and methods are being carried out
ii. Readings are taken at regular intervals
iii. Results are recorded on the manual reading sheet, which are then identical to those in the data acquisition system

Errors Example in Gas Measurement

In the Gulf of Mexico (GOM) it was common for the well testing companies to assume the gas gravity of the produced gas. The reasoning behind it:

We have been producing the gas from this region for years and we know what the gravity and contaminants are.

On such a well test, where no gas physical measurements were made by the well testing company, the operating company had employed a separate PVT sampling specialists' company to gather samples. This company had a mini lab and were able to measure gas gravity and its contaminants.

The standard gravity used by the well testers in all calculations and reports for years was 0.657 with no H_2S or CO_2; this apparently had not been checked or verified for years.

Example of Gas Measured

Table 10.1 represents gas gravities taken using a Ranarex gravitometer and contaminant measurements from reaction tubes. While not disclosing the full data or flow

DOI: 10.1201/9781032623689-10

TABLE 10.1

Example Gas Measurement Results

dt Hrs	Choke 64ths	Gas Gravity	CO_2 %	H_2S ppm
1.50	14		3.1	0
2.50	16	0.728		
10.50	8		2.9	
12.00	12	0.700		
15.00	14	0.706	2.9	0
18.50	12	0.708	2.7	
24.25	18	0.707	2.9	4
37.50	18	0.704	3.4	6

(dt = time since well flowed)

rates, this mistake, not taking the gas gravity and its associated contaminants, causes a calculation error of over 2.5% based on the flow rates at the time.

However, a lot of well testers will fall back on a gas gravity of 0.6 with no contaminants, which then yields a calculation error of over 6% of the flow rates at the time.

This is based on the AGA 3 calculations where the major components affecting gas flow rates are:

- Gas gravity factor (F_g)
- Gas supercompressibility (F_{pv})

Most well test calculation packages incorporate a Wichert-Aziz type correlation which compensates for these non-hydrocarbons in the compressibility calculations so these readings should be taken and used.

It is recommended that the operating company representative overseeing the well test, employee, reservoir engineer, production, etc. witness and verify the specific gravities, BS&W, and contaminants of gasses and liquids from sample collection to measured results for at least one of the readings.

10.1 ASTM STANDARD FOR GAS GRAVITY

ASTM International Designation D 1070–03
Standard Test Methods for Relative Density of Gaseous Fuels

There is a standard for taking gas gravity readings published by the ASTMS, D1070–03, which can be applied to different methods for gas gravity measurements. The most common of these methods in well testing is the Ranarex gas gravitometer, which is covered in detail later on in this chapter.

10.2 GAS SPECIFIC GRAVITY

Gas specific gravity is commonly referred to as *s.g.* (or *sg*) and is the measurement of the produced gas gravity (when compared to air), while including the contaminant gas components often produced with natural gas. There is a connection between the gas mixture of elements and its specific gravity.

It is important for the operator to know what the gravity of the produced gas is for commercial reasons but it is critical to get a correct reading in order to calculate an accurate gas rate.

The gas gravity measurement is a standard part of well testing and is carried out on both gas and oil wells, but the significance and the effect on the flow calculations from true and accurate readings is not fully understood. There is an ASTM standard that covers the measurement of gas gravity, ASTM D1070; therefore if this is in use, the ASTM procedures should always take priority.

10.2.1 GAS SAMPLING

Unless carrying out a check on the contaminants being produced from the well during the clean-up or early phases of production, all gas sampling should be taken from the highest point on the separator gas line, shown in Figure 10.1.

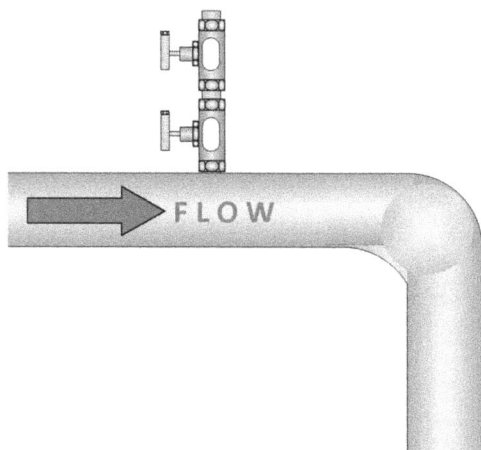

FIGURE 10.1   Gas Sampling Position on Separator.

The gas sample must always be taken at the top of the gas line to try to ensure that only a dry, clean gas sample is taken.

10.2.1.1 Gas Sampling Safety

All sampling must be carried out in accordance with the safety practices and structure by all personnel while on the location.

Before attempting to obtain a sample from any designated sampling point ensure that the operator is wearing the full and proper PPE with special attention to H_2S, safety glasses, and gloves.

Until the well has been tested and proved not to have dangerous contaminants, H_2S, acid, etc. the sampling should be conducted with extreme caution. Therefore, all sampling should be assumed to contain hazardous components and gasses; all monitors and precautions must be in place while adhering to safety procedures.

Drip trays or suitable buckets should always be placed beneath all the sampling points to contain spillage.

10.3   GAS GRAVITY MEASUREMENT

When well testing, two methods are normally adopted in the determination of the produced gas gravity of a well:

    i. Gas gravitometer
   ii. Chromatograph

The dominant method employed in well testing is the gas gravitometer, which is a simple and portable device. This is normally part of the standard equipment of a well testing company.

The gas chromatograph, although capable of giving a much more detailed and precise make-up of the gas, tends to be a specialist device and used more with the mud logging companies and in underbalanced operations.

10.3.1   Ranarex™ Gas Gravitometer

The most common form of gas Gravitometer is the Ranarex, illustrated in Figure 10.2; however most well test operators do not treat it with the care and respect it deserves.

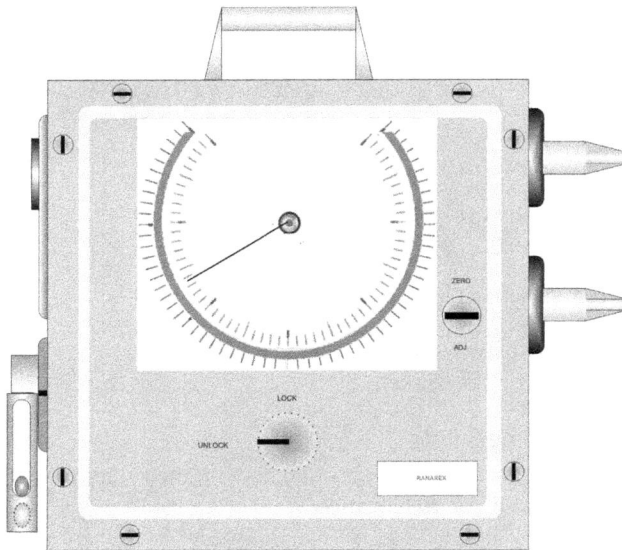

FIGURE 10.2   Ranarex™ Front View.

In fact, it is an accurate device if properly looked after and maintained and complies with the ASTM Procedure D1070 Standard Test Methods for Relative Density of Gaseous Fuels.

ASTM International Designation D 1070
Standard Test Methods for Relative Density of Gaseous Fuels

The principle behind the operation of the Ranarex gas gravitometer and its design is simple. (Refer to the figures in this section to aid in understanding the operation of the Ranarex.)

- Within the instrument there are two cylindrical, gas-tight measuring chambers, one for the measured gas and the other for the reference gas: air. Each measuring chamber has separate inlet and outlet vents, one for gas and the other for the reference air.

NOTE: *The chambers are referred to as "gas tight" as they are low-pressure chambers and not to be confused with the higher pressure vessels used in well testing.*

In each chamber there is an impeller and an impulse wheel, both with axial vanes as shown in Figure 10.3.

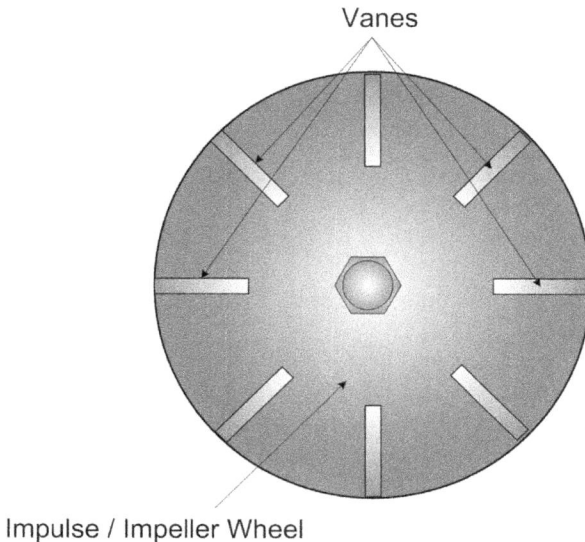

FIGURE 10.3 Impulse and Impeller Wheel Construction.

Both the impulse and impeller wheels are of identical manufacture and are mounted on independent and separate axles facing each other, but they make no physical contact in a gas-tight chamber as shown in Figure 10.4.

Both the air and gas chambers are manufactured to the same dimensions.

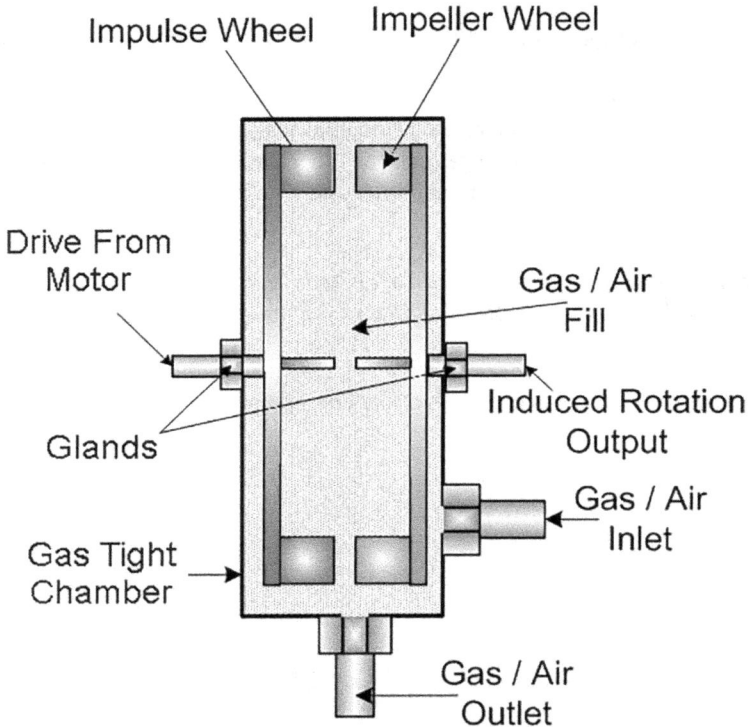

FIGURE 10.4   Impulse and Impeller Wheels Mounted in Gas-Tight Chamber.

A non-sparking electric motor is used to drive both individual impulse wheels in the separate chambers.

Gas impulse wheel
Air impulse wheel

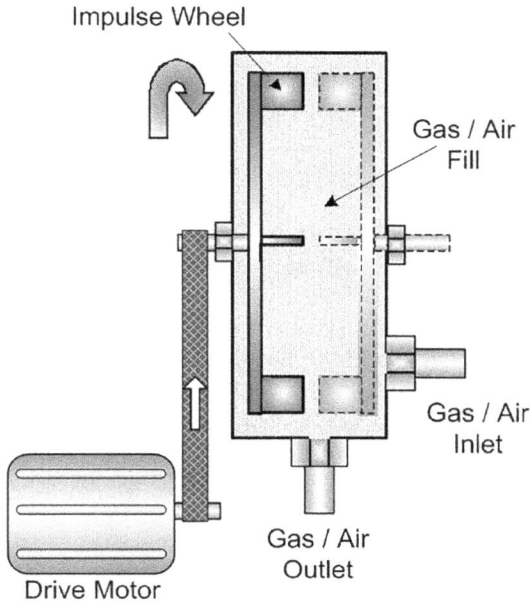

FIGURE 10.5   Impulse Wheels Driven by an Electric Motor.

Only the impulse wheels are driven by a belt from the motor. The impeller wheel is not connected to or driven by the electric motor, as shown in Figure 10.5.

- The rotation of the impeller wheels draws the sample gas into the upper chamber and air as a reference into the lower chamber
- While there is power to the Ranarex, impulse wheels, reference air, and measured gas are consequently rotating. The rotation of the impulse wheels causes the gas and air in their respective chambers to rotate in the same direction as shown in Figure 10.6
- The powered rotation of the impulse wheel causes the gas inside the chamber to rotate.
- This aids in the input and exhaust of the sample gas/air.
- As the two impellor wheels are tethered together it effectively puts a torque on the axis. Shown in Figure 10.7.
- The rotation of the gas or air by the impellor wheel forces a similar rotation in the facing impeller wheels in each of the chambers

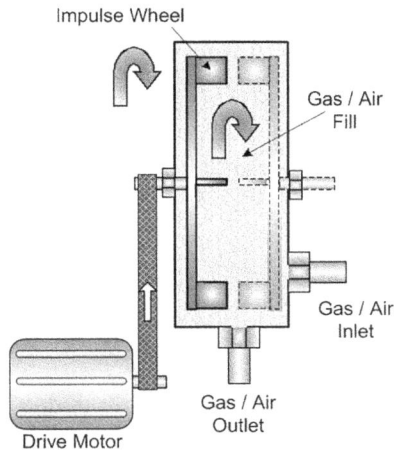

FIGURE 10.6   Impulse Wheels Causing the Gas/Air to Rotate.

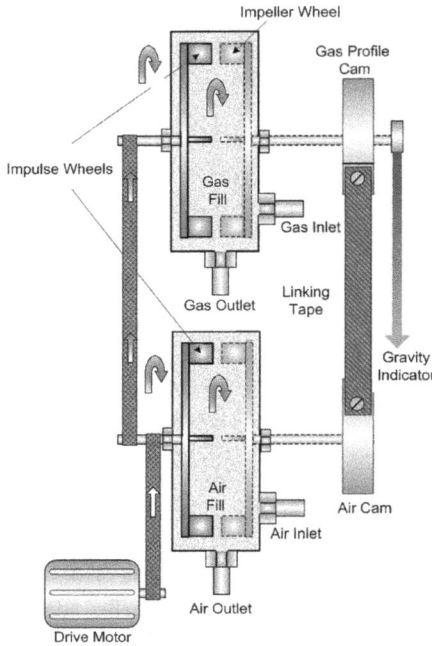

FIGURE 10.7   Connected Chambers.

- The impeller wheel's axles have the following:
- Reference gas impeller – standard drive cam
- Measured gas impeller – shaped cam wheel

These two are connected together using a tape as shown in Figure 10.7.

Figure 10.7 shows the two chambers with the individual impellor wheels joined by the linking tape.

The two impellor wheels are not free to rotate and are joined, shown in Figure 10.8.

The rotation of the gasses in the chambers now causes a torque to be generated at the output cams.

Joining the two cams causes the torque from the gas sample to be deducted from the torque of the standard air torque.

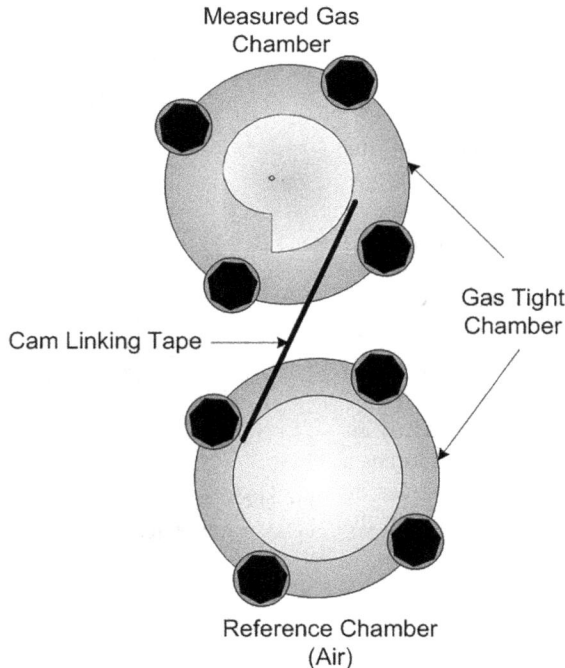

FIGURE 10.8   Air and Gas Impellors Connected by Flexible Tape.

This is measured by a calibrated profile cam on the gas chamber which gives the torque difference in specific gravity.

- The calibrated cam on the measuring gas is made to correspond with the torque generated by the measured gas
- The torque is directly proportional to the density of the gas the impulse wheel exports
- The individual torques from both chambers are connected to the two cams which effectively measure the difference in torque between the reference gas (air) and the measured gas
- The resultant torque is then measured using the customised cam on the measuring impeller which is connected to an arrow on the Ranarex to give the gas gravity, depicted in Figure 10.9

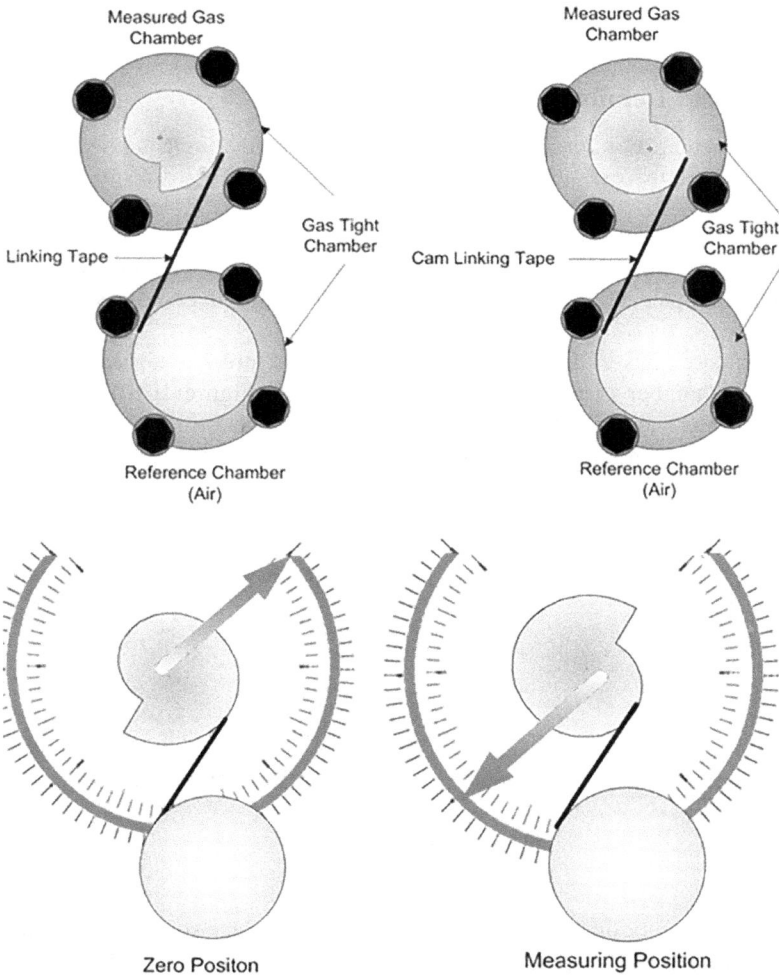

FIGURE 10.9   Position of Cams Showing Zero and Gas Measuring Positions.

Since this instrument is fundamentally a mechanical comparison between sampled gas and air, the results achieved depend on any external variables to both fluids being common: temperature, atmospheric pressure, and even humidity/wetness. If these variables are NOT eliminated or controlled equally for both measured elements, then one would have an undesirable influence over the other.

The Ranarex gravitometer will measure the relative density of the sample gas to that of air in accordance with the following specification:

> The ratio of the density of the gas, under the observed conditions of pressure and temperature to the density of dry air at the same pressure and temperature.

The pressure of the sample must be limited to the operational limits and if the temperature of the gas sample is significantly different than the atmospheric conditions it too must be processed.

Full Ranarex Details:

www.ametekpi.com/products/brands/chandler-engineering/gas-gravitometers-portable-ranarex

A simple way to cool the gas is to utilise a coil of stainless-steel liner before the input to the Ranarex. However, the addition of extra liner increases space for the volume of gas and there may be an insufficient sample and pressure to operate the Ranarex.

The Ranarex has to have dry air reference. Therefore, the air sample is passed through a chamber containing silica gel, shown in Figure 10.10, to absorb the potential water vapour in the air. Unless the silica gel is bright blue, the gel is not fully dry and will not absorb the water it is intended to and incorrect readings can result.

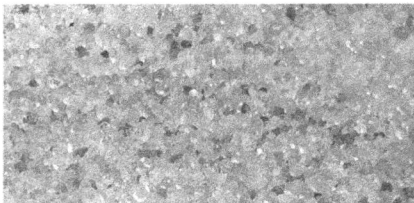

Pink / White Colour - Unsuitable   **Full Blue Colour - Suitable**

FIGURE 10.10   Ranarex™ Silica Gel Status.

To recondition or dry the silica gel it can be put in a heat source, such as an oven, for 15–30 minutes so the absorbed water is evaporated and removed; this is evident when the silica gel returns to bright blue.

10.3.1.1 Ranarex Basic Specifications

NOTE: *Provided that the Ranarex is in good condition and the silica gel is fully dry and in good condition.*

Accuracy:	± 0.5% of actual value
Ambient air reference dryer:	silica gel
Drive motor:	115 VAC 60 Hz
Sample flow rate:	10–15 scf/hour
Sample pressure:	20 psig maximum
Specific gravity range:	dual scale
	0.52 to 1.03
	0.97 to 1.90
Ambient temperature rating:	
Saturated gas and air:	40 to 120 degrees F2 (4–50 °C).
Normal gas and dry air:	0 to 130 degrees F2 (-18–54 °C)

The Ranarex should be level when mounted and placed in a clean, dry area.

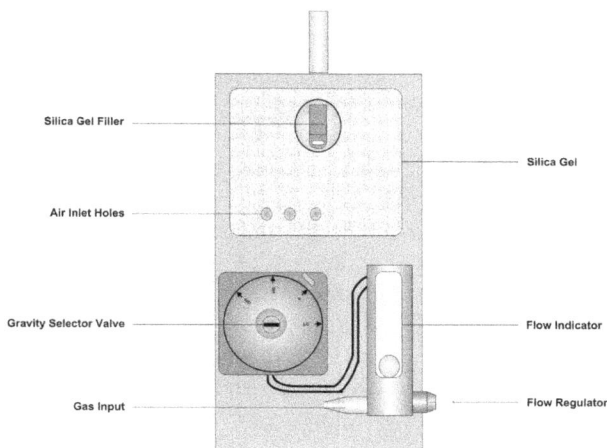

FIGURE 10.11 Ranarex™ Side View.

10.3.1.2 Ranarex Operational Considerations

As the Ranarex has no certification for hazardous areas and there is not any metal certification plate for hazardous area use, the Ranarex is a safe-area-only device. Any safe-area devices cannot be mounted on the separator or within any zoned areas as it is a potential source of explosion. The standard practice is to bring a small limited pressured sample of gas to the Ranarex while it is in a safe area.

Unfortunately, some operators are extending a hose from the separator to the Ranarex at the safe area. This brings live gas to a safe area, which effectively breaches the safe area classification so it has to be reclassified as a Zone 2 area. This means that all non-certified equipment has to be turned off. This includes the Ranarex.

Some operators are taking the gas contaminant readings of H_2S, CO_2 etc. from the gas outlet of the Ranarex. The thinking behind this is that the gas is passed through the silica gel crystals and hence will give a more accurate reading which is completely in error and only measures dry air.

To clarify some of the misconceptions:

i. The gas sample is not passed through the silica gel – only the reference air is

ii. The presence of H_2S or other hazardous gasses in the Ranarex area is an indication that the device is not being adequately ventilated

iii. The Ranarex™ is a mains electricity operated device; it has **no** electrical certification

iv. The Ranarex™ cannot be connected via a hose to the separator as it makes the area around the Ranarex Zone 2 hazardous area. The Ranarex does not have the following:

   a. Pressure certification and is not rated above pressures above 20 psig

   b. Any electrical certification for hazardous areas

There are lots of arguments that a hose can be connected to a Ranarex direct from the separator gas line, but, in addition to the aforementioned misconceptions, the following also need to be evaluated before allowing the practice.

i. Is the regulator used on the separator capable of handling the range of pressures in the separator? or is it just a needle valve slightly cracked open?

ii. Is the regulator certified for use with natural gas and possibly hazardous contaminants?

iii. What is the length of the hose used to connect to the Ranarex? The hose's volume must be purged before a true gas gravity reading can be taken

iv. What is the rating of the hose used, with a test certificate?

v. Is the hose rated for use with natural gas and hazardous contaminants?

vi. How is the flow shut off from the separator?

These questions should be asked before piping a Ranarex directly into a separator gas line because if there is an accident or incident, the investigation will be researching these questions and why it was permitted.

NOTE: *This practice has led to an explosion in a lab cabin.*

10.3.1.3 Gas Sampling

The Ranarex has a maximum pressure rating of 20 psi, therefore the use of a hose from a separator should be considered suspect regardless of electrical certification issues.

It also has a limited flow rate. The flow rate is controlled by a ball float regulator on the side of the Ranarex – again the pressure comes into play and can result in erroneous readings.

The safest way to bring the gas sample to a Ranarex is to use the inner rubber bladder of a football/soccer ball, as shown in Figure 10.12. This gives a sufficient sample at a reduced pressure, thus minimising any accidents in carrying the sample to the Ranarex.

FIGURE 10.12 Sample Gas Bladder.

As mentioned earlier all gas samples for analysis should be taken from the highest point of the gas line on the separator.

10.3.1.4 Ranarex Gas Filtering

If the sampled gas contains suspended particles larger than 25 microns, they must be removed or filtered before supplying the gas to the Ranarex. This is can be achieved using either a commercial gas filter or a manufactured one made from steel or glass wool in a pressure-bearing tube.

Whichever method is employed it should be positioned at the Ranarex gas inlet.

10.3.1.5 Locking and Unlocking the Ranarex Pointer

The Ranarex utilises a pointer lock mounted on the front of the instrument. The pointer lock must be engaged when there is no power applied to the Ranarex and the internal motor is not running.

The selector valve should be set to off and the power removed before any locking action.

The primary purpose of this mechanical lock is to exert tension on the flexible tape that interconnects the measuring wheels; the applied tension helps to maintain an accurate "tracking" of the tape when the instrument is not in use.

When moving the Ranarex the pointer assembly should be locked; this is a simple mechanical lever that prevents movement while the instrument is in transit. This should not be forced locked. After the Ranarex has been mounted for use, the pointer assembly should be unlocked and the Ranarex zeroed. If the pointer does not move then the drive tape inside the Ranarex has probably come off the wheels. It is a simple process to open the front of the Ranarex and put the tape back into place. All the wheels freely rotate and there should be no force required. Do not try to open the front of the Ranarex with the lock in place as it will bend and damage the lock mechanism. When locking the Ranarex, first zero the instrument to purge any trapped gas.

10.3.1.6 Zeroing the Ranarex

There is a mechanical zero on the right-hand side of the Ranarex, but this is only effective if the silica gel is in good condition as it references dry, clean air. Before zeroing the Ranarex, the selector valve, Figure 10.13, must be switched to the zero position and the air inlet ports cleared.

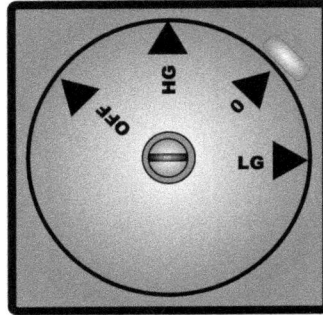

FIGURE 10.13   Ranarex™ Gas Gravity Select Switch.

The Ranarex should be left to run with the gas gravity selector switch in the zero position for several minutes before zeroing from the front screw adjuster.

10.3.1.7   Ranarex Dual Range

This is an option to allow for switching the Ranarex functionality between two specific gravity ranges, as defined in the basic specifications of the instrument. The dual scale defines the SG for the heavier and lighter type of gases.

This dual range is not automatic and has to be selected manually by the user. What is not widely realised is that there are two vents from the Ranarex as shown in Figure 10.14, for heavy and light gasses. Often the venting for one gas range is not accommodated for and in the presence of high toxic gasses this can cause operational issues.

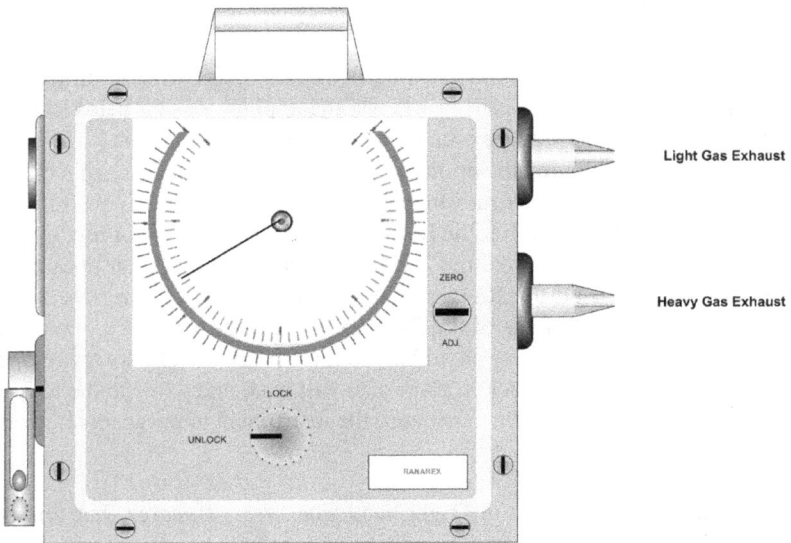

FIGURE 10.14   Ranarex™ Outlets.

10.3.1.8   Ranarex Flow Adjustment

The flow adjustment on a Ranarex is a simple device where the operator opens the input valve until the ball floats at approximately the specific gravity of the gas to be measured. If the gas gravity is unknown it can be estimated, a first reading taken, and the measurement repeated after the first reading for correction.

10.3.1.9   Ranarex Response Times

When determining the sample gas gravity at the recommended flow rate of 12 SCFH, the response time for the Ranarex to stabilise is normally 40–45 seconds.

However, increasing the flow rate by adjusting the inlet valve so the float reads higher than the actual gas specific gravity can reduce this time.

The inlet valve should **NOT** be opened fully to the degree flow is maximised and the float is lodged at the stop at the top of its flow tube.

10.3.1.10   Verifying the Ranarex Results

There is a very simple way to verify the functionality of the Ranarex and its results, while on a rig or location, by using gases with known specific gravities through the Ranarex and checking the result.

The welder will have at least acetylene and oxygen; Propane is usually used on burner ignitions. If there is a mud logging unit on the location then there may be other known gravity gasses available. Care is to be paid to the scale range of the Ranarex and the tested gas with respect to its gravity:

Gas Type	Specific Gravity
Acetylene	0.899
O_2	1.105
Propane	1.522

Other available gas types can be looked up on the web.

Unfortunately, there is no other real field calibration of the Ranarex than the zero. Therefore, if the readings are in error the only thing to look at is the pressure and flow rate of the sample fed into the Ranarex.

10.3.2   Gas Gravities Using a Chromatograph

When using a gas chromatograph the gas sample is injected into an internal gas stream which then moves the gas sample into a separation tube, referred to as the "column" of absorbent material.

The inert gasses helium or nitrogen are normally used as the carrier gasses to move the gas sample across the column where the various components are separated by the absorbing material and measured.

With underbalanced drilling and other types of operation where nitrogen is used to control the flow this can cause issues as it is an inert gas and normal "mud logging" chromatographs often cannot measure it. Therefore, the need to measure nitrogen needs to be specified at meetings like the TWOP.

For safety reasons Nitrogen/Helium remains the most commonly used carrier gases. Hydrogen is popular for improved separation of gas components but is far more volatile.

The readings from the chromatograph are then transmitted electronically by several standard methods, WITS, MODBUS, etc. to recording, analytical, and controlling SCADA instruments.

The instrumentation companies have developed a "chromatograph type" sensor that can be used in a hazardous area, provided all safety and wiring standards are applied, which automatically samples the gas stream. This device should be considered a simple type of chromatograph with all the restrictions and errors.

10.3.2.1   Chromatograph Problems

What should be stressed regarding chromatographs is that they do not always have the capability to detect the full range of gasses and hydrocarbon ranges. Therefore, it is critical to know what gasses the chromatograph will and will not detect.

As nitrogen can be used as a carrier gas, a chromatograph using nitrogen cannot detect nitrogen.

10.3.2.2   Chromatograph Gas Gravity

The chromatograph does not output the mixed gas gravity but rather the percentage of each individual gas constituting the mix. So, from the quantity and gravity of each component we can calculate the total/collective gas gravity.

In order to calculate the total gas gravity the specific gravity of all the gasses' components must be known. With access to the internet this is easily determined. If no access, Table 10.2 can be used.

TABLE 10.2

Component Gas Gravities

Hydrocarbon Gas Names Commonly Used	Gas Fraction Name	Ideal Gravity
Methane	C1 Fraction	0.553888
Ethane	C2 Fraction	1.038175
Propane	C3 Fraction	1.5222462
Iso-Butane	IC4 Fraction	2.006750
Nor-Butane	NC4 Fraction	2.006750
Iso-Pentane	IC5 Fraction	2.491047
Nor-Pentane	NC5 Fraction	2.491037
Neo-Pentane	EC5 Fraction	2.491037
Iso-Hexane	IC6 Fraction	2.975324
Nor-Hexane	NC6 Fraction	2.975324

TABLE 10.2 (Continued)

Contaminant Gas Names Commonly Used	Gas Contaminant Name	Ideal Gravity
Carbon Dioxide	CO_2	1.59450
Nitrogen	N_2	0.967167
Oxygen	O_2	1.104763
Hydrogen Sulphide	H_2S	1.176615

10.3.2.3 Gas Gravity Calculation

The measured data from a chromatograph will be exported in the parts per million for each of the measured component parts, as shown in Table 10.3.

The data in Table 10.3 is an exaggerated example of readings from the chromatograph.

TABLE 10.3
Example Component Gas Gravities

Chromatograph Gasses by Fraction	Parts per Million (ppm)
Gas C1 Fraction	560,000
Gas C2 Fraction	82,000
Gas C3 Fraction	41,000
Gas IC4 Fraction	10,000
Gas NC4 Fraction	6,000
Gas IC5 Fraction	2,000
Gas NC5 Fraction	900
Gas EC5 Fraction	860
Gas IC6 Fraction	240
Gas NC6 Fraction	0
CO_2	0
N_2	240,000
O_2	0
H_2S	1,600
Totals	944,600

What the most common error, with most field chromatographs, is that they actually *do not* measure the full mixtures of gasses present in the gas stream. This is due to the restrictions on the chromatograph performance as not all the mixed gasses are actually measured. Nitrogen for example can be a problem as it is an inert gas and does not easily react as other gasses do.

The issue is that the chromatograph can give results that "look like" full data has been measured but it has been mathematically extrapolated.

It is critical to know the true results and the types of gasses that are not measured by the chromatograph as this affects the gas gravity calculation result.

10.3.2.4   Gas Gravity Calculation from Chromatograph Results

The example Table 10.4 gives the gas readings in both parts per million and percentage. If the percentage is not given it can easily be calculated using:

$$\text{Gas Fraction \%} = \frac{ppm}{10,000}$$

TABLE 10.4

Example Component Percentages

Chromatograph Gasses by Fraction	Parts per Million (ppm)	Percentage (%)
Gas C1 Fraction	560,000	56.00
Gas C2 Fraction	82,000	8.20
Gas C3 Fraction	41,000	4.10
Gas IC4 Fraction	10,000	1.00
Gas NC4 Fraction	6,000	0.60
Gas IC5 Fraction	2,000	0.20
Gas NC5 Fraction	900	0.09
Gas EC5 Fraction	860	0.09
Gas IC6 Fraction	240	0.02
Gas NC6 Fraction	0	0.00
CO_2	0	0.00
N_2	240,000	24.00
O_2	0	0.00
H_2S	1,600	0.16
Totals	**944,600ppm**	**94.46%**

From Table 10.4 it is possible to integrate the actual gas gravity fraction from the published tables of ideal gas gravities.

The following procedure has been adapted from the Geoservices' mud logger's quick method of determining actual gravity from chromatograph readings.

Gas Fraction Gravity = (Gas Fraction In ppm/10^6) * Gas Gravity

The final gas gravity from the sum of all the individual gas fraction gravities, in Table 10.5, is highlighted in bold at 0.73391.

However, from Table 10.5 it is obvious that the chromatograph has not measured all the gasses from the gas sample as only 94.46%, highlighted, are accounted for; there is a discrepancy of 5.56% of unmeasured gasses.

Some manufacturers of chromatographs automatically do mathematical corrections for the error generated by missing component gasses, by expanding the actual measured percentages of the gasses measured to fill the error.

TABLE 10.5

Example Component Gas Gravity Calculation

Chromatograph Gasses by Fraction	Measured Parts per Million (ppm)	Measured Percentage (%)	Ideal Gas Gravity from Table	Gas Fraction Calculated Gravity
Gas C1 Fraction	560,000	56.00	0.553888	0.31018
Gas C2 Fraction	82,000	8.20	1.038175	0.08513
Gas C3 Fraction	41,000	4.10	1.522246	0.06241
Gas IC4 Fraction	10,000	1.00	2.00675	0.02007
Gas NC4 Fraction	6,000	0.60	2.00675	0.01204
Gas IC5 Fraction	2,000	0.20	2.491047	0.00498
Gas NC5 Fraction	900	0.09	2.491037	0.00224
Gas EC5 Fraction	860	0.09	2.491037	0.00214
Gas IC6 Fraction	240	0.02	2.975324	0.00071
Gas NC6 Fraction	0	0.00	2.975324	0.00000
CO_2	0	0.00	1.5945	0.00000
N_2	240,000	24.00	0.967167	0.23212
O_2	0	0.00	1.104763	0.00000
H_2S	1,600	0.16	1.176615	0.00188
Totals	944,600 ppm	94.46%		0.73391

From Table 10.5, the total percentage of gasses that the chromatograph measured was 94.46 %; mathematically correcting the error to zero or deducing results of 100% measured gasses, the following equations are used

$$\text{Chromatograph Gravity Correction Factor} = \frac{100}{94.46} = \mathbf{1.058649}$$

The mathematically corrected gas gravity is then calculated by:

Mathematically corrected chromatograph gas gravity = original gas gravity * Chromatograph gravity correction factor

Mathematically corrected chromatograph gas gravity = 0.73391 * 1.058649 = **0.77695**

This compensation for unmeasured gasses gravities may appear a good fix, but there is a fundamental discrepancy in the true measurement of the gravity that has been mathematically corrected with no regard to what the missing gas is or its effects.

If part of the missing gasses is oxygen which is a heavier gravity gas, at 5% the results become as shown in Table 10.6:

It is essential to know which gasses the chromatograph can read and more importantly which gasses it cannot. If possible, the gravity result with the Ranarex should be compared with the chromatograph, the main components of the gas can be measured with the chemical reaction tubes. If the missing gas capability is known then a

TABLE 10.6

Example Component Gas Gravities Correction

Chromatograph Gasses by Fraction	Measured Parts per Million (ppm)	Measured Percentage (%)	Ideal Gas Gravity from Table	Gas Fraction Calculated Gravity
Gas C1 Fraction	560,000	56.00	0.553888	0.31018
Gas C2 Fraction	82,000	8.20	1.038175	0.08513
Gas C3 Fraction	41,000	4.10	1.522246	0.06241
Gas IC4 Fraction	10,000	1.00	2.00675	0.02007
Gas NC4 Fraction	6,000	0.60	2.00675	0.01204
Gas IC5 Fraction	2,000	0.20	2.491047	0.00498
Gas NC5 Fraction	900	0.09	2.491037	0.00224
Gas EC5 Fraction	860	0.09	2.491037	0.00214
Gas IC6 Fraction	240	0.02	2.975324	0.00071
Gas NC6 Fraction	0	0.00	2.975324	0.00000
CO_2	0	0.00	1.5945	0.00000
N_2	240,000	24.00	0.967167	0.23212
O_2	50,000	5.00	1.104763	0.05524
H_2S	1,600	0.16	1.176615	0.00188
Totals	**994,600 ppm**	**99.46 %**		**0.78915**

The gravity now becomes **0.78915** instead of a corrected value of **0.77695**

specific chemical reaction tube can be used and the data can be entered into the table to correct the chromatograph for the missing component(s).

What should always be realised is that if the chromatograph does not measure a toxic gas and it is assumed that it is not present based on its readings, this becomes a potential hazard.

10.4   GAS CONTAMINANTS

A gas contaminant simply put is the presence of a non-hydrocarbon-based gas.

H_2S & CO_2 are the two most commonly measured contaminants although there are others that are undetected or not tested for. These components are usually detected in the PVT lab analysis where a comprehensive gas chromatograph is used to analyse the true composition of the gas. These contaminants are normally compounds of sulphur, nitrogen, and oxygen. The most commonly found and tested for non-hydrocarbons in well testing are:

- CO_2
- H_2S
- N_2

All these contaminants must be taken into account when calculating the gas rate as it affects the gas orifice calculation factors, predominantly supercompressability (F_{pv}).

There are other gasses such as helium and argon that are not regularly tested for. The more dangerous components that are not often tested for, often found across the entire range of natural gases, are the mercaptans. These like H_2S are highly toxic and corrosive.

The well testing companies rely on the use of the chemical reaction tube for detection of the non-hydrocarbon components, the contaminants. The majority of "mud logging" companies utilise a gas chromatograph which will identify all the components of the gas stream.

The service company should declare in advance which tubes will be available on the location and in what quantity. In case the operating company has any prior knowledge of the well or field this information should be passed to the testing company so that the correct range of tubes are on location for more accurate results. Unless specified in the contracts the well test company should supply reaction tubes across a spread of ranges from very low range to high ranges.

There is a standard for taking H_2S readings published by the ASTMS, D4810–06, which can be applied to all chemical reaction tube measurements.

ASTM International
Designation D4810–06 (Reapproved 2011)
Standard Test Method for Hydrogen Sulfide in Natural Gas Using Length-of-Stain Detector Tubes

10.4.1 GAS DETECTION TUBES – BASIC PRINCIPLE OF OPERATION

Gas detection tubes operate on a chemical reaction between the vapour phase compound and a liquid or solid detecting reagent, which is supported on an inert matrix. The tubes are length-of-stain types. In the specific tubes, the reaction of the specified gas with the chemical reagent is usually responsive and fast

Figure 10.15 shows a new unused tube with the ends capped in glass. After usage, the chemicals inside will change colour if the pertinent gas is present as shown in Figure 10.16. It should be noted that the colour change is not always the same for different manufacturer's tubes or gas detected.

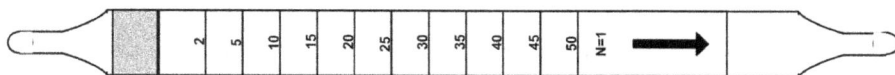

FIGURE 10.15 New Reaction Tube.

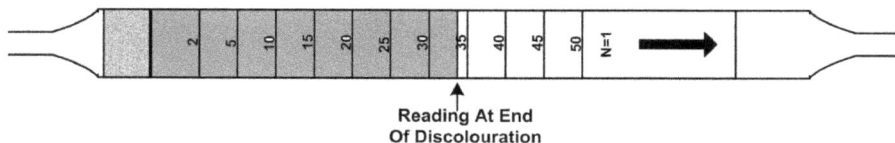

Reading At End
Of Discolouration

FIGURE 10.16 Used Reaction Tube.

There is dependence on the rate at which the gas is sampled for a correct reading with these tubes. A very high flow rate can cause some errors on a high reading. Conversely, low-flow rates are less likely to be affected. To combat these errors a fixed amount of gas is drawn through the tube using a dedicated pump at a fixed rate. The pump is used to draw the correct volume of gas at the specific flow through the reaction tube for optimum results.

There are several manufacturers of the gas detection tubes, each with its own different style of sampling pump. The sampling pump can vary from a syringe/pump type to a bellows mechanism. Each is calibrated to sample the correct amount of gas for that brand's tube and they should not be mixed with other brands.

The arrow marked on the tube will point in the direction that the gas passes through the tube or which end is placed into the pump. To be clear the pump is a suction pump so all gasses flow towards it.

10.4.2 REACTION TUBE OPERATION

The operators *must* always refer to the instruction sheets that come with the tubes to the number of strokes required for a correct reading. Often this is also written on the tube itself, highlighted in Figure 10.17, as N = X; where X is the number of strokes required for a valid sample.

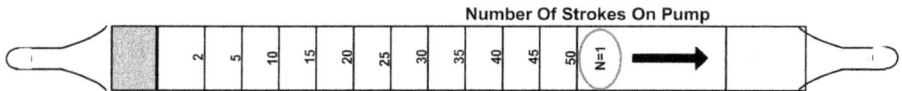

FIGURE 10.17   Number of Strokes on a Reaction Tube.

Chemical reaction tubes have a use by date; this is normally printed on the box, and before taking the sample the tubes used should be validated against the validity time on the box. Practically tubes up to six months past their date have been used and little difference to the readings were noticed, but valid date tubes should replace outdated tubes as soon as possible – first in first out.

Before the tube is prepared for use by snapping the glass caps off the end, the efficiency of the pump should be tested. This involves placing the intact tube in the sampling pump – shown in Figure 10.18 – and carrying out a sampling cycle.

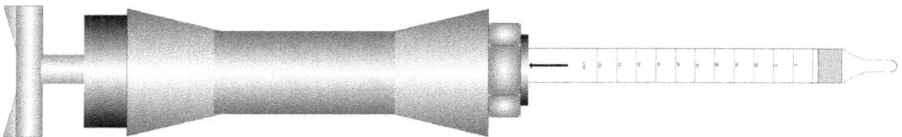

FIGURE 10.18   Example of a Reaction Tube Pump.

The previous figure shows a version of a reaction tube pump; other models exist from differing companies but the tests specified should still apply.

If the pump is fully functional and all the seals are correct the plunger/bellows should not expand as they are holding a vacuum. If the pump does not hold the vacuum and draws in air then if it is used it will not pull the complete volume through the reaction tube; air will also be drawn in through unfit seals, which will result in erroneous readings.

10.4.3 USING THE REACTION TUBES ON A SEPARATOR

Before using a tube, the instructions on using the tube should be read and understood; there can be a release of harmful chemicals when the tube seals are broken. The reaction tube should be left unbroken until it is ready to be used to prevent any ingress of fluids or contamination to the internal chemicals.

All gas sampling must be carried out from the highest point of the separator gas line to try to ensure dry gas is sampled. The layout shown in Figure 10.19 is the standard as per the ASTMS document. The bottle is specified as a polypropylene

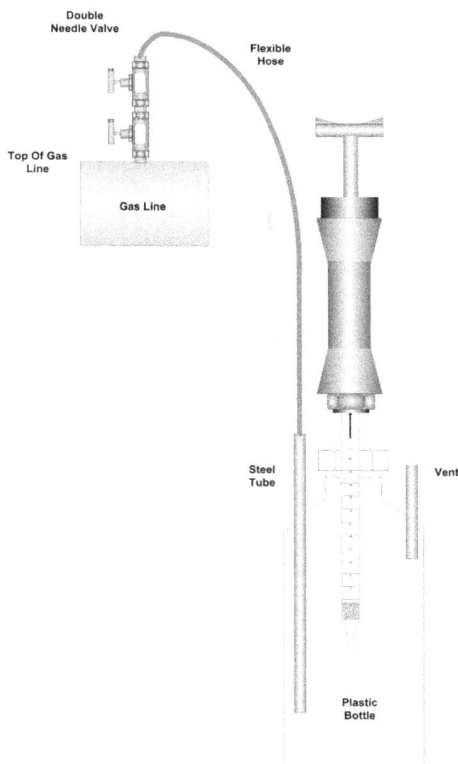

FIGURE 10.19 Taking a Sample with Reaction Tube.

construction but the most important thing is the flow through the bottle and potential hazards associated with it. The proscribed assembly can be easily constructed on location; there should be no excuses for the wrong procedures. The vent pipe is optional and is only necessary if the hole in the top is not suitably large enough to allow the escape of gasses.

10.4.4 TAKING A GAS SAMPLE FOR REACTION TUBE PROCEDURE

Ensure all HSE precautions are taken especially if H_2S is present or suspected. With the reaction tube and sampling pump removed, as shown in Figure 10.20;

i. Slowly turn on the gas needle valves so a slow steady low-pressure flow of gas is achieved through the bottle.
ii. Wait 2 to 3 minutes for the system to purge.
iii. Carefully break the seals on the reaction tube at both ends and insert the tube into the pump.
iv. Insert the assembly into the top of the bottle, support and wait a further minute in case air has been introduced.
v. Holding the pump, take a sample as described in the reaction tube instructions.
vi. Wait for the pump to fully return to its default condition before removing the assembly.
vii. Turn off the gas flow.
viii. Read the reaction tube.

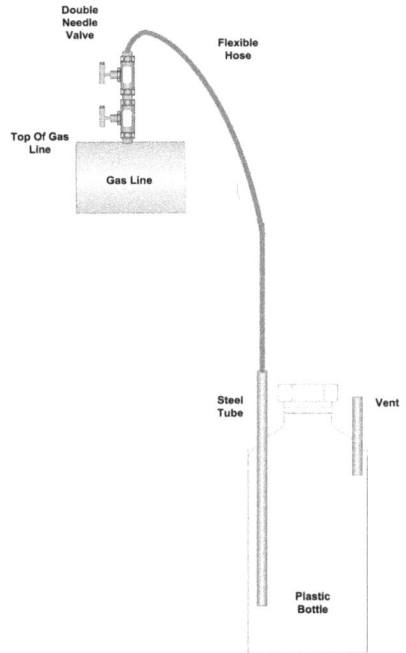

FIGURE 10.20 Illustration on Gas Sampling Setup Purging for Reaction Tube Sampling.

In the case of a wet gas flow, a piece of tissue or absorbent clean paper can be placed at the bottom of the plastic bottle, shown in Figure 10.21, which helps absorb any liquid entrained with the gas.

This technique is not included in the standard paper but has been adopted by the sampling companies and does not interfere with the gas sample in any manner. The tissue paper or equivalent should be removed after each sample to prevent any cross contamination.

There are instances where the gas cannot be brought in a liner to the sampling apparatus, distance, and access issues. In this case the sample gas is caught in a football or soccer bladder, similar to the Ranarex sample. The use of a pressurised sample bottle should be avoided due to safety issues involved with the carrying or of the transporting of the sample.

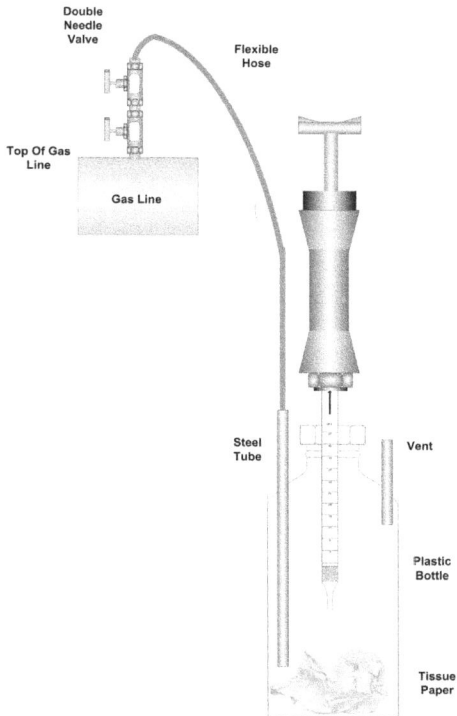

FIGURE 10.21 Taking a Sample of a Wet Gas with a Reaction Tube.

Figure 10.22 shows the wrong way gas contaminants are measured from gas held in a bladder. This causes erroneous readings due to pressure in the bladder and potential gas separation; this technique *should not* be used. The gas in the bladder should be vented into the gas sampling bottle and measured there, as shown in Figure 10.23.

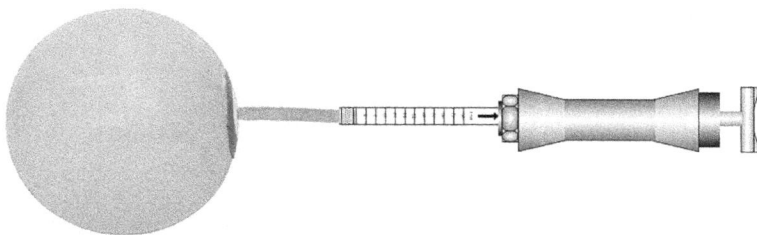

FIGURE 10.22 Gas Sampling Error.

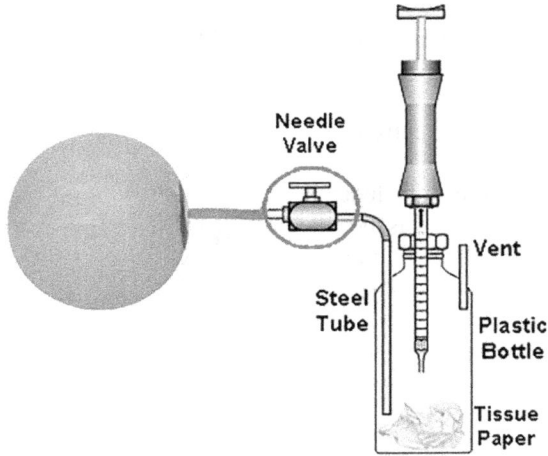

FIGURE 10.23 Correct Sampling with a Bladder.

The diagram shown in Figure 10.23 shows a better way of sampling gas in a bladder for contaminants. It is essential that a needle valve – or other form of flow restriction – is placed in the line from the bladder to control flow out of the bladder, otherwise the bladder will empty immediately. With the needle valve closed, connect the bladder and insert the reaction tube assembly. Slowly open the needle valve and let the gas purge through the bottle before taking the sample through the reaction tube.

NOTE: *It is better to get a full bladder rather than multiples for this procedure.*

10.4.5 READING THE REACTION TUBES

Reading the reaction tubes is fairly straight forward, a simple matter of reading off the length of the discolouration.

The tube shown in Figure 10.24 has a reading of 30. The units or percentages are dependent on the reaction tube used. However, the reaction is not always as clear cut as illustrated.

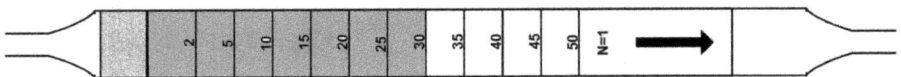

FIGURE 10.24 Reaction Tube with a Clear Response.

The reaction tube shown in Figure 10.25 has a slow discolouration fade back to the background. In this case the reading is taken at the end of the discolouration before any fading commences. In some cases, the flow through the reaction tube was not laminar and this results in a decline, shown in Figure 10.26, rather than an edge.

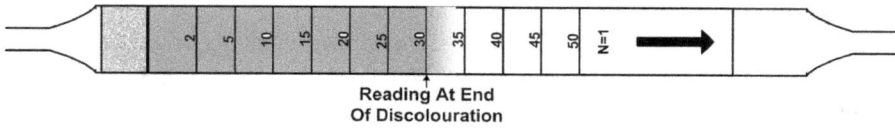

FIGURE 10.25   Reaction Tube with a Faded Response.

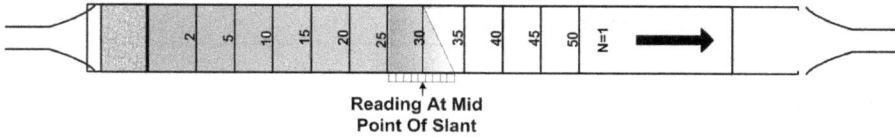

FIGURE 10.26   Correct Reaction Tube Reading.

In this case the reading is taken at the midpoint, expanded view shown in Figure 10.27.

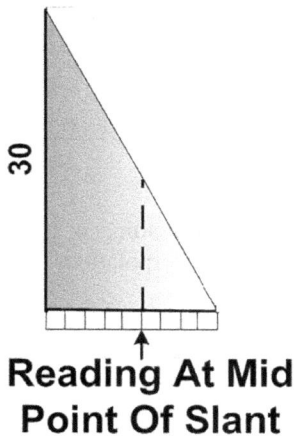

Reading At Mid Point Of Slant

FIGURE 10.27   Expanded Correct Reaction Tube Reading.

10.4.6   Gas Sampling Pressure

There is a lot of confusion with regard to the conditions of taking the sample through the gas tube; however, if the ASTMS procedures are followed this should not be an issue. All sampling by reaction tubes must be carried out with the gas at atmospheric conditions otherwise a correction should be carried out. This can vary from tube to tube depending upon the gas sampled and the manufacturer's recommendations. The general equation for correcting one manufacturer's gas tubes for pressure is:

$$True\, Concentration = Tube\, Reading \times \frac{1013}{Pressure\ in\ hPa}$$

Where hPa is the abbreviation for Hecto Pascal.

1 hPa = 0.01450377 psi   Or   1psi = 68.9476 hPa

Other manufacturers' correction procedures may vary.

Table 10.7 shows the variance with pressure on the % reading of a reaction tube, with the tube reading along the top and the sampling pressure in psig on the side of the table.

TABLE 10.7

Reaction Tube Variance with Pressure Table

Press	Tube %									
Psig	10.0	20	30	40	50	60	70	80	90	100
0	10.0	20.0	30.0	40.0	50.0	60.0	70.0	80.0	90.0	100.0
1	9.4	18.7	28.1	37.5	46.8	56.2	65.5	74.9	84.3	93.6
2	8.8	17.6	26.4	35.2	44.0	52.8	61.6	70.4	79.2	88.0
3	8.3	16.6	24.9	33.2	41.5	49.8	58.2	66.5	74.8	83.1
4	7.9	15.7	23.6	31.5	39.3	47.2	55.1	62.9	70.8	78.6
5	7.5	14.9	22.4	29.9	37.3	44.8	52.3	59.7	67.2	74.7
6	7.1	14.2	21.3	28.4	35.5	42.6	49.7	56.8	64.0	71.1
7	6.8	13.6	20.3	27.1	33.9	40.7	47.5	54.2	61.0	67.8
8	6.5	13.0	19.4	25.9	32.4	38.9	45.4	51.8	58.3	64.8
9	6.2	12.4	18.6	24.8	31.0	37.2	43.5	49.7	55.9	62.1
10	6.0	11.9	17.9	23.8	29.8	35.7	41.7	47.7	53.6	59.6

Referencing Table 10.7; if the tube reading is 25% at 3 psig, then the corrected reading is obtained by using a linear interpolation.

Press	Tube (%)	
(Psig)	20	30
3	16.6	24.9

((Reading at 30% – Reading at 20%)/2) + Reading at 20%

((24.9–16.6)/2) + 16.6 = **20.75**

Differing manufacturers of reaction tube can have differing pressure variants so it is necessary to verify the figures.

10.4.7   GAS CONTAMINANT READING AT CHOKE MANIFOLD

When flowing a well and gas reaches the surface it is important to identify any hazardous components as soon as possible, to organise operational safety. It is therefore necessary to take gas contaminant readings at the choke manifold as the earliest point of contact. Again, all these readings should be recorded on the manual readings sheet.

This type of sample does not conform to set procedures but is critical to determine the presence of any hazardous gasses. Full PPE should be employed with breathing apparatus if H_2S is suspected. As preferred standard, the sample must be taken downstream of the choke manifold, a lower pressure source and the gas will break out of the flow easier than with a higher pressure upstream of the choke manifold. Ensure all safety practices are enforced and full PPE is worn especially if it is a new well or field.

NOTE: *This procedure is only for determination of hazardous gasses and NOT for use in gas calculations. All gas calculation readings are from the separator gas line only.*

Steps

    i. Using a measuring cylinder, as shown in Figure 10.28, position the goose-neck from the sampling point so the flow is forced to flow round the inside of the measuring cylinder which will help to separate the entrained gasses for sampling

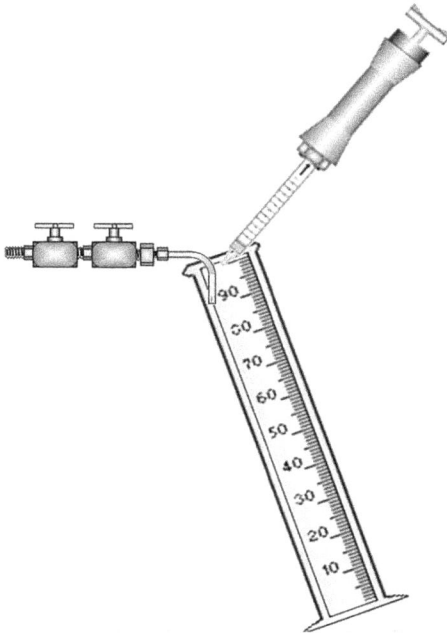

FIGURE 10.28 Taking a Reaction Tube Reading at the Choke Manifold.

NOTE: *This is an estimation for safety purposes; the correct readings are taken from the separator with the correct procedure.*

    ii. With the fluid spinning round the inside of the measuring cylinder it in effect aids the gas to break out of solution and exit the top of the measuring cylinder
   iii. Wait for sufficient gas to escape to ensure that the sample taken is gas and not mainly air and take a sample from the highest point of the cylinder
   iv. The reaction tube should not be put into the cylinder itself as fluid can be drawn into the tube and affect the results

v. It is recommended that the test for H_2S is carried out first for safety reasons before other contaminants

vi. Ensure that the measurements made from the choke manifold are labelled correctly on the report as there can be discrepancies with the separator readings due to the nature of the sampling and the changes in the constituents as the well effectively cleans up

10.4.8 EXTENDING THE REACTION TUBE RANGE

As the contaminants produced by a new well or field are often not fully known, it is possible that the correct range reaction tubes are not available on location. There is a method of extending the reaction tubes range, but it is not always applicable to all gasses and reaction tube brands. The manufacturer's instructions should be consulted to see if it is applicable. This measurement does not result in a precise/accurate result but rather a "good" indication of the contaminant until the correct reaction tubes are sourced. This technique involves the use of a 100 ml plastic syringe.

Steps

i. The syringe is used in place of the reaction tube and similar steps are followed with the configuration illustrated in Figure 10.29

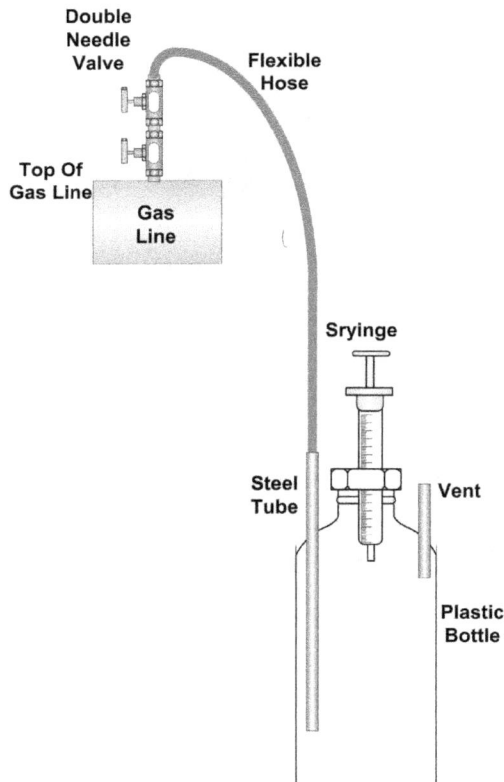

FIGURE 10.29   Taking a Gas Sample Using a Syringe.

ii. When the operator is confident that the gas sample is valid and representative within the bottle or measuring cylinder as depicted in Figure 10.30, 50 ml of gas only is drawn into the syringe

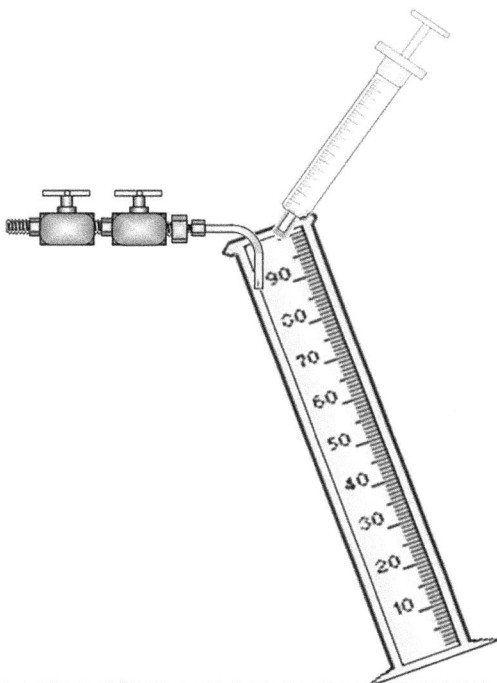

FIGURE 10.30   Taking a Gas Sample at the Choke Manifold Using a Syringe.

iii. The syringe is then removed from the sampling section and 50 ml of **clean air** is drawn into the syringe making it a total sample volume of 100 ml. Agitate the syringe to ensure mixing of the gas and air. In principle the gas sample from flow stream has been diluted by 50% by adding the equal volume of air into it

iv. Insert the reaction tube into the syringe and perform a standard sampling cycle, as illustrated in Figure 10.31. Be prepared to inject the gas from the syringe simultaneously as necessary without applying too much pressure

v. After completion the reaction tube should be read in the normal manner, but in this case the reading on the reaction tube should be *doubled*

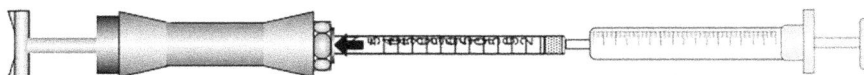

FIGURE 10.31   Transferring a Gas Sample from a Syringe.

The tube shown in Figure 10.32, has a reading of 30. The units or percentages are dependent on the reaction tube used, but as the gas mixture has been diluted by 50% air then this reading is actually 60.

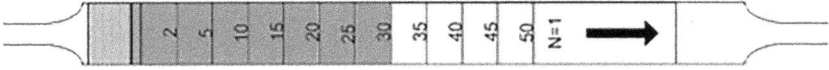

FIGURE 10.32 Reading a Mixed Gas Sample on a Reaction Tube.

It is important that these readings are declared as estimates and mixtures following this type of technique on both the manual reading sheets and the final reports; there could be discrepancies when the correct reaction tubes are used and the reasons for such difference made clear.

NOTE: *This technique should be used as a last resort and the correct reaction tubes ordered as soon as possible. Not all reaction tubes may work with this technique.*

11 Flow Data

It is often considered that the well testing staff should just measure the data from the well. This is not true; they should also keep records of other companies' activities that affect the performance, flow, and pressures involved with testing the well.

What this means is that while the well is live the well testers should act as a central point for collecting and collating all the relevant data. Therefore, a well tester should be located on the well test equipment at all times and in the vicinity of ongoing operations under safety guidelines.

The following operations should be monitored and all events, pressures, depths, and activities recorded using the standardised timing on the manual reading sheets.

This may involve the operating company personnel instructing the other service companies to set their watches to a standard time and recording all their events which should be passed to the well test personnel on regular intervals.

This includes:

- E-line
- Slickline
- Coiled tubing
- Any pumping operations
- Any sub-sea operations

11.1 CLEAN-UP PERIOD

If it is a new well and freshly perforated then there is a necessary clean-up period to remove all mud and debris before the well can be put through the separator.

Rather than flow the well through the burners the initial flow should be to tanks. Flowing to a stock or non-pressurised tank is used but as soon as any hydrocarbons are present then it is possible for pressure to build up in the tanks.

Therefore, the stock tanks must have a vent system of large bore pipework to a safe location. Pressure relief valves should be installed on all tanks as well as sight glasses to calculate tank volumes.

A well tester should be assigned to monitor and switch the tanks as required.

Unless cleaning up after stimulation operations, the volume of introduced fluids to the well that will be returned during the clean-up can be estimated by calculating the volume of the tubing in the well, often called bottoms up. When the well is flowed to the tanks the volume can be monitored to determine when the tubing contents have been flowed to the surface.

The level in the tanks, pressure, and time should be recorded by the data acquisition system and on the manual sheets so flow rates can be determined. As the higher gravity liquids are flowed out the flow rate to the tank should increase.

DOI: 10.1201/9781032623689-11

In the case of the suspected presence of H_2S, to minimise human interaction, a simple modification to a differential cell can be used to estimate the level in the tanks, as shown in Figure 11.1. This is not an accurate system but an indication.

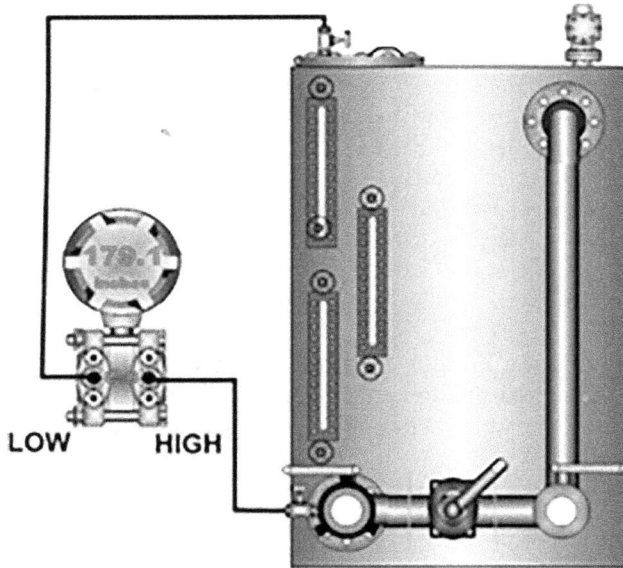

FIGURE 11.1   Tank Level Monitoring.

- A Standard Differential Transducer Range is 0–400 inches of water
- The Specific Gravity of the Tubing Fluid Contents = 0.85
  - the range of the transducer is 400/0.85 = 471.6 inches of fluid

The 471.6 inches of fluid is the corrected number that is put into the data acquisition system and is a measurement of the fluid height in the tank. With this fluid height and an equation for tank volume, the volume of fluid within the tank is easily calculated.

Measuring liquid levels with a differential transducer only works if the fluid in the tank is all the same density. If there is an interface of two levels, oil and water, then the method is inaccurate.

11.2   SEPARATOR FLOW REGIMES

As a recap, within the well testing scenario there are two major types of flow that are measured using a three-phase separator.

Three-Phase Flow

This is where there are three distinct phases of the produced fluids: oil, water, and gas. In this scenario all three phases are measured by independent meters on the separator.

BS&W must be taken from the separator oil line for related calculations. The measured water and oil flow rates are then corrected for the water escaping through the oil flowline.

Two-Phase Flow

In this scenario there are only two phases considered as produced fluids, liquid, and gas. In this case the liquid mixture of oil and water is measured as a single entity with the gas being measured separately as normal.

This is the case; when there is a liquid flow stream where the oil and water phases are not separable into individual phases using standard well testing techniques, and a single liquid phase measurement is the only option.

In this case the individual liquids, oil and water, components of the flow, are calculated from the measured BS&W taken at the choke manifold or before the test separator.

From the aforementioned flow regimes, it shows just how important an accurate and reliable BS&W measurement is and where it is taken from. There is a simple guideline to determine which mode the test separator should be run in:

If there is measurable water and oil then the separator should be run in three phases.

Co-Mingling

Using a standard test separator running in three-phase operation, it is possible to combine or co-mingle the oil and water flows into a single line to achieve a two-phase output from the separator. Verify this capability on the outlet manifold of the separator.

11.3   LIQUID FLOW MEASUREMENT

11.3.1   FLOW METER RANGES

Regardless of the type of meter installed on the liquid lines, the range of the meter must be known. Flow rate errors are often caused by flowing a liquid through the meter at a rate outside the meter's capacity. The well test crew should know the ranges of the liquid meters and be prepared to change them as required.

It is important not to confuse flowline bore size with turbine meter size, with common practice of installing a 1.5-inch turbine meter on a 2-inch flowline. The turbine meter will have its parameters stamped on a metal plate attached to its body. Any issues with the range of a turbine meter should have been eliminated with flow tests from maintenance and last major inspection. This is where water at varying rates was pumped through the system to verify meter functionality and to get a water meter factor.

NOTE: *All meter factors used on a test should be carried out using the fluid from the well.*

11.3.1.1 Turbine Meter Ranges

Table 11.1 is an example of the nominal operating range of common turbine meters used on a separator. The separators will have different meters installed depending upon its operating parameters; these may be different flow ranges to the ones in the table.

TABLE 11.1
Standard Turbine Meter Nominal Ranges

Meter ID Size Inches	Low Rate Bbl/D	High Rate Bbl/D
1	170	1,700
2	1,300	13,000
3	2,750	27,500
4	3,400	41,000

It is only common to have two or three different sized turbine meters on the separator oil flowline. The water flowline normally has one flow meter, sometimes two, occasionally the same size and maybe different. This gives a number of different configurations of turbine meters and combinations of them on each flow outlet.

Meters overlap in ranges which makes it possible to use different flow meters for the same flow rate. For instance, if the flow rate is 10,000 Bbls/D, then this in theory can be measured by either the 2-inch or 3-inch turbine, defined in Table 11.2. This may lead to minor differences in the reading between the two different sized turbine meters. This should be corrected by use of a meter factor at the conditions of the flow.

TABLE 11.2
Turbine Meters Centre Points

Meter Size Inches	Low Rate Bbl/D	High Rate Bbl/D	Centre Point Bbl/D
2	1,300	13,000	**7,150**
10,000 Bbl/D			
3	2,750	27,500	**15,125**

The general rule of flowing through a turbine is to consider its flow range centre point (50% of its full range); the turbine meter with a flow range centre point closest to the expected flow rate would be the best fit and least error generating. Avoid using the bottom and top ranges of the turbine meters.

With a separator, the option of different sized output flowlines, each with a differently sized turbine, allows a wider range of flow rates to be metered accurately and with the flexibility of selecting which turbine size would best suit the particular flow as a preferred option.

Manipulating the configuration of turbine sizes against expected flow rates for the full well test, even for the well test campaign, ensures smooth and efficient operations of fluid measurement. If any of the turbine meters are replaced or maintained, then a separate meter factor must be performed for that meter, for calculations and to confirm functionality. If the separator does not have the capacity to measure a high flow rate through a single installed meter, then two meters, in parallel, can be used to extend the maximum flow permissible.

It should be kept in mind that although flowing using two parallel turbine meters, the internal diameter of the individual liquid flow meters and the liquid level valves will also have restrictions on the flow rates. If the liquid meters, flowlines, and valves are unable to cope with the flow rate from the well, then, the flow rate capacity should be limited to the capabilities of the of the flow meters.

In this case the well flow has to be limited by reducing the choke size that suits the operating range of the separator/flow meters, or a larger separator is brought in to replace the existing one. In some cases, a second separator could be brought in and both separators run in parallel. In both these scenarios the export lines from the separator(s) may have to be increased in size to cope with the fluid flows.

NOTE: *This scenario also applies to high gas flow rates.*

11.3.2 METERING TIMING ERRORS

As previously mentioned where manual readings are being used there can be a large discrepancy in the reported fluid flow rate due to the readings taken at incorrect or different times; this isn't applicable in data acquisition systems. Therefore, in order to get a consistent flow rate, the time intervals between readings must be consistent and using the unified standard test time previously described.

11.3.2.1 Shift Changes

Over an extended period, it is often necessary to have multiple shifts of personnel. When these "shifts" change over it is common for the reading at the time of change-over to show flow rate discrepancies, due to mismatches in time between different shifts. The time should be set as a unified standard for all personnel to correspond with the data acquisition system time from the first shift change.

11.3.3 METER FLOW RATE ERRORS

With a data acquisition system measuring liquid flow from a separator using turbine or positive displacement meters there are two ways of measuring the flow:

Analogue Flow Meters
This is normally a 4–20 mA output that measures the rate at the instance of reading.

Pulse Flow Meters

This is where the rotation of the meter is counted in individual pulses in a manner similar to a car odometer.

Where a stable flow is involved, there is very little difference between the two measurement principles, but, where unstable or surging flow is encountered, the pulse system has the advantage.

If we consider the flow period shown in Figure 11.2, the following applies:

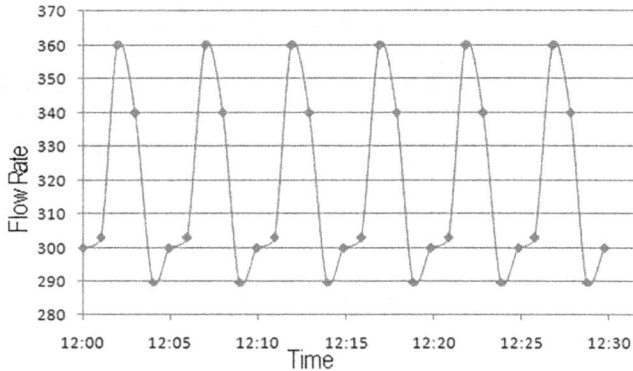

FIGURE 11.2   Analogue Meter Output.

Analogue

If the readings are taken every 15 minutes, then the flow rate will be reported as 300 Bbl/D, which is the averaged reading over 15 minutes.

If connected to a data acquisition system (DAS) there are options for instantaneous reading at the time of acquisition or average based on a timeframe of previous readings. The functionality is also dependent on the instrument used, speed, and mode of communication.

Pulse

If the readings are taken by counting the pulses, then all the instantaneous flow rates are taken into account and an average flowrate is reported of 318.6 Bbl/D.

This is a more accurate measurement of the cumulative flow through the meter.

When measuring fluid flow, it is advisable not to measure the flow at short intervals; in some cases it is possible that what seems like a fluctuating flow rate is actually due to the position of the control valve.

NOTE: *One minute sampling rate is usually the fastest DAS sample rate for turbine type Flow Meters because if it is taken faster, it records the increases and decreases in flow due to the position of the level valves. This gives the impression of flow instability where in fact it is measuring the position of the valves rather than flow rates.*

11.3.4 GOR2

GOR2 is the amount of gas dissolved or entrained in the oil flow as it leaves the separator. The higher the operating pressure of the separator the higher the GOR2, the higher the entrained gas in the oil leaving the separator.

If we could measure the amount of gas that is released from the surge tank then we can determine the GOR2 and compensate for its effects. It is not possible to measure the entrained gas in the discharged oil in field operations, since initial high gas escape (flash) is from the light end hydrocarbons (methane predominantly) and then the heavier end hydrocarbon escape is considerably slower, often hours. There is not normally a meter on the tank capable of reading the vast ranges of escaping gas.

Example

Many years ago, a testing company used a commercial gas meter on a tank in conjunction with a pitot tube in order to try to get a compound reading of escaping gas to measure GOR2. The pitot tube was for the fast-escaping gas and the gas meter for the slow. In theory it would work but operating pressures were overlooked and the commercial gas meter exploded. No one was hurt.

This is included to try to explain the difficulty in measuring escaping gas from a tank.

True GOR2 can only really be determined in the laboratory using PVT samples, which is a relatively expensive process, and therefore graphs were originally used to give an estimation of GOR2. But with the advent of computers equations that generated the original graphs have been used.

GOR2 Estimation Equation

$$GOR2 = \frac{10^4}{4.67} \times \left[\frac{1}{(1-S) + 2.1 \times ((1-S) - 0.81)^2 \times (1 - (1-S))^{0.8}} - 1 \right]$$

Where:
S = shrinkage factor corrected to standard conditions (60 degrees F)
GOR2 units SCF/Bbl (standard cubic feet per barrel)

This equation is commonly used in the oilfield and is taken from the book *Prediction of Shrinkage of Crude Oil* by D.L. Kattz. This GOR2 equation uses shrinkage at its core, which would require using a shrinkage tester in order to determine the actual oil shrinkage rather than relying on a correlation. Shrinkage should be taken every time there is a change in the oil properties from pressure and flow rates. Like all estimations there are limits for the shrinkage range that the GOR2 equation is valid for:

$$(1-S) \text{ between } 0.74-1$$

- If the shrinkage is outside of the valid range and the data is used then this must be declared on the field and final reports, to keep all those involved aware of the limitation affects
- If estimations of the shrinkage is used to derive GOR2 then this *must* be declared in the reporting

11.3.4.1 High Values of GOR2

The units of GOR2 is SCF/BBL; therefore if the GOR2 result is multiplied by the oil flow it will yield the flow of gas in the oil line. This can then be added to calculated gas results. The derived flow from the GOR2 is in SCF but the flow rate is MMSCF so the units have to be matched. This should be in a separate column in the report if used either as standalone value or correction of the gas flow.

It should once again be highlighted that GOR2 is often a compounded estimation.

11.3.5 SHRINKAGE

The wrong shrinkage reading will have effects not only on the oil calculations but also the GOR2 estimations, so particular emphasis should be placed on ensuring the correct factor is used.

Shrinkage is generally carried out on the oil phase alone and it is often impractical to acquire with two-phase testing, since water and oil would constitute the total volume of the tested liquid. Further difficulties are faced with heavier oils and physically measuring shrinkage, so correlation will have to be used. To assist in judging the circumstances, previous test records of existing wells in the same field should be referred to.

11.3.5.1 Shrinkage Estimation

Where well testing units are employed for routine well checks (production monitoring tests) or similar, it is common not to use a mechanical shrinkage tester but to rely on predictive methods of either formulae or graphs.

The graphs were originally published by Flopetrol in the 1970s but with current computers a formula, usually in Excel, is used. The shrinkage estimation equation is usually a computer-based program that functions within a manual based reporting package but is generally not declared by the well testing company. A common form of shrinkage prediction is given by the equation:

Shrinkage Estimation Equation

1) $R_S = \left(\dfrac{1.797}{g} - 1.838 \right) P$

2) $R_S = pgd \left[\left(\dfrac{P+14.73}{18.2} + 1.4 \right) 10^{(0.0125 \times °API - 0.00091T)} \right] \Big/ 0.83$

3) $pgd = °API \left(0.02 - 3.57 \times 10^{-6} R_S \right)$

4) $\dfrac{1}{1 - SHR} = 0.9759 + 12 \times 10^{-5} \times F^{1.2}$

With $F = R_S \left(\dfrac{pgd}{g} \right)^{0.5} + 1.25T$

Where:
P = pressure in psig
API = API gravity at 60 degrees F
F = temperature in degrees F
1-SHR = shrinkage

This equation is taken from the Prediction of Shrinkage in HP41C Petroleum Fluids Pack Manual and sites the following as references.

(1) and (3): "Two-phase flow metering with two fluxi Flow Meters" – P. Vigneaux = RND/1680.

(2) and (4): "Volumetric and phase behaviour of oilfield hydrocarbon systems" – Standing (1977 Ed.)

There are limits of validity to this equation which are often ignored or unknown to the user and the receiver of the data. This makes it essential that if estimations are used in the report calculations then a declaration of the estimations used must be clearly made with all relevant equations and information.

The Validity Ranges Are:

- Temperature range 60–260 degrees F
- Oil gravity range 16–70 API
- Pressure 500–7,000 psi

If the measured parameters are outside of the range, it is necessary to consider the following:

i. Continuing with the equation, then it must be declared in the report as such
ii. A shrinkage tester should be used
iii. Combined meter/shrinkage flow calculations should be used

In all cases the practice of leaving a sample in a one-litre measuring cylinder should be employed to give a more definitive result that can be compared with any of the aforementioned points.

It is important and must be made clear that the shrinkage is now derived from an equation that uses manual readings at its core. This calculated shrinkage *estimation* is then used in the GOR2 equation.

Therefore, it is critical to know that these two equations are significantly affected by manual readings and is one of the reasons for this document because this "estimation" is not made clear in operations or reports. While a GOR2 measurement may not be physically possible, checks on shrinkage are feasible and should be carried out.

When shrinkage is calculated it directly bases GOR2 and oil calculations on its estimation, so it would be necessary to confirm the estimated value and its dependent values by using a shrinkage tester. The measurement of shrinkage should be addressed before the job is mobilised; then, if necessary, the mechanical shrinkage tester is mounted on the separator.

11.3.5.2 Sample Shrinkage Measurement

If a shrinkage tester is not available or cannot be used in two phase or the oil viscosity is too thick to use one, the well testers can resort to measuring cylinder shrinkage estimation.

If a 1,000 ml measuring cylinder, for the specific purpose of shrinkage measurement, is filled with oil at the start of the main test and left with no additives or demulsifiers then a value of shrinkage can be *physically estimated* and used to validate the shrinkage calculation estimation equation result. For this comparative purpose with the shrinkage equation, the oil sample taken here MUST be from the separator oil line.

If the oil in the measuring cylinder is left for a long enough duration it will give an indication of the amount of shrinkage that can be expected. Taking the final temperature would allow for volume correction using the volume correction factor figures. Refer to Figure 11.3.

A 1,000 mL sized measuring cylinder should be used and not a smaller one (500 mL); a plastic bottle shouldn't be used either. The cross-sectional area of the one-litre

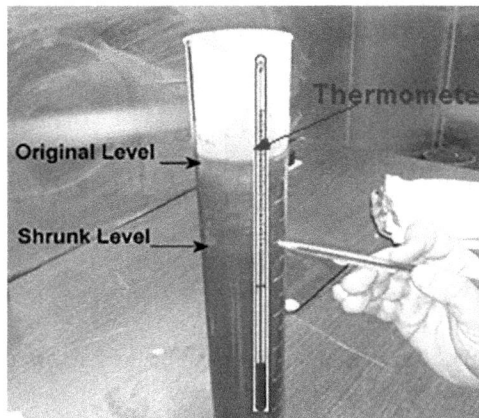

FIGURE 11.3 Basic Shrinkage for Verification of Estimation.

measuring cylinder plays an important role in de-gassing with a larger liquid surface area; the bigger the better, as the smaller cylinders will restrict the escape of gas from the liquid.

All this data and sequence of events should be written on the manual readings sheet at the start of the measurement and at the end, including the final temperature.

11.3.5.3 Shrinkage Tester

As the use of the shrinkage tester involves vented gasses, all safety precautions and PPE must be in place and must be utilised before any operations are carried out. Shrinkage testers are used predominantly in three-phase tests on separated oil that is measured due to gas escaping the volume; this does not apply to water.

Water generally speaking does not shrink except in the case of high CO_2 content. CO_2 is soluble in water as in the majority of fizzy/soft drinks. There is no real rule here except to try to see if it shrinks. One indication could be that the water is slightly acidic due to the formation of carbonic acid where the water has mixed with the CO_2.

The shrinkage tester should also be pressure tested as part of the separator; it is often left out by being isolated and could be a safety issue when handling the equipment.

The physical construction of the shrinkage tester vessel has not changed since it was first introduced by the Metrol Corporation as shown in Figure 11.4.

The calibration is simple a fluid volume of 3,000 cc put into the vessel body and its height noted on the sight glass. This is the 25% shrinkage mark.

The fluid volume is increased to 4,000 cc and this is noted on the sight glass. This is the 0% shrinkage mark.

The distance between the two points is then subdivided to give a scale. Traditionally this is every 5% but can be increased to give a better reading.

The mounting of the shrinkage tester should be below the oil level in the separator; the oil level is taken from the separator "oil" sight glass. It is often common to use the gas line from the top of the oil line sight glass. The setup is illustrated in Figure 11.5.

Like all sight glasses on the separator the sight glass on the shrinkage tester must be kept clean and it is recommended that it is cleaned after every use to prevent oil "settling" on the glass obscuring the view.

11.3.5.3.1 Operation of the Shrinkage Tester

Full PPE must be used especially if any hazardous gasses are present in the gas stream.

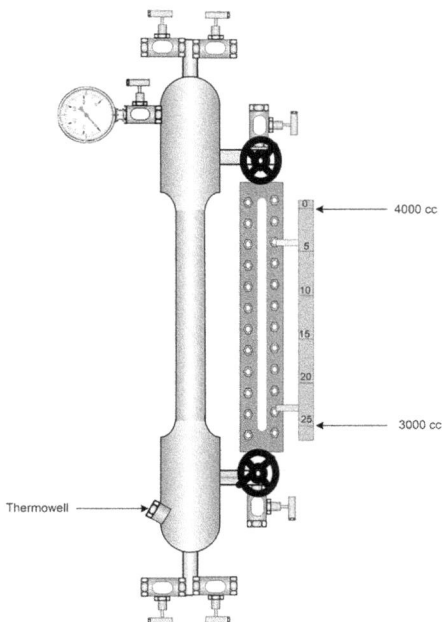

FIGURE 11.4 Shrinkage Tester.

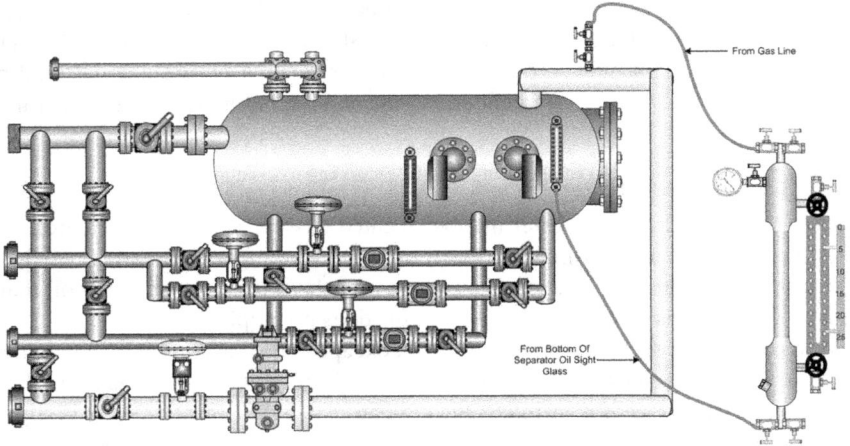

FIGURE 11.5   Separator with Shrinkage Tester Connector.

With the shrinkage tester connected to the separator as shown in Figure 11.5, configure the shrinkage tester with the valves as depicted in the following steps:

$$\mathbf{X} = \text{valve closed}$$
$$\mathbf{O} = \text{valve open.}$$

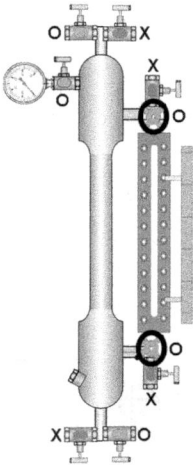

11.3.5.3.2   Step 1 – Purge

Using the gas line from the separator, purge the shrinkage tester of all fluids and residues. This includes the sight glass (Figure 11.6).

FIGURE   11.6   Purging the Shrinkage Tester with Separator Gas.

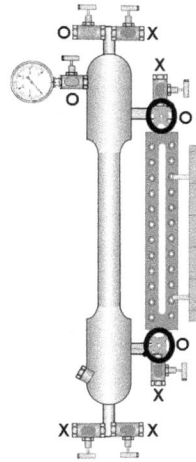

11.3.5.3.2   Step 2 – Equalise Pressure

Using the gas line from the separator equalise the pressure from the separator to the Shrinkage Tester.

The pressure gauge reading on the shrinkage tester should match the gauge on the separator.

There should be no difference in pressure between the two vessels. (Figure 11.7)

FIGURE 11.7 Equalising the Shrinkage Tester.

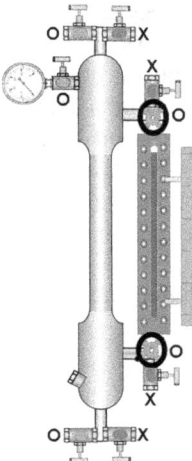

11.3.5.3.2 Step 3 – Fill Shrinkage Tester

Open the valve from the bottom of the separator oil line sight glass and allow the oil level to fill to the 0% mark.

It is important that the gas line be left open during this operation and the pressure inside the shrinkage tester be constant with the separator. (Figure 11.8)

FIGURE 11.8 Filling the Shrinkage Tester.

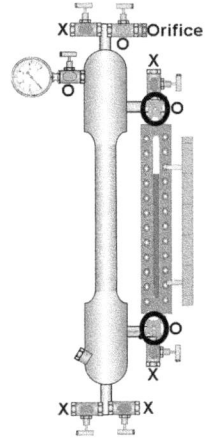

11.3.5.3.2 Step 4 – Shrinkage Operation

With the valves all closed, install an orifice no larger than 1/4" into the vent line detailed in Figure 11.9.

Slowly open the vent line valve and allow the gas to escape through the orifice.

The orifice is in place to regulate the rate of release of gas from the shrinkage tester so the gas does not "flash" from the shrinkage tester (Figure 11.9).

FIGURE 11.9 Measuring Shrinkage.

The shrinkage tester should be allowed to settle for a period of no less than 20 minutes, longer if possible.

One of the errors often involved with a shrinkage tester is omitting the orifice in the vent line. Without the orifice in the vent line, the escape of gas will be uncontrolled and considered as a "flash" release; the lighter-end hydrocarbons (C_1, C_2, etc.) will be quickly released to the atmosphere, leaving behind the heavier hydrocarbon ends. A controlled release of gas stops the "flash" effect and allows the lighter end hydrocarbons to assist in "lifting" the heavier ends to atmosphere; otherwise the heavier ends are left behind in the solution and a true shrinkage is not performed.

Upon completion of shrinkage, where the level of the oil in the shrinkage tester no longer reduces over time, the amount should be read from the sight glass scale while taking a temperature reading from the thermowell at the bottom of the shrinkage tester.

The reading from the shrinkage tester must now be corrected using the volume correction tables to get a shrinkage at 60 degrees F. Finally, the corrected reading should be compared to that given by the shrinkage algorithm for accuracy checks.

11.3.6 FLUID FLOW METERS

On most well test separators the liquid flows are measured with a turbine type flow meter although the Coriolis meters are gaining more favour. Coriolis meters require an electric stimulation, therefore any Coriolis must have an electrical safety certificate. These have a standard in API 5. The meters are subject to calibrations, as well as the analogue sensors and fluid tanks.

The stock/surge tanks have a mathematical calibration based on the dimensions of the tank. A certificate is issued or a metal plate on the tank defines the barrels per centimetre/inch of tank height. This is critical in the case of tanks with semi hemispherical construction. The tank should have a fixed scale next to all sight glasses and the corresponding tank calibration is critical to fluid flow measurement.

Because the normal liquid flow meters are a type of displacement meter, the properties of the liquid passing through them should be taken into account within the calibration. The liquid passing through a meter will change its performance according to the pressure, temperature, gravity, and amount of dissolved gas in the fluid.

These changes primarily affect the viscosity of the fluid which will affect the response from the flow meter. To correct for these changes a calibration is required for each flow rate at differing pressures and temperatures.

It is common place to actually do a "supposed" calibration with water. Even water with different salinities will have a slight effect on the flow meter's response. This is not a calibration or meter factor but a function check. The calibration or meter factor must be with the specific liquid flowing through the meter at the flowline's pressure and temperature. This practice of just using a water calibration is common with trailer mount packages as a tank is not often included in the package.

Considering temperature alone as an example of effect on the turbine meter functionality; there is expansion on the turbine blades and on the meter body and as these are dissimilar metals they expand and contract in different rates. This assists in slippage where fluid passes beneath the rotor blades without affecting the rotation speeds. The degree of slippage is also dependant on the fluid viscosity.

11.3.7 BASIC OIL FLOW EQUATION

$$Q_o = V_s \times M_f \times (1 - Shr) * (1 - BS\&W) \times k$$

Where:

Q_o = corrected fluid flow
V_s = uncorrected fluid flow
M_f = meter factor at 60 degrees F
Shr = shrinkage at 60 degrees F
BS&W = BS&W
k = volume correction factor.

From the formulae we find a number of variables that would influence the oil flow rates.

11.3.7.1 Meter Factor

The meter factor is one of the most important factors in well testing as it is an effective calibration of the liquid measuring meter, especially with turbine meters. If the meter is not calibrated correctly, it will significantly affect the reported liquid flows. The meter factor varies with the following factors:

i. Fluid type
ii. Flow rate
iii. Viscosity
iv. Pressure
v. Temperature
vi. Entrained gas
vii. Compressibility
viii. Slippage past rotor blades

These factors can change at differing flow rates and choke sizes when flowing a well; therefore the only way to get a true meter factor is to measure and flow the liquid to a calibrated tank.

11.3.7.2 Shrinkage

Shrinkage is also a primary factor in the calculation of fluid flows but it can be difficult to measure accurately therefore the practice of carrying out a combined meter shrinkage factor had been adopted to resolve the difficulty. Shrinkage can be determined by:

i. The shrinkage tester mounted on the separator
ii. Using a pressurised stock or surge tank
iii. Mathematical estimation

Shrinkage is defined as

$$\frac{\textbf{Volume Measured By Meter}}{\textbf{Actual Volume Flowed}} \times 100$$

The actual volume flowed is the measured volume in the calibrated tank.

- A pressurised tank is used for **meter factor** alone
- An atmospheric stock tank is used for **combined meter shrinkage factor**

Meter or combined meter factors when on a live well test are usually co-ordinated with the data acquisition and the well testers using time or volume as a reference.

Before beginning a meter or combined meter factor, a stable measure volume in the tank is necessary. The initial level and its corresponding volume should be recorded on the manual readings sheet. The respective manual reading from the Flow Meter and also from the DAS system should be recorded.

If using a pressurised tank (Figure 11.10) to measure a meter factor then the tank should be pressurised first using the gas from the separator line so that both equipment are at equal pressure. This is to stop any gas breakout and errors due to shrinkage.

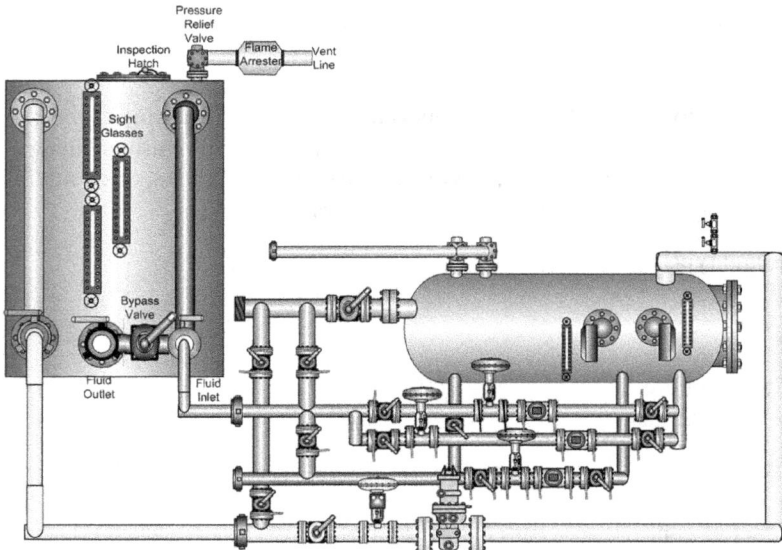

FIGURE 11.10 Separator with Surge Tank.

However, if the same process is carried out using a stock or non-pressurised tank (Figure 11.11) then the reduction of pressure from separator conditions to tank conditions will allow any entrained gas in the oil to break out, flashing, as the light ends of the gas composition flash off immediately whereas the heavy ends take longer to break out.

Either at a set time or at a signal from the well test crew, the oil flow from the separator is switched to either tank type depending on the factor measured. Recording the start time of the factoring flow period is important.

The tank and the separator must be correctly grounded in order to prevent any static electricity build-up from escaping gas that may result in an explosion.

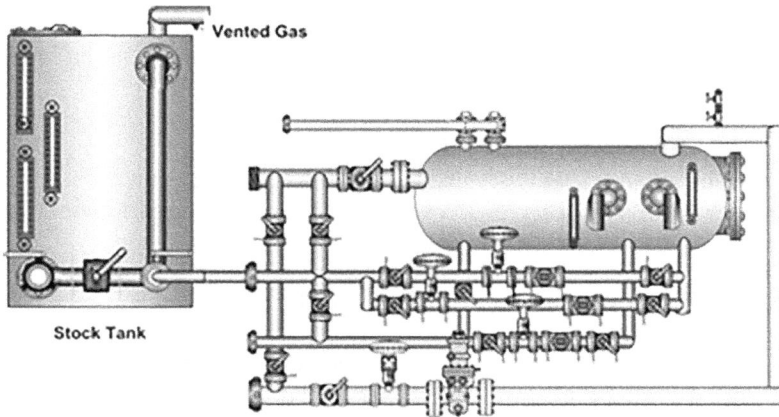

FIGURE 11.11 Separator with Stock Tank.

Upon completion all the final readings must be recorded.

- Time and date
- Final tank volume
- Final flow meter reading
- Final DAS reading
- Flowline and tank temperatures (to correct the volumes to standard conditions)

If meter factor alone is required the final tank volume is then used.

$$\frac{\text{Final Volume Measured By Meter} - \text{Initial Volume Measured By Meter}}{\text{Final Tank Volume} - \text{Initial Tank Volume}}$$

When a combined meter shrinkage factor is measured then the final volume in the atmospheric tank must be left to complete shrinking. This is not a quick process and a minimum of an hour is suggested as the minimum period although monitoring every 15 minutes is carried out to ensure complete shrinkage, until there is no further decrease in the volume of the oil.

The temperature in the tank will be used to correct the meter factors to the standard 60 degrees F using volume correction factor tables or equation.

11.3.7.1   Meter Factor Procedures

i. Ensure an initial level is present in the intended test tank, this level should be taken and recorded on manual reading sheets from the sight glass. Tank sight glasses should have the same level of maintenance and cleaning as separator sight glasses

   a. If using a stock tank, time should be allowed for the dissolved gas to escape from the solution

   b. Because of the volatility of condensate already present in a stock tank, there should be ample time for the fluid to reach its static state (total shrinkage)

   c. Ensure that the liquid level is over any spherical base or other designs of the tank so that any liquid levels would be linearly proportional to the measured level

   The initial level in the tank must be clearly recorded as the initial or starting level of the tank on the manual readings sheet

ii. Within the oil chamber of the separator, sufficient oil volume/level should be present to carry out the meter factor and to prevent any gas flow through the oil line which could cause false measurements of fluid through the turbine meter

iii. BS&W from the oil leg of the separators should be taken to confirm no water is present prior to diverting flow to the surge tank

iv. Oil should be flowed until a measurable volume has accumulated in the tank, nominally no less than 25% of the tank volume depending on operational considerations

v. Volume of oil transferred to the tank is measured by the centimetres/inches taken on the test tank. In case of magnetic floats and coloured flaps, ONLY fully turned flaps should be taken into consideration and slightly flipped flaps ignored as a standard to avoid discrepancies

vi. The final level in the tank must be clearly recorded as the final un-shrunk volume/level of the tank on the manual readings sheet

vii. The final un-shrunk volume must be left for a period of time to shrink. This is not a quick process and ample time should be allocated. Again, the final shrunk reading should be clearly recorded as final shrunk volume on the manual readings sheet

11.3.7.2   Combined Meter Factor Procedure

This ultimately results in the oil shrinking due to the escape of the entrained gas components. This meter factor is then of a twofold effect:

- The effect of the meter factor error
- Shrinkage

As these cannot be individually separated this effect has come to be known as combined meter/shrinkage factor.

All vented gasses from a tank should go through a flame arrester which allows the vented gas to exit the tank but if there is a fire the flame cannot return back to the tank. The flame arrester should be matched to the equipment it is installed on.

The vent from the flame arrester is usually an expandable plastic line. This MUST be a separate line that is not joined to any other vent lines, including from other tanks. Also, it must be vented way from any burners or source of ignition. Wind direction should carry away any vented gasses and in the event of wind direction change then the vent lines should be re-positioned.

During the following steps, ensure all safety precautions are taken especially when using surge tanks:

i. Switch the oil/condensate flow from the separator oil line to the tank and simultaneously take the meter reading. Record the meter reading as an initial reading and the timing accurately on the manual readings sheet

ii. Oil should be flowed until a measurable volume has accumulated in the tank, nominally no less than 25% of the tank volume, but as we are going to allow the tank volume to shrink the volume should be increased for high GOR rates

iii. When there has been sufficient flow, either from the meter or tank volume readings, switch the flow back from the separator oil line to its original destination. The final reading from the fluid meter should be noted on the manual readings sheet, at the same time the inlet valve to the test tank is closed

iv. The tank should remain isolated for a minimum of one hour, allowing the tank volume to vent off any entrained gas; this level reading is recorded. To be sure the volume is stable the level is checked again after 15 minutes to see if any change has occurred

v. Once the tank volume has stabilised, the final level reading is recorded; timing for this event is to be recorded accurately as it is used for calculation

vi. The final temperature of the fluid in the tank should be taken and recorded

NOTE: *In a combined meter factor the temperature in the tank must be taken at the time of recording the final shrunk volume.*

As many well test packages do not utilise a surge tank connected to the separator, the flow equation when used with a stock tank now becomes:

Combined Meter Factor Flow Calculation

$$Q_o = V_s \times C_{mf} \times (1 - BS\&W) \times k$$

Where
Q_o = corrected fluid flow
V_s = uncorrected fluid flow
C_{mf} = combined meter factor at 60 degrees F
k = volume correction factor

The shrinkage now is not used independently but factored into the combined meter factor.

Whatever method employed it is critical that meter factors or combined meter factors are carried out to match the flow conditions as the performance of the flow meter changes: viscosity, pressure, temperature, etc.

11.3.7.3 Gravity Correction to 60 Degrees F

Because oil expands or contracts with temperature the calculated oil flow has to be corrected for the temperature of the flowing oil. This is done using a volume correction factor which is also known in well testing as a k factor.

NOTE: *A k factor is sometimes used with flow meters; this is not the same as the k for volume correction.*

Before the advent of computers this had to be looked up in a reference book; however, with the advent of computers, volume correction can be calculated. The first step is to correct the oil specific gravity to the standard 60 degrees F.

$$SG_{60} = SG_T + (0.00069 - 0.000372\, SG_T)\,(T - 60)$$

Where
  SG_{60} = corrected gravity to 60 degrees F
  SG_T = specific gravity at temperature T
  T = observed temperature
  Limits of Validity
  T = 0–110 degrees F
  SG_T = 0.68–0.92

11.3.7.4   Volume Correction Factor (k)

$$k = 1 - \left(0.66\,API_{60} + 2.75\right) x \left(T - 60\right) x 10^{-4}$$

Where
  k = volume correction factor (k)
  API_{60} = specific gravity in units API at 60 degrees F
  T = observed temperature
  Limits of Validity
  T = 0–150 degrees F
  SG_{60} = 0.68–0.92
  API_{60} = 76–22.3

11.3.8   Oil Calculation with a Three-Phase Separator

The following measured parameters need to be taken in order to undertake the oil flow calculation for both data acquisition and manual systems. There can be minor differences between the DAS and manual reading calculations arising from minor differences in the readings, but these differences should not be excessive. Ideally there should be no differences between the two systems. Timing differences between readings become a major source of error.

Parameter Measurement

- Oil line temperature (degrees F)
- Cumulative reading from oil fluid meters (Bbls)
- Oil gravity (API or s.g.)
- Oil temperature (degrees F)
- BS&W taken from oil line (%)
- *Shrinkage (%)
- *Oil meter factor

(*) Depending upon the configuration of the well test package, shrinkage and oil meter factor are often expressed as a combined meter factor especially if a tank is used. In other configurations shrinkage is determined from a correlation using pre-set meter factors.

Physical Measurement

- Oil meters sizes (inch)
- Number of meters

These are necessary in order to determine the meter factors are available to carry out the calculations. Referring to the previously detailed equations:

Standard Oil Flow Calculation

$$Q_o = V_s \times Mf \times (1 - SHR) \times (1 - BS\&W) \times k$$

Combined Meter Factor Oil Flow Calculation

$$Q_o = V_s \times C_{mf} \times (1 - BS\&W) \times k$$

Where
  Q_o = corrected oil flow
  V_s = uncorrected oil flow
  Mf = meter factor
  C_{mf} = combined meter factor at 60 degrees F
  SHR = shrinkage
  k = correction factor.

11.3.9   WATER CALCULATION WITH A THREE-PHASE SEPARATOR

The following measured parameters need to be taken in order to undertake the water flow calculation for both data acquisition and manual systems.

Parameter Measurement

- Cumulative reading from oil fluid meters (Bbls)
- Cumulative reading from water fluid meters (Bbls)
- BS&W taken from oil line (%)
- Water meter factor (%)
- Salinity (for reference)

Physical Measurement

- Water meter sizes (inch)
- Number of water meters

Standard Water Flow Calculation

This is the calculation if water is flowing through the water turbine alone and no water is in the oil line, BS&W = 0. This normally only occurs with gas wells with associated water.

$$Q_w = V_w \times Mf_w$$

Where
Q_w = corrected water flow
V_w = uncorrected water flow
Mf_w = meter factor for water meter – this is usually 1.

Water Flow Calculation Using BS&W

This is the calculation if water is flowing through oil turbine as an emulsion like flow which is normal for most oil wells

$$Q_w = V_s \times BS\&W$$

Where
Q_w = corrected water flow
V_s = uncorrected oil flow
BS&W = BS& from oil line

Water Flow Calculation Using Meter and BS&W

This is the calculation if water is flowing through a water and water in oil flow oil which is normal for most oil wells

$$Q_w = \text{water from turbine} + \text{water from BS\&W}$$

$$Q_w = (V_w \times Mf_w) + (V_s \times BS\&W)$$

Where
Q_w = corrected water flow
V_w = uncorrected water flow
V_s = uncorrected oil flow
BS&W = BS&W from oil line
Mf_w = meter factor for water meter – this is usually 1.

11.3.10 OIL CALCULATION WITH A TWO-PHASE SEPARATOR

The following measured parameters need to be taken in order to undertake the oil flow calculation for both data acquisition and manual systems.

Parameter Measurement

- Oil line temperature (degrees F)
- Cumulative reading from the oil fluid meters (Bbls)

- Oil gravity
- Oil temperature (degrees F)
- BS&W taken from before separator (%)
- *Combined meter shrinkage

(*) In two-phase operations shrinkage and oil meter factor are often expressed as a combined meter factor as a tank has to be used to determinate the meter factor

Physical Measurement

- Oil meter sizes (inch)
- Number of oil meters

Two-Phase Combined Meter Factor Oil Flow Calculation

$$Q_o = V_s \times C_{mf} \times (1 - BS\&W) \times k$$

Two-Phase BS&W Water Flow Calculation

$$Q_w = V_s \times BS\&W$$

Where
Q_o = corrected oil flow
Q_w = corrected water flow
V_s = uncorrected oil line flow
C_{mf} = combined meter factor at 60 degrees F
BS&W = BS&W from before the separator
k = correction factor.

11.3.11 WATER CALCULATION WITH A TWO-PHASE SEPARATOR

The following measured parameters need to be taken in order to undertake the water flow calculation for both data acquisition and manual systems:

- Cumulative reading from oil fluid meter (Bbls)
- BS&W taken from before separator (%)

Two-Phase BS&W Water Flow Calculation

$$Q_w = V_s \times BS\&W$$

Where
Q_w = corrected water flow
V_s = uncorrected oil line flow
BS&W = BS&W from before the separator (normally choke manifold)

All these factors must be either known or accurately measured in order to achieve an accurate flow rate calculation for all the phases from the separator.

11.4 GAS FLOW MEASUREMENTS

It is standard practise to measure the gas flow through a known orifice plate although the use of Coriolis meters to measure gas flow is becoming more popular.

With the Daniel changeable orifice systems using the orifice plate has one real advantage: rangeability. This is ability of the orifice system to be configured to meet the flow requirements by changing the orifice plate, whereas metering systems like the Coriolis have a fixed range requiring multiple meters to be installed to cover the separators full flow range.

11.4.1 GAS METERING BY ORIFICE VARIABLES

11.4.1.1 Gas Line Pressure (Static Pressure)

The gas calculation uses the gas line pressure, often referred to as the static pressure; this is not the same as the separator pressure but a measurement of the pressure of the flowing gas. With the orifice plate this pressure can be taken either upstream or downstream of the orifice, as illustrated in Figure 11.12.

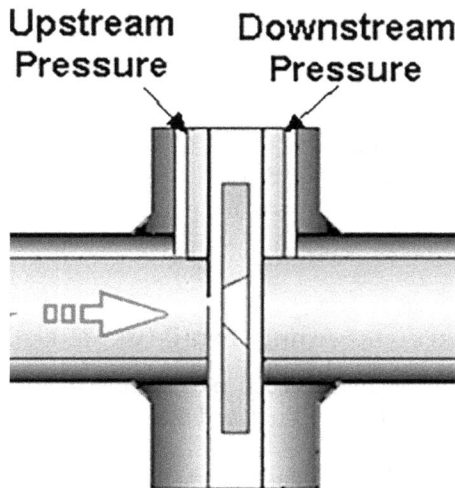

FIGURE 11.12   Orifice Plate Mounting with Flange Taps.

This is sometimes difficult to determine as the liners run behind the Barton and orifice plate holder, but the well testers must know which is being used; there is a small difference that influences calculations.

11.4.1.2 Differential Pressure

Traditionally Barton chart recorders are used for the supply of static and differential pressure readings. With the advent of modern data acquisition systems the original use of the Barton differential meter has been superseded with individual or digital sensors. The Barton is still used as a mechanical source of data as redundancy and to verify the electronic data acquisition system. This comparison should be carried out regularly and written on the manual readings sheet.

Mounting the Barton Meter

The Barton Meter along with the electronic differential meter must be mounted vertically with knockout pots to ensure there is no build-up of fluids within the lines or differential cell that would affect readings.

Figure 11.13 shows a typical mounting of the Barton differential meter with its associated fittings and transducers.

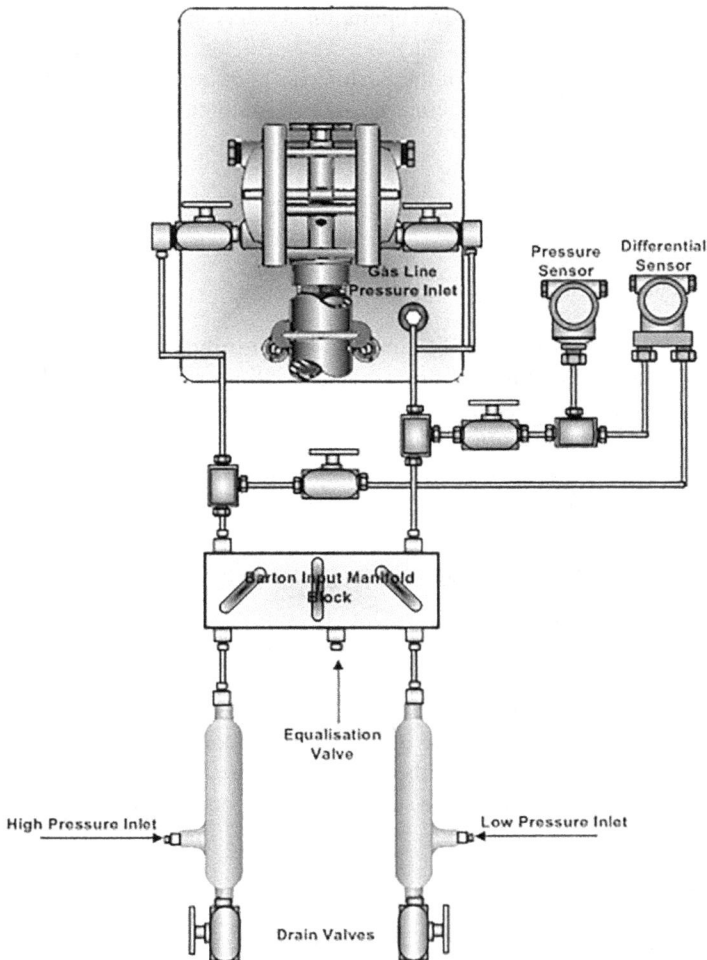

FIGURE 11.13   Barton Meter Mounting.

In this case it is configured as downstream pressure taps as the Barton pressure sensor is connected to the low-pressure inlet which corresponds to the downstream tapping.

As the separator is not 100% efficient there will always be a small amount of liquid carried over or condensed/coalesced in the gas line. This liquid will tend to congregate in the bottom of the orifice plate holder and can be passed through to the Barton lines.

Hence the use of knockout pots, and it is recommended that these lines and knock-out pots are bled down on regular intervals, every two to three hours, to ensure that the liquid does not reach the differential cell.

Bleeding off the differential cells will cause a change in the differential readings and will appear on the Barton chart as well as the DAS. To avoid the bleed off operations appearing on the Barton chart and transducers, they can be isolated by the needle valves.

It is important that the bleed down operation is noted on the manual reading sheets and in the sequence of events.

It is important that the centre of the Barton chart be completed with this basic data added onto a Barton chart as necessary. The information should be entered using a black pen:

• Well name
• Date
• Time chart started
• Time chart finished

Upon the completion of the test the chart should be fully annotated with the orifice plate, choke sizes, etc. that match the change in readings as in the sequence of events. This data should be taken from the manual readings sheets.

The readings from the Barton Chart should be used and recorded on the manual readings sheet; this means that the supervisor should physically check the Barton chart during the flow period and compare it with the data acquisition readings.

11.4.1.3   Gas Line Temperature

Similar to the gas line pressure the gas line temperature is not the same as the temperature in the separator or the oil line. The gas line temperature has to be mounted downstream of the orifice plate, shown in Figure 11.14, and as there is a temperature drop associated with gas flowing through an orifice, the gas line temperature is lower than the rest of the separator. Some Barton Chart Recorders still have the temperature option in place; if present this also should be utilised.

There should be two temperature sensors, at a minimum, on a separator:

• Gas line temperature
• Oil line temperature

Using the wrong temperature – or even separator body temperature – will result in errors in fluid flow calculations. The position of the gas line temperature with respect to the orifice line pressure is set by the American Gas Association (AGA) reports.

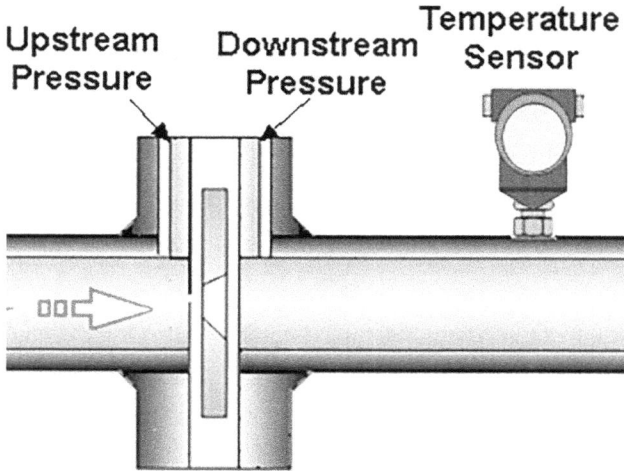

FIGURE 11.14   Gas Line Temperature Mounting.

11.4.1.4   Orifice Plates

Unlike the liquid flow meters the orifice meters have an interchangeable orifice plate system. This involves the changing of an orifice plate during flow and under pressure. This is what the Daniel Senior orifice plate holders are designed for, and they work well with this advantage, if they are properly maintained and greased.

Since there is gas under pressure, it will have to be vented and care must be taken when operating the plate holders especially if gasses similar to H_2S are present. It is always better to have a two-person operation similar to the "buddy system."

11.4.2   GAS CALCULATION BY ORIFICE PLATE

The most common calculation method in use with well testing is the AGA Report Number 3 method from the American Gas Association (AGA). This report has been updated several times since it was originally published in the 1950s. Other methods such as the AGA8 and NX-19 reports are standard but most of the factors of AGA3 apply.

The AGA 3 report procedure is given here as it can be used without a computer and accurate gas flow rates calculated by hand as in well tests in the 70s. The AGA 3 Report is constituted as a set of look up tables, when used for calculation purposes, enabling the operator to use a simple calculator to determine gas rates.

NOTE: *The well test supervisor should carry a copy of a manual method in case of power failures so flow rates can be determined until power is restored.*

There are several computer "applications" that have been written by individuals that will calculate gas and oil rates. While laudable and often useful there often is no traceability in the equations and functions used. These "applications" should not be used unless authorised and vetted by the service company.

11.4.2.1 AGA 3 Gas Calculation Equation

Using the standard American Gas Association (AGA) equations (AGA Report #3) the determination of the gas flow across an orifice plate is shown in the following equations.

American Gas Association
AGA Report No. 3
Orifice Metering of Natural Gas and Other Related Hydrocarbon Fluids

i. Formula 1 – Gas Calculation

$$Q_h = \sqrt{h_w p_f} \times F_b \times F_{pb} \times F_g \times F_{tf} \times F_r \times Y \times F_{pv}$$

The calculated gas flow (Q) is expressed in MMSCF/D where the M in MSCF/day stands for thousands; MM in MMSCF/day stands for millions. Although the AGA 3 report is normally used in imperial units the metric versions can be similarly used.

This equation is often split into two parts where the calculated – or looked up – factors are separated into the orifice flow rate constant (C') and the acquired measurements.

Therefore, the gas flow equation is often simplified as:

$$Q_h = C' \sqrt{h_w \times p_f}$$

Where:

- $Q_h = Quantity\,of\,flow\,at\,base\,condition - cubic\,ft\,/\,hr$

  This is then usually expressed in MMSCF/day. SCF is standard cubic feet per day which means the flow rates have been corrected to 14.73 psi and 60 degrees F.

- $h_w = Differential\,Pressure\,in\,inches\,of\,water\,at\,60\,Deg\,F$ –

  its abbreviation comes from height of water, which the differential pressure is measured in. Therefore is the actual measurement of the differential pressure across the orifice plate from either the Barton or digital sensor.

- $p_f = Absolute\,Static\,Pressure\,in\,psi$ –

  stands for flowing pressure and is the pressure measurement at the orifice plate, either upstream or downstream, and it is taken in absolute pressure units, normally psia.

- $C' = Orifice\,Flow\,Constant$

  (psia is considered as **psia = psig + 14.73**)

<u>NOTE:</u> *If the test is being performed at elevated locations, such as mountains, then*

psia = psig + Atmospheric Pressure

In certain countries the standard atmospheric pressure is not considered to be 14.73 psia. The standard pressure needs to be verified and declared in the final report.

ii.   Formula 2 – Orifice Flow Constant

The Calculation of the Orifice flow constant C' standard equation becomes:

$$C' = F_b \times F_{pb} \times F_{tb} \times F_g \times F_{tf} \times F_r \times Y \times F_{pv}$$

Where:

F_b = Basic Orifice factor

also called the hourly basic orifice factor whose value is dependent upon the line bore and the installed orifice plate as well as a dependence on the type of tapings used as described earlier. With a Daniel orifice meter the taps are normally flanged on a well test separator.

The line bore is the line bore of the orifice plate meter and not the gas pipe line. The orifice line bore is normally given on a metal plate at the top of the meter.

F_{pb} = Pressure base factor

Pressure base factor is a correction for the pressure base if the pressure base is not the standard 14.73 psi.

In the southern states of the USA differing base pressures are used; these should be declared before the test starts and included in the final report.

Normally this is fixed at 1 for standard operations.

F_{tb} = Temperature base factor

Temperature base factor is a correction for the temperature base if the temperature base is not the standard 60 degrees F.

F_g = Specific gravity factor

Specific gravity factor corrects the gas gravity of the measured gas in the gas line.

F_g determines the ratio of the density of the gas to that of dry air when both the pressure and temperature of the gas and air are at standard conditions to give a factor known as the real specific gravity which is then used in the calculations.

This is one of the factors that is directly affected by incorrect manual readings, gas gravity, and contaminants.

F_{tf} = Flowing temperature factor

used to correct the measured gas temperature to standard conditions of 60 degrees F.

F_r = Reynolds number factor

used to correct the effect of turbulent flow in calculating the gas rates.

Y = Expansion factor

used to correct the gas flow calculation for the expansion of the gas. Y is based on the ratios of orifice plate to meter run and hw to Pf.

F_{pv} = *Supercrompressibility factor*

used to correct for the compression of the gas at different pressures, temperatures, gravities, and non-hydrocarbon contaminants.

$$\text{Often expressed as } F_{pv} = \frac{1}{\sqrt{Z}}$$

This is one of the factors that is directly affected by incorrect manual readings, gas gravity, and contaminants.

Given all the previous factors it can be seen that the accuracy of the calculated gas flow is dependent on the accuracy of the manual information and readings from the separator.

NOTE: *This is the standard equation although there are advanced equations which take into account abnormal atmospheric conditions and counter corrections. Advanced equations should only be used under guidance and with proper expertise.*

iii.   AGA 3 Gas Calculation Factors

There are several factors involved in the gas calculation and it is usually a simple chore to get the data acquisition system to re-calculate the flow rates but, by the time it has been discovered, the erroneous gas flow data has been transmitted, sent to town, and entered into a "reservoir model." Having to later admit the data is wrong often causes questions to be asked about the competency of the personnel on location.

The most common error with an orifice plate is the incorrect size being entered into the records and calculations. This is easily prevented by having a second crew member verifying the size before it is installed. The plate size is normally etched into the plate, most often at the edge as shown in Figure 11.15.

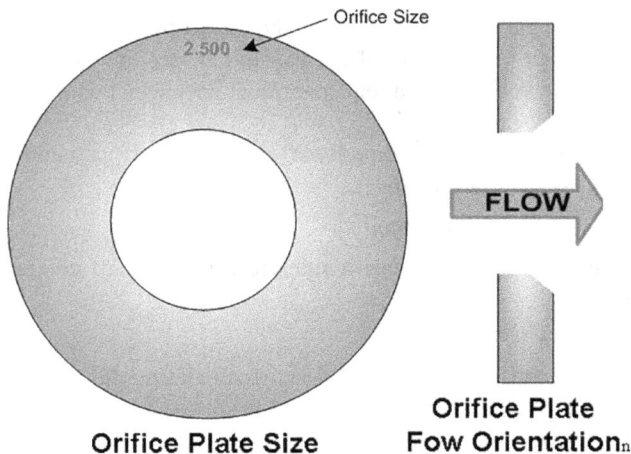

Orifice Plate Size   **Orifice Plate Fow Orientationn**

FIGURE 11.15   Orifice Plate.

Another common error arises from the way the orifice plate is entered if there is a bevel on the plate; it should be entered as shown in the following illustration, as well as the mechanical condition of the plate. The plate orientation in the orifice box should also be verified by a second crew member.

There are options within the AGA 3 gas calculation and the most important is the selection of the tap type. The majority of well test separators have a Daniel senior orifice box fitted. This usually means that flange taps should be used.

Occasionally mistakes are made with the type of pressure tapping used on a separator gas line. This will make a very big difference in the calculations if not correctly selected. Figure 11.16 shows the basic differences. There is a simple rule of thumb:

If the gas line uses a Daniel senior orifice plate holder, then it is flange taps.

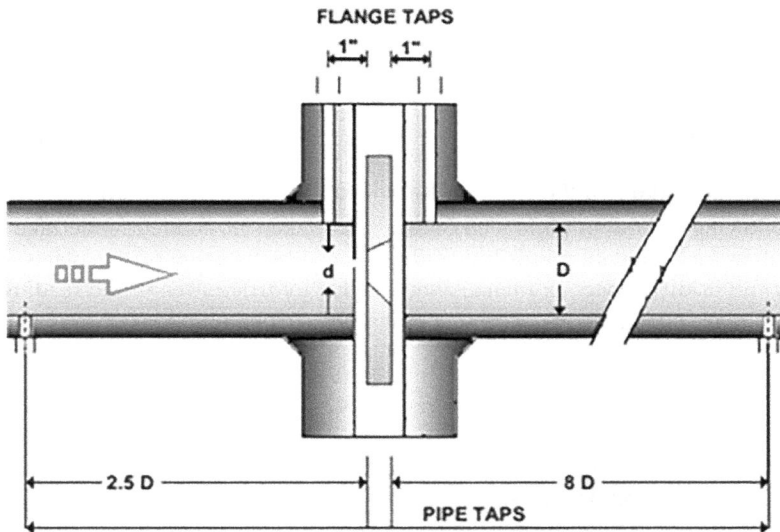

FIGURE 11.16   Orifice Plate Spacing/Mounting.

There is also a limit to the size of the orifice plate that is installed in the gas line which is expressed as a ratio between the orifice plate size and the line bore diameter (d/D) shown in Figure 11.12. This is known as the beta ratio (ß), and orifice plates whose ratio is outside the limits should not be used if possible.

AGA 3 states that:

- For meters with flange taps ß shall be between 0.15 and 0.70
- For meters with pipe taps ß shall be between 0.20 and 0.67

Different calculation systems have slightly different values so it is advisable to check with the calculation method. If not available then these figures give a good guide.

As the calculation changes with the type of pressure tapping used, it is important to know how the pressure is taken at the orifice plate.

All the calculation, conditions, and respective errors are detailed in the American Gas Association (AGA) Reports which should be on the location for reference. The majority of today's well testing units have installed electronic data acquisition systems and automatic calculation of the factors still can be cross checked from the AGA Manuals.

iv.   Estimation of Gas Flow through an Orifice

The gas flow can be estimated from the following equation.

$$Q = \sqrt{h_w p_f} \times F_b \times Time\ Interval$$

The remainder of the listed factors can be considered as the "correction factors for the gas flow to standard conditions" as most of them have little effect.

The time interval is included to correct the hourly orifice factor effect from hours to more standard days.

NOTE: *this estimated flow rate has not been corrected to standard cubic feet.*

11.4.2.2   Summary of Parameters Required for Gas Calculation

The following measured parameters need to be taken in order to undertake a gas calculation for both data acquisition and manual systems using orifice plates. There can be minor differences in the calculations due to differences in the readings but this difference should not be excessive as it shows errors in the system.

Parameter Measurement

- Average gas line pressure (psig)
- Average gas line temperature (degrees F)
- Average differential pressure across the orifice plate ("wg (Inches of Water))
- Gas gravity at time of measurement
- Gas contaminants (H_2S, CO_2, N_2 etc.) at time of measurement

Physical Measurement

- Gas line bore (inches)
- Gas orifice size (inches)
- Pressure tapping

11.5   NEW TECHNOLOGY FLOW METERS

11.5.1   CORIOLIS METERS

Coriolis meters are today replacing conventional metering in both oil and gas lines.

The modern Coriolis meters are based on the work of a French physicist Gustave Coriolis (1792–1843) who published a paper in 1835 on the energy yield (force) of equipment and machines with rotating parts.

Coriolis observed that all moving bodies tend to drift sideways because of the rotation of the planet. This changes depending upon which hemisphere the body is in; in the Northern Hemisphere the deflection is to the right of the motion and in the southern it is to the left. The Coriolis drift plays a major part in the tidal activity of the oceans and the weather of the entire planet. Coriolis called this the "compound centrifugal force" but by the 1920s it was renamed to the now familiar "Coriolis Force."

The first Coriolis meter was not released by MicroMotion until 1977. Today's Coriolis flow meters basically work by inducing a Coriolis acceleration a fluid flow, then measure the angular momentum resulting from the fluid flow.

The fluid flowing in a tube within the Coriolis meter is subjected to what is called a "Coriolis acceleration" by transducers that mechanically oscillate or vibrate the tube which imitates the same effect as tube rotation. The force generated by the Coriolis effect will be a measure of the mass flow rate of the liquid flowing through it, calculated as a function of the Coriolis force.

The mechanical oscillation is commonly induced by having the flow tube fixed at the two end points enclosed within the meter and then vibrated by vibrating components.

While Coriolis meters offer an accurate form of metering there are drawbacks that can often make it impractical to use in modern well testing. It is recommended that the mechanical metering is left in place in case of Coriolis failures.

Power

Coriolis meters rely on electrical power to produce the vibration or oscillation of the flow tubes and accurate configuration of the instrument. Therefore, if there is a loss of power then the meter ceases to function.

The actual Coriolis meter and the associated cabling all need a certification in order to work in hazardous areas.

Repair

A Coriolis cannot be easily repaired in the field; with no real moving parts it is down to replacement of parts or the complete meter.

Safety factors also come into this as there can be a breach of electrical safety if the meter is opened while the well is flowing through the separator.

Calibration

What is not realised is that there are calibration factors associated with the flow tubes; if the tube is damaged or eroded then the calibration changes.

Pressure Drop

With several versions of the Coriolis meter there is a pressure drop across the flow tube where the tube is bent. This can cause issues especially in underbalanced applications.

Some Coriolis meters exist with little or no pressure drop, depending on the model selected.

Meter Size

Like the turbine meters the Coriolis meters have limited flow ranges, then multiple ranges must be equally available to suit the potential capability of the separator.

Coriolis meter manufacturers only tend to produce meters that are greater than 1/2" (15 mm). There are problems with balancing the tube and thin wall flow tubes that meet the necessary tolerances, making them difficult to manufacture.

With the continuing development of Coriolis technology the measurement of both liquid and gas rates are now being offered by some manufacturers. The liquid is a mass flow of the total liquid but if independent liquid densities or BS&W are taken and applied to the flow reading then both oil and waters can be measured.

One solution is to have a separate Coriolis meter skid that connects to the outlet of the separator retaining the original meters but with the added accuracy of a Coriolis.

11.5.2 MULTI-PHASE FLOW METERS (MPFM)

The Multi-Phase Flow Meter (MPFM) developments began in the 1980s centred in Norway with the aim to replace expensive separation technologies and to reduce the size of the equipment necessary to carry out a well test.

This initial development has expanded over time to encompass myriad different technologies.

- Nuclear
- Infrared
- Microwave
- Pressure drop
- Flow segregation, etc.

To go into the operation and issues with each different method and technique would be a book in itself, therefore the operational aspects will be considered.

Most of the MPFMs require details of the composition of the flowing fluid(s), either full composition or basic gravities. What this now reduces to is the reliance on manual data in the sampling and analysis of the well. In the case of a known producer the gravities are available but in an exploration environment the errors in sampling procedures mentioned earlier come into play. These include manual readings of:

- Gravity
- Salinity
- BS&W

Some MPFM's also require PVT samples for a correlation to calculate the flow rates and associated parameters.

The accuracy and resolution of the sensors have an influence on the accuracy of the calculated results; a lot of MPFM companies do not declare the resolution of the physical data acquisition system. For instance a standard well test data acquisition system uses 16 bit; it can be 24 bit, but some of the MPFM data acquisition systems are 14 bit and lower.

From the instrumentation stage we know that the resolution of the transducer is the smallest part a sensor can measure, Table 11.3 details the measured resolution of a transducer, at maximum range when measured by analogue to digital converters in data acquisition systems with differing capabilities.

TABLE 11.3
Transducer Resolution Table

	Transducer Pressure Range					
	1,000	**2,000**	**5,000**	**10,000**	**15,000**	**20,000**
8 Bit	4.878049	9.756098	24.39024	48.78049	73.17073	97.56098
12 Bit	0.30525	0.610501	1.526252	3.052503	4.578755	6.105006
16 Bit	0.019074	0.038148	0.095369	0.190738	0.286107	0.381476
24 Bit	0.00007	0.000149	0.000373	0.000745	0.001118	0.00149

Most data acquisition systems (DAS) in use in well testing are 16 bit devices so the maximum resolution from a 5,000 psi transducer is 0.04, but if the DAS system is 24 bits, then the maximum resolution is 0.002 psi.

This is based on the Table 11.4 where the resolution is one part in the maximum range of the associated electronics.

TABLE 11.4
Individual Bit Transducer Resolution

Resolution	1 Part In
8 Bit	256
12 Bit	4,096
16 Bit	65,536
24 Bit	16,777,216

Therefore, in order to maximise the accuracy of a MPFM it stands to reason that a high accuracy measuring system is used. This is not always the case in some MPFM systems; the pressure response from the MPFM is artificially dampened by using pressure liners full of a high-viscosity fluid like silicon oil.

While not wanting to criticise all MPFMs as the technology is still developing, the user should first be sure that the intended MPFM will accurately measure the well with no additional measurement systems. The best way to do this is to carry out a series of trials with the intended MPFM and a test separator in parallel and independently of each other.

What is important is that the data from the well test separator is isolated and kept separate from the MPFM data otherwise the data from the well test can be used as reference – or seed – for the MPFM data. The object is to ensure that the MPFM will accurately measure data from a well as if a separator is not present. Using on-site data which will not be available for future tests must not be allowed to contaminate a comparison test.

With regard to comparing data from a MPFM with a separator-based measurement, if the separator is run properly with all the required data collected in a timely and efficient manner, then the data from the separator should be taken as an established benchmark for a true comparison. The separator will be measuring a physical quantity, proven over the years, in comparison to a system that predicts the response, often with empirical means, which may not fit the mathematical model used by the MPFM.

One of the areas an MPFM comes into its own is in clean-ups, especially where acid and frac fluids are present, as this can affect and damage a separator. In this scenario it is advisable that checks on the MPFM flow rates are carried out using a tank at least during the flow periods. Tanks have their safety issues, which should be addressed, but in a clean-up or flowback the major hazard, gas, is not present in large quantities and reduced risks.

Another option is to place a Coriolis meter with gas measurement capability downstream of the MPFM for qualitative and quantitative purposes, if applicable.

There are well test reports from an MPFM that utilise a GOR2 reading; this should not be included. GOR2 is only applicable when the fluid leaving the vessel has undergone a pressure drop. This is not the case with MPFMs; there inlet pressure is the same as the outlet pressure and hence the GOR measured is the total GOR (GOR tot). In MPFMs GOR2 is often used as a fudge factor to correct the results.

Often the approach that it is "new technology" so it is right – or simply accepting that the results are a from computer so it has to be right – this approach of accepting data can lead to errors. Check and verify.

11.5.3 WATER CUT METERS

The use of water cut meters, sometimes known as a BS&W meter, is becoming more common within well testing. There are multiple different technologies employed and it can also be a product of other instruments such as MPFMs and Coriolis meters.

The more commercial water cut meters are used for export lines and as such do not have natural gasses entrained in the flow. Consequently, they only work accurately with "dead oil" so before placing them in a well test environment, check if gas is included in the compensations.

What is not widely acknowledged is that there are two sub divisions of the meter which can cause issues if misunderstood or mixed:

Water oil meter – This is where the amount of water in the sample is less or equal to 50 %.
Oil in water meter – This is where the amount of water exceeds 50% so there is more water than oil.

NOTE: *These characteristics can affect the types of sensors to the degree that they do not work.*

While it is not the purpose of this document to extol the virtues and performance of water cut meters, the issues involved when using them on well testing units will be highlighted.

BS&W Manual Sampling

A lot of water cut meters use a manual BS&W reading as their reference and to check their performance.

The earlier chapter concentrated on the importance of procedures to take and process of BS&W sampling. It is important that the crew all follow the same BS&W procedure, otherwise errors can result.

BS&W Sample Correction

Some water cut meters use the manual BS&W result as a seed or calibration point for the water cut meter. This can improve results but then it cannot be considered a water cut meter, since it is more likely measuring deviation from a pre-determined set point configured by the manual BS&W result.

What some systems do is allow the operator to put the manual reading into the host software and then write it to the water cut meter which then is accurate only to the time it was entered. If this approach is to be used then an average or a table should be used.

However, the method of writing to the water cut meter is usually through a serial computer link which would still be subject to the hazardous area rules and regulations. This means that the water meter housing must be certified and the computer cable either armoured or run in a certified armoured conduit. Without these standards the link between the computer and the meter cannot be established. (There are also distance limitations on RS232 cabling systems.)

Once a correctional system as such is used, then the meter must be either updated or checked regularly.

BS&W Sampling Frequency

With the BS&W sample correction the meter is only as good as its sample frequency, especially in clean-up operations.

Often the BS&W correction is disguised by the operators either seeding the system with data from previous wells or a site visit the day before to obtain a BS&W reading. In this case if the well test package is not present then the equipment to determine a true BS&W will not be truly representative and an average BS&W is often utilised.

It is important for the operating company to know what is actually happening and that any BS&W seeding should be recorded on the manual readings sheet, well test report, or daily logs to act as future reference, making it clear that BS&W seeding is taking place and the interval of the seeding.

12 Data Instrumentation

With today's technology and its subsequent advances, the well testing and associated equipment is covered with digital and electronic sensors and transmitters.

It should always be remembered that just because they are new technology, digital and electronic, it does not mean that they are always better in terms of performance or for the intended application.

By having a digital display, they can't be considered more accurate. It is not always the case as all sensors are subject to calibration data entry, drift, signal processing, and other processes with which a digital display is just as likely to be in error as an analogue.

This is why manual readings and verifications are necessary in all instances.

Most errors detected are from the most popular pressure transducers; the standard error is that the sensor-transmitted data to the data acquisition system shows one value while the digital display on the sensor shows another. There is a very simple solution to determine where the error lies; using the traditional and mechanical deadweight tester for a field calibration and recording the results.

NOTE: *A deadweight tester must be supplied connected and ready to use with every well test package*

With the transducer mounted on the deadweight tester as reference we would record at the same instance and compare:

  i. The applied pressure from the deadweight tester
 ii. The reading from the digital display of the transducer
iii. The reading from that data acquisition system

This simple comparison will highlight where the error lies. In the case of the data acquisition system it may simply be the entered calibration that is in error, as often it is not updated and left as is from the previous job despite changes made with sensors. Regular field calibrations will detect any source of error and the necessary solution can be applied accordingly.

Importance of Sensor Selection

The importance of the actual sensor is often overlooked when carrying out a data acquisition job. There are several factors that constitute good data from a sensor:

- Suitability/range
- Calibration
- Response
- Maintenance

                                          DOI: 10.1201/9781032623689-12

- Type of mounting
- Method of measurement
- Safety classification of the sensor

12.1 DEFINING PRESSURE

To begin with we must understand the concept of pressure. Pressure, like light and current flow, cannot be seen or measured. A strange concept, but true. We can only measure the effect of pressure, which in the vast majority of cases relies on the deformation of a material. We can look at a basic dial pressure sensor, Figure 12.1, which is a dial gauge using a Bourdon tube.

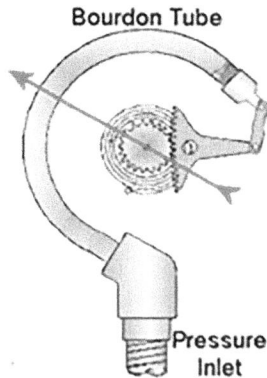

FIGURE 12.1 Internals of a Dial Gauge.

Application of pressure to the inlet port will cause the internal Bourdon tube to expand and effect a change on the indicator mechanism. As shown in Figure 12.2, the material has deformed with the application of pressure to give us a reading of how much pressure it has taken to achieve the resultant deformation.

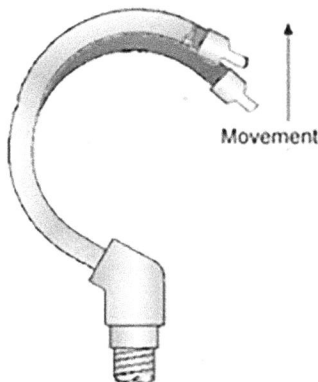

FIGURE 12.2 Bourdon Tube Deformation.

This deformation of material is the basis of most of the pressure sensors and transducers; we are not actually measuring the pressure but the degree of deformation the applied pressure has caused.

If we look at the majority of the electronic sensors or transducers used in data acquisition surface measurement, they can be considered a diaphragm type of sensor. There are numerous types of sensors, too many to detail, so generalities will be dealt with.

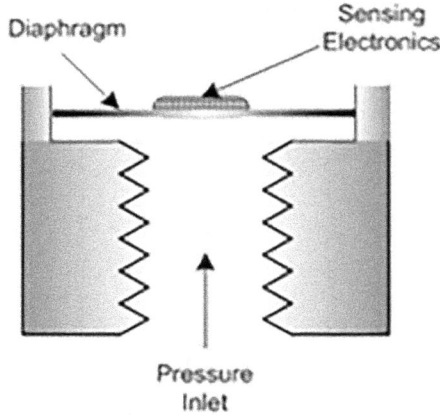

FIGURE 12.3   Basic Pressure Gauge Construction.

Figure 12.3 shows a basic diaphragm sensor with no pressure applied. The type of sensor again is generalised as "Sensing Electronics" to cover all the differing options. When pressure is applied to the device its response is as shown in Figure 12.4.

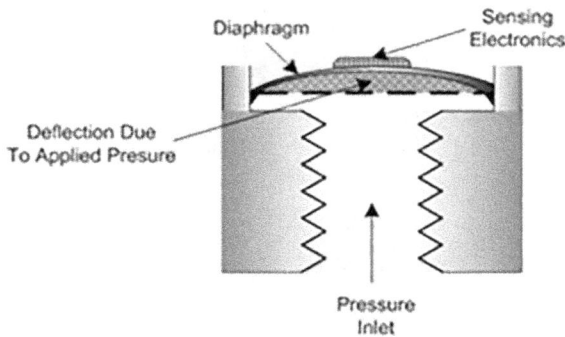

FIGURE 12.4   Basic Pressure Gauge with Pressure Applied.

The applied pressure causes the diaphragm to deflect or bend according to the amount of pressure applied. The deflection is a function of several factors.

- The thickness of the material
- The type of material
- The physical shape of the diaphragm
- The amount of pressure applied
- The modulus of elasticity of the material

The modulus of elasticity is a number that measures an object or substance's resistance to being deformed elastically (non-permanently) when a stress, in our case pressure, is applied to it.

The curved line in Figure 12.5 shows the non-linear response of most pressure transducers, but most pressure sensors are used in DAS with a two-point linear type calibration. What actually happens is that there is often a linearity table built into the transducer that converts the non-linear movement into a linear one, shown as a straight line in Figure 12.5.

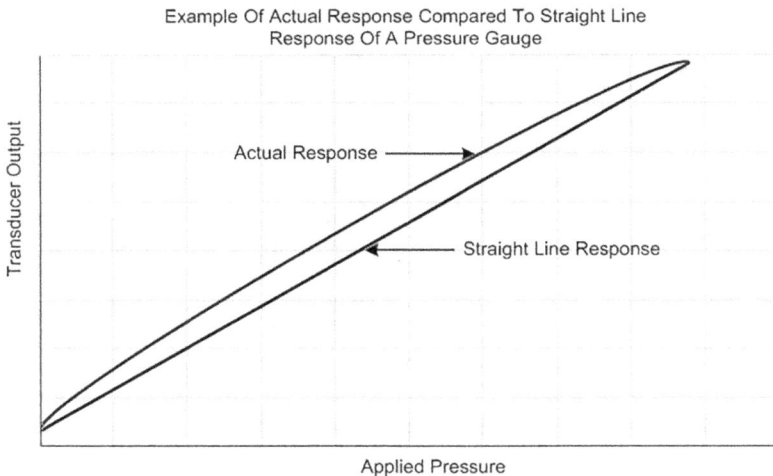

Example Of Actual Response Compared To Straight Line
Response Of A Pressure Gauge

Transducer Output

Actual Response ⟶

Straight Line Response ⟵

Applied Pressure

FIGURE 12.5   Actual Response of a Pressure Gauge.

Some data acquisition systems actually offer this linearisation table as a calibration option to increase the accuracy of the transducer by entering the true output against a known pressure rather than rely on a generic internal table.

Figure 12.5 graph shows the actual curved response, and the resulting lin-
earized response is shown as a straight line. This can lead to errors depending
upon the response of the transducer and the number of pressure points used in the
linearisation.

The only way to get the true readings from the transducer is to actually use the
response or calibration points and not a straight line fit. This is borne out if a transducer does not respond according to its manufacturer's specification. This is where
the diaphragm is damaged out of a number of reasons.

Figure 12.6 graph shows the response of a transducer that has either been used at
a limited pressure range for long periods or has been physically damaged.

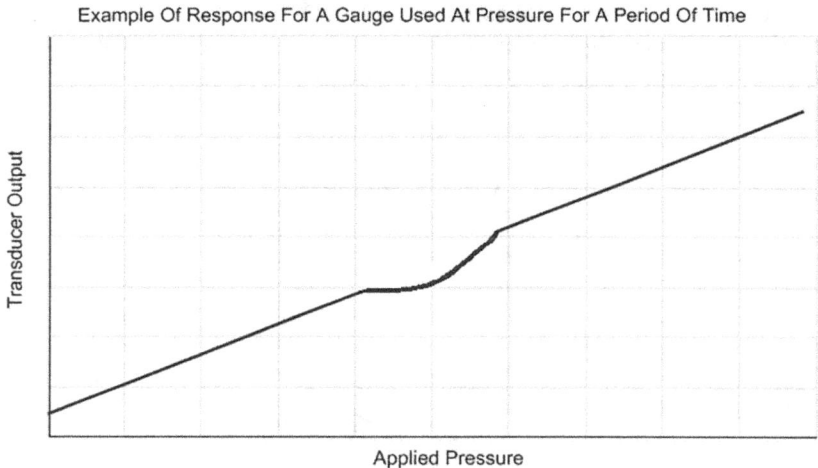

Example Of Response For A Gauge Used At Pressure For A Period Of Time

Transducer Output

Applied Pressure

FIGURE 12.6   Response for a Pressure Gauge Used at Pressure for a Period of Time.

The graph in Figure 12.7 shows the linearity response that can be applied. As can
be seen there are two responses shown: linear regression and point to point calibration, which yield different results if used in the field. The point-to-point calibration
and the result of a linear regression curve fit, and both are valid.

*It is this simple approach of assuming everything is a straight line that causes a lot of
data errors in the field.*

What is often not realised is that the pressure transducers are only normally calibrated with increasing pressure points. Downhole sensors, however, are normally
calibrated in both increasing and decreasing steps. This is due to an effect called
hysteresis.

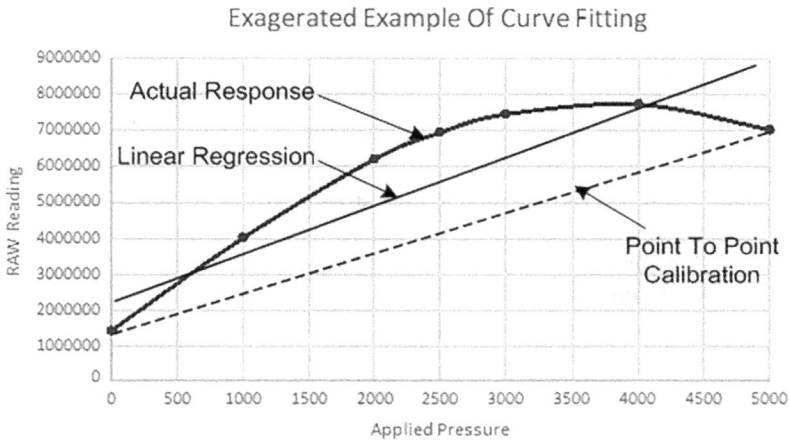

FIGURE 12.7   Response for Pressure Gauge on Point-to-Point Calibration.

Hysteresis is generally defined as:

The maximum difference in output at any measurement value within the sensor's defined range when approaching the point first with increasing pressure and then with decreasing pressure is known as the hysteresis error of a pressure sensor.

This is better illustrated by the graph in Figure 12.8.

The pressure error between the two curves is known as the hysteresis error and will vary from transducer to transducer.

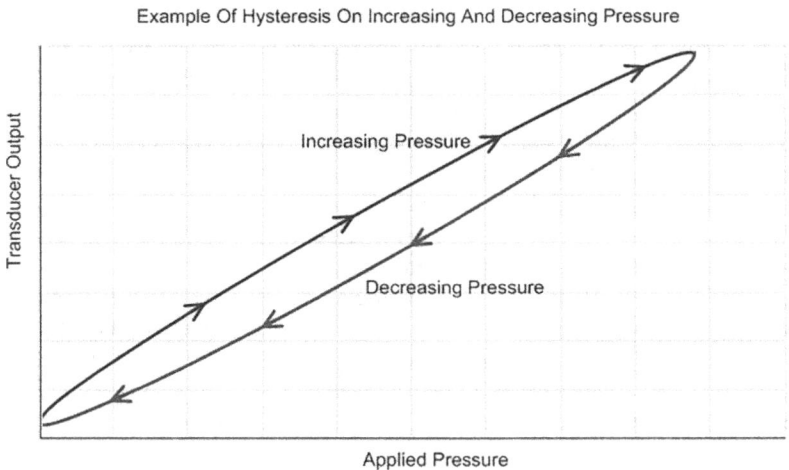

FIGURE 12.8   Hysteresis on Increasing and Decreasing Pressure.

What is happening across well testing is that the oil companies are demanding a calibration from a third-party calibration house to prove the sensors are calibrated. However, the actual calibration conducted is not checked, just assumed as correct.

The newer "smart" transducers have a capability built in to them called HART – highway addressable remote transducer – or a variant. Without going into details, it means that there is a serial communication stream built into the transducers output that enables the operator to change the parameters of the transducer. So, no matter what the third-party calibration certificate says, the operator can manipulate the calibration of the transducer while it is in place and in operation. There are other manufacturers' communication protocols similar to HART.

HART communicator can be used as a stand-alone instrument or be built into a computer.

Another means of manipulation is to change the calibration within the data acquisition system itself to achieve the same results. This is a major reason why all manual readings are required and all data is to be recorded on the manual readings sheet, stored, and saved; then they are available to the client as proof of data integrity.

12.2   CALIBRATIONS

The majority of the sensors on a well test are surface-mounted 4–20 mA transmitters, and before we can understand them it is necessary to understand the terms and technology behind them.

There is also an increase in use of wireless systems that do not rely on power or signals being transmitted through cabling. However, the concepts of calibrations and operations still hold true for all transducers.

Many operating companies insist on the transducers being calibrated by an independent, also referred as a third-party, calibration house. These independent calibrations are normally only carried out once annually, where repeat well tests, PGORs, are involved these transducers and are used daily; therefore the instance of calibrations should be higher. This does not have to be an independent calibration but one carried out onsite, with a suitable deadweight tester and associated equipment, to verify the transducer's performance.

In many cases a single calibration is issued, whereas if the transducer has a digital display and a data output, such as a current signal, then a calibration should be submitted for both displayed and transmitted data.

Site representatives do not often check the calibration that has been entered into the data acquisition system, and the calibration from a previous job/test can often be left in the system. Operators do not set up the data acquisition's software system configuration from scratch on every test but rather just restart an existing one with modifications if needed. In many cases the previous calibration is set up in system configuration, which is no longer valid with the sensors installed on the new test, and it is left undiscovered until the end of the test.

It is important that the entered data acquisition calibrations are verified by the company representatives at the beginning of the job. If in doubt calibration verification can be carried out by using:

- Pressure – deadweight tester
- Temperature – stick thermometer
- Differential – air line/gas backpressure controller
- Flow – stock tank

12.2.1 CALIBRATION ISSUES

As previously mentioned, many of the smart transducers use HART protocol. This is often used in the calibration of the transducer; the HART communication will transmit the calculated pressure. This is one of HART's many functions.

The HART transmitted pressure data is based on a 24 bit analogue to digital (A/D) converter and is reliable. However, most of the data acquisition systems do not use the HART formatted transmission but the standard 4–20mA current loop.

The 4–20mA current loop system has been in use for many years and is used with PLC and SCADA systems with some drawbacks. We have mentioned the resolution drawbacks but there is an issue with measuring current.

Currents like pressure and light cannot be measured directly. The effect of current is measured across a known resistor as a voltage, as illustrated in the setup of Figure 12.9. The known resistor is normally 250 ohms (Ω), because it gives a convenient voltage drop across the resistor of 1–5 volts.

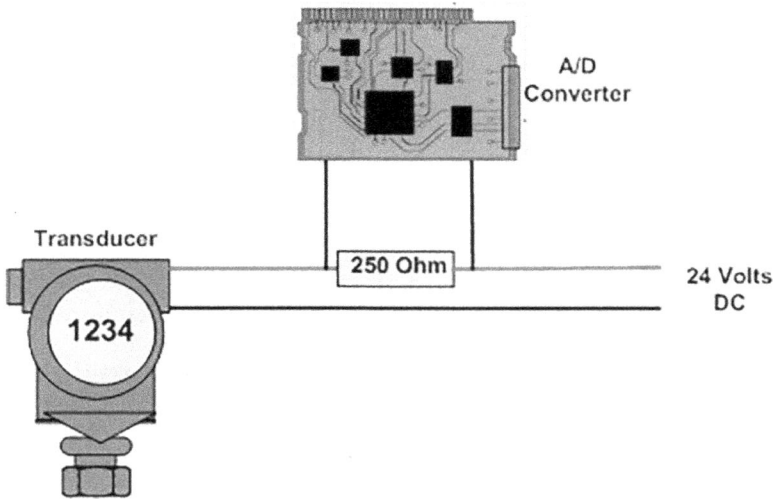

FIGURE 12.9   Current Loop Measurement.

Ohms Law

$$V = I \times R$$

Where
  V = voltage in volts
  I = current in amperes (A)
  R = resistance in ohms (Ω)

At 4mA the voltage across the 250 Ω resistor is

$$V = 4x\,10^{-3}\,x\,250 = 1\;\text{Volt}$$

At 20mA the voltage across the 250 Ω resistor is

$$V = 20\times10^{-3}\times250 = 5\;\text{Volt}$$

The calibration issue arises from tolerance. Everything manufactured has a tolerance, as do resistors; therefore, the actual resistance of the 250Ω can vary. Normally due to sensitivity these are chosen to be high tolerance resistors with better than $\pm0.1\%$ of the value.

Therefore, the calculated pressure of the sensor can vary $\pm0.1\%$

Table 12.1 shows the variance of pressure associated with just the variance in the resistor alone using a tolerance of 0.1%.

TABLE 12.1
Pressure Variance with Resistor Tolerance

Pressure	Minimum Pressure	Maximum Pressure	Error Range
1,000	999.0	1,001.0	2.0
5,000	4,995.0	5,005.0	10.0
10,000	9,990.0	10,010.0	20.0
15,000	14,985.0	15,015.0	30.0

The relevance of equipment tolerance errors is faced in well testing data acquisition; we should consider the calibrations by a third party that do not take these electrical tolerances/errors into account. The third-party equipment will have different resistor tolerances than the intended data acquisition system or use a HART communicator which bypasses the 4–20 measurement system.

The third-party calibrations are correct within the system they have been calibrated against but with different data acquisition electrical specifications the errors will not be constant.

Basically, if a transducer reading 5,000 psi on a channel with a low resistor tolerance is now plugged into a channel with a high resistor tolerance, the value can increase by up to 10 psi. This effectively negates the concept of third-party calibrations as it is obvious the sensors have to be calibrated and checked on the channel they are going to be used on, otherwise errors will occur.

There is no reason the third-party calibrations can be carried out as long as a secondary calibration is carried out with the sensor in its allocated channel on the data acquisition and verified in its operational environment.

12.2.2 FIELD CALIBRATION

In the cases where the sensor is damaged or simply out of calibration most sensors have a method of calibrating them if the HART system is not available on the location. These are the manual zero and span buttons that are often included with the majority of sensors. Consult the manufacturer's documentation for availability and location.

As the preferred sensor manufacturer in well testing seems to be Rosemount the probable location is shown in Figure 12.10. Other manufacturers often use the same method but the transducer manuals should be referenced.

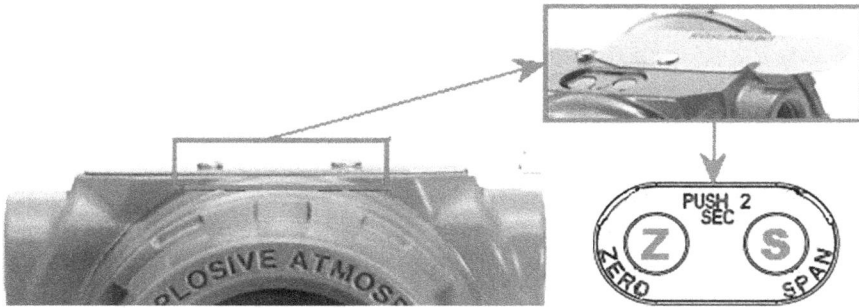

FIGURE 12.10   Reset and Span Buttons on Transducer.

Slide open the plate on the top of the transducer by loosening the screw (do not remove it). There will be two buttons or screws shown depending on the transducer make and model.

The deadweight tester now comes into play as a calibrated pressure source.

In the case of the push button option:

i. With NO pressure on the transducer push and hold the Z (zero) button for 2 (two) seconds
ii. Apply maximum/required pressure to the sensor
iii. With the pressure constant push and hold the S (span) buttons for 2 (two) seconds
iv. Bleed off the pressure and carry out a spot check to verify the calibration has been applied

In the case of the screw option:

i. With NO pressure on the transducer push adjust the Z (zero) screw until zero is displayed on the display or DAS system
ii. Apply maximum/required pressure to the sensor

iii. With the pressure constant on the transducer push adjust the S (span) screw until the correct pressure is displayed on the display or DAS system

iv. Bleed off the pressure and carry out a spot check to verify the calibration has been applied

NOTE: *If the maximum pressure of the transducer has been changed the transducer must be marked accordingly. This information must be included on the manual readings sheet.*

If possible it is recommended that the field calibrated transducer be in an area where the quality of the data is not critical, such as annulus pressure or if a transducer is damaged, needs re-calibration, and has not got a traceable calibration certification.

12.2.2.1 Checking a Pressure Transducer

Use the deadweight tester as a calibrated pressure source, with the transducer connected and purged through with hydraulic oil, as illustrated in Figure 12.11. All steps must be documented and noted, especially applied pressures and data acquisition system responses.

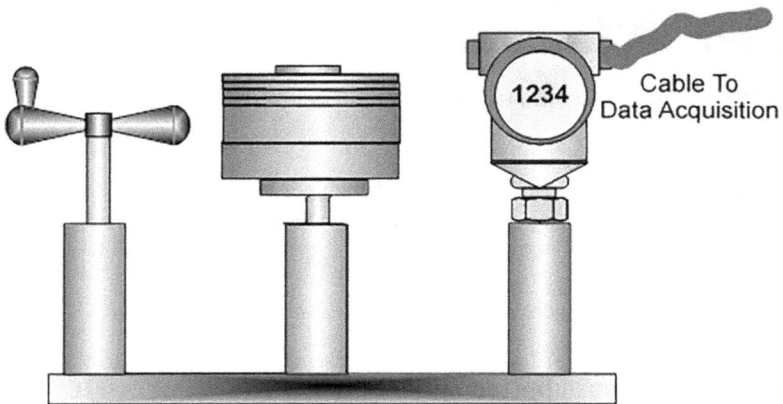

FIGURE 12.11   Field Deadweight Tester Used to Check/Calibrate a Sensor.

The transducer is mounted on the deadweight and connected to the correct channel on the data acquisition system. The deadweight tester's pump and isolation valves are opened so there is no pressure on the transducer and left for several minutes to stabilise.

In the case of the push button option:

i. With NO pressure on the transducer push and hold the Z (zero) button for 2 (two) seconds

ii. Apply maximum/required pressure to the sensor

iii. With the pressure constant, push and hold the S (Span) buttons for 2 (two) seconds

   iv. Bleed off the pressure and carry out a spot check to verify the calibration has been applied on both the display and DAS
   v. The transducer now should be field calibrated to the applied pressure

In the case of the screw option:

   i. With NO pressure on the transducer push adjust the Z (zero) screw until zero is displayed on the display or DAS system
   ii. Apply maximum/required pressure to the sensor
   iii. With the pressure constant on the transducer push and adjust the S (span) screw until the correct pressure is displayed on the display or DAS system
   iv. Bleed off the pressure and carry out a spot check to verify the calibration has been applied on both the display and DAS
   v. The transducer now should be field calibrated to the applied pressure

NOTE: *If the maximum pressure of the transducer has been changed the transducer must be marked accordingly. This information must be included on the manual readings sheet along with the transducers identification.*

The applied pressure does not have to be a standard range. You can choose whatever span necessary for the application. Carry out a calibration check across the new range with the deadweight tester to ensure the transducer is working.

12.2.3 DEADWEIGHT TESTER

A deadweight tester is a piston-cylinder type measuring device that functions in accordance with the basic principle that:

Pressure = Force/Area

In a deadweight tester the applied pressure acts on a piston assembly with a known cross-sectional area. Weights that correspond to known pressures are then added to the piston assembly until the system is in balance. The value of the sum of the weights on the piston equate to the value of the pressure applied.

A typical Deadweight Tester consists of the same components comprising a Deadweight Gauge, in addition to a hand-operated, self-contained pump for the generation of pressure. Included are a piston, cylinder, table, weights, associated plumbing, and oil reservoir, illustrated in Figure 12.12.

Tester components are mounted on a base and are protected by a carrying case. Connecting adapters, pointer drivers, and a hand jack (for working on gauges) accompany each instrument.

The deadweight tester's resolution is generally 1 psi, and its accuracy depends on the manufacturer, the quantity, and quality of the weights used.

NOTE: *Deadweight physical weights must not under any circumstances be cleaned using any abrasive substances such as a wire wheel and abrasive paper. Any deadweight showing these symptoms should be replaced.*

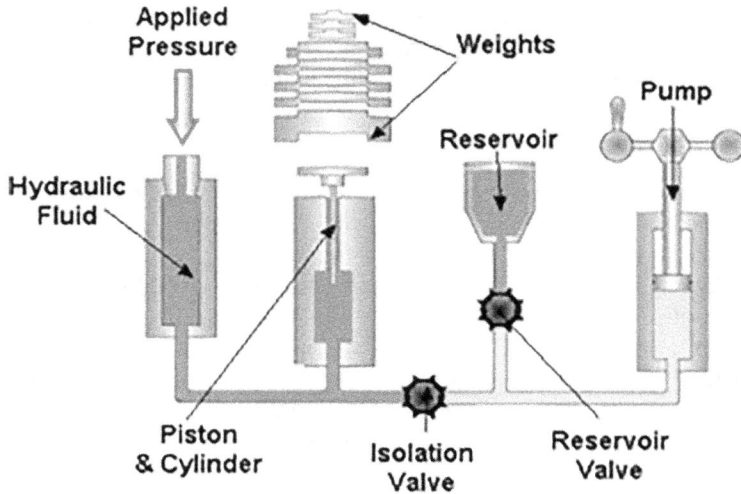

FIGURE 12.12 Schematic of Basic Portable Deadweight Tester.

A deadweight tester has two functions:

  i. As a pressure measuring device
 ii. As a pressure source where it can generate pressure

The deadweight tester when used to generate pressure has a high-pressure/low volume-pump which enables hydraulic fluid to be applied at the pressure dictated by the weights on the piston assembly. Therefore, it can be used to generate the pressure necessary to calibrate a pressure transducer.

This is a mechanical instrument to check the calibration of pressure instruments, a simple Pressure vs Force with no deformation errors to take into consideration. High specification deadweight testers, referred to as master-class, are used for calibrations of high precision instruments like downhole gauges.

12.2.3.1 Using the Deadweight

Before the job starts, critical sensors should be checked and witnessed by a company representative; these checks should include.

  • Area certification (ATEX or equivalent) where applicable
  • Pressure calibration certificate and date of issue
  • Calibration entered into the data acquisition system
  • Checks with deadweight tester

If the transducer has not got a steel – or equivalent – certification plate attached to the device, then the device must be considered uncertified and cannot be used in a hazardous area.

The well testing operators will try to avoid using the deadweight tester but it should be connected to the data header upstream of the choke manifold at all times. The well testers should take a reading from the deadweight tester at regular intervals; every 30 minutes is the standard time interval. This can then be used as a qualitative check of the digital pressure sensor performance. Standard Bourdon tube gauges should not be relied on instead of a deadweight tester, at least on the upstream of the choke manifold with higher pressures, since they are prone to the mechanical deformation errors and failures.

The "primary" excuse nowadays for not using the deadweight tester is that is has been replaced by newer digital gauges. These gauges are still subject to errors due to physical damage. The standalone digital gauges that are a resonant quartz are not normally used except in calibrations and generally are not robust enough for well testing operations.

The purpose of the deadweight tester is to introduce a mechanical instrument which is more reliable, accurate in its operation, and does not have the mechanical deformation errors. All errors on a deadweight tester are easily detected by the user, whether it is a hydraulic leak or any broken parts, and as long as it is not damaged physically its principal function is unquestionable.

This is why it is the most popular device used to calibrate all other pressure sensors, mechanical or electrical.

A master class deadweight is the primary calibration source for ALL pressure sensors, including down-hole quartz sensors, so the deadweight tester should not be simply fobbed off as old technology; it is the pressure reference used in calibration labs.

Under all circumstances a manual reading sheet should be kept by the well testers for all the readings from wellhead, separator, and other associated equipment.

These **MUST NOT** be readings from the digital transducers but the deadweight tester, dial gauges, thermometers, and chart recorders. Again, to act as qualitative check for all the DAS sensors on the well test.

The completed manual readings sheet should be kept with the final report in its hand-written form; a scanned copy is almost always necessary in final reporting; a computer-generated manual reading sheet is not accepted. It should be available to the operating company with the final report if requested. Its prime function is to verify the readings and sequence of events of the final report.

12.2.3.2   Measuring Pressure

If the deadweight tester is to be used to measure pressure the deadweight, internal lines must be first flushed through with hydraulic fluid before applying pressure. The attached reservoir is normally filled with oil.

i. The isolation valve is closed and reservoir valve is opened to draw the reservoir oil into the pump

ii. With the reservoir valve closed and isolation valve open it is now possible to apply pressure using the pump

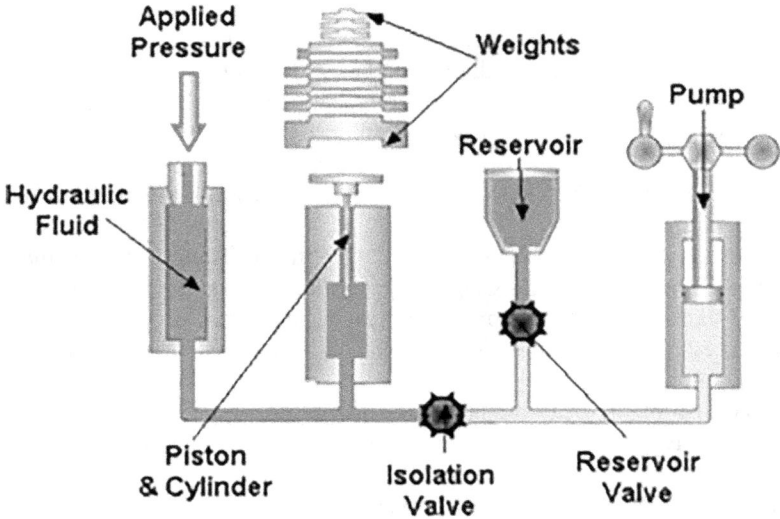

FIGURE 12.13 Schematic of Basic Portable Deadweight Tester.

   iii. The specific individual weights for that deadweight are placed on the piston assembly, shown in Figure 12.14, which is slowly spun clockwise until it floats between two marks on the piston

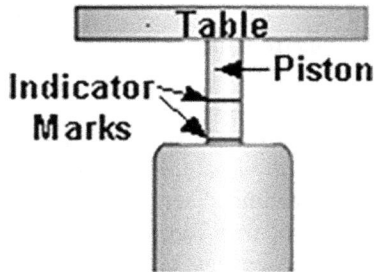

FIGURE 12.14 Deadweight Piston Assembly.

   iv. The deadweight table, with weights, should *not* be spun anticlockwise as there is a risk that the table assembly will unscrew
    v. The relevant weights are slowly added to the table until the table and weights begin to float between the two lines
   vi. With the table balanced between the two indicator marks, the pressure is then determined by adding the pressures stamped on the weights plus the pressure of the table to calculate the applied pressure

12.2.4 Transducer A/D Response

As the most common analogue card in well testing data acquisition is a 16-bit device, it can be seen in Table 12.2 that the resolution decreases as the range of the transducer increases. However, a lot of industrial data acquisition systems are also used and care should be taken with the lower resolution.

TABLE 12.2
Analogue to Digital (A/D) Digital Converter Response

	Transducer Pressure Range					
	1,000	**2,000**	**5,000**	**10,000**	**15,000**	**20,000**
8 Bit	4.878049	9.756098	24.39024	48.78049	73.17073	97.56098
12 Bit	0.30525	0.610501	1.526252	3.052503	4.578755	6.105006
16 Bit	0.019074	0.038148	0.095369	0.190738	0.286107	0.381476
24 Bit	0.00007	0.000149	0.000373	0.000745	0.001118	0.00149

The output from the measuring devices is usually in a digital unprocessed form, raw readings, abbreviated as RAW.

Full details of A/D resolution are in Table 12.2. A 400 pressure transducer range has been included as it is a common range used with differential pressure cells for orifice calculations. If we take a 5,000 psi transducer and measure it with a 16-bit analogue card the resolution is 0.095369/psi from the A/D converter. This is in direct conflict with the normally displayed resolution of 0.01 on the logger screen. This is because it is a mathematical generated number on the display and the true resolution should be considered.

With the introduction of higher range transducers the true resolution decreases.

Table 12.3 where it is one part in the maximum range of the analogue to digital converter in the DAS electronics.

TABLE 12.3
Smallest Amount Measured by
an Analogue to Digital Converter

	1 Part In
8 Bit	256
12 Bit	4,096
16 Bit	65,536
24 Bit	16,777,216

As the most common analogue card in well testing data acquisition is a 16-bit device it is apparent that the resolution decreases as the range of the transducer increases. However, a lot of industrial data acquisition systems are also used and care should be taken as their resolution is often lower.

If we take a 5,000 psi transducer and measure it with a 16-bit analogue card the resolution is 0.095369 psi from the A/D converter. This is noticeably in direct conflict with the normally displayed resolution of 0.01 on the logger screen. This is simply because it is a mathematically generated number on the display when the true resolution should be considered.

With the introduction of higher range transducers the true resolution decreases. It is important to realise this, because it explains why the transducers often seem unstable. It is a function of the resolution limits. For example, if a 10,000 psi transducer is used with a 16-bit A/D then its true resolution is 0.19. When used with the data acquisition system the decimal digits are not constant and jump up or down in the magnitude of this bit resolution value of 0.19.

Therefore, there is better resolution using a 24-bit resolution board. It becomes a more accurate reading with less fluctuation using a 24-bit resolution board. With the same 10,000 psi transducer and a 24-bit A/D the true resolution becomes 0.000745 so the sensor will be seen to be stable without erratic jumps up and down.

With wells getting deeper and deeper, higher pressures are encountered so higher range transducers are required. Therefore, the data acquisition system capability should meet these requirements.

The previous details apply to all sensor types that utilise a 4–20 mA output.

- Pressure
- Temperature
- Differential
- Flow

Normally the resolution is expressed as

$$\frac{\text{Device Range}}{\text{Measuring Instrument Range}}$$

With a 4–20mA instruments this is not the case; since the full range of the A/D converter is not used in the measurement we only use 4 to 20mA part of the A/D. Therefore, if we calculate the actual measuring range of 16 mA:

TABLE 12.4

Actual Measuring Range

A/D Range	20 mA	4 mA	20–4 mA (16mA)
8 Bit	256	51.2	205
12 Bit	4,096	819.2	3,277
16 Bit	65,536	13,107.2	52,429
24 Bit	16,777,216	3,355,443	13,421,773

With a 10,000 psi transducer and the 16 mA span the resolution is:

$$\frac{10,000}{52,429}$$

A resolution of 0.190735 psi

Using this method all the transducers' resolution can be calculated.

12.2.4.1   Transducer Linearisation Tables

Some transducer companies utilise a linearisation table approach to calibrations but the majority use a two-point calibration when using 4–20 mA transducer.

The linearisation table is where the data points from a stepped calibration check at set points are entered into a table, applied pressure and its corresponding raw data. The raw data in this case is the reading direct from the analogue to digital converter based on the number of bits.

In electronic terms an analogue-to-digital cnverter (A/D) is a system that converts an analogue voltage into its digital equivalent. The A/D comes in myriad formats varying from voltage input to output resolution, which is covered in more detail later in this chapter.

Figure 12.15 is a simplified explanation of how an A/D functions in a computerised system; the information can now be stored digitally. Using the raw value reading from the data acquisition system rather than the actual milli-ampere (mA) current output value it makes the numbers easier to handle without using long strings of decimal points.

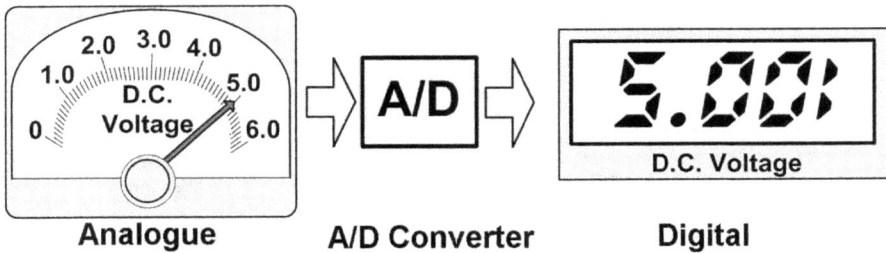

FIGURE 12.15   Example of Voltage Measurement.

Based on a 5,000 psi transducer, Table 12.5, and an output of 4–20 mA, which is a span of 16mA, this gives an output of 0.0032 mA per psi change, as calculated by:

$$\frac{\text{Current Span}}{\text{Maximum Pressure}} \qquad \frac{16\text{mA}}{5000\text{ psig}}$$

For every increase or decrease of 1 psi the current output is of a value of 0.0032 milli-amperes, which is a difficult number to deal with because of the number of decimal points.

If the raw data from the analogue to digital converter is used the increment value is 2,802 raw counts per psi from the A/D and easier to deal with, as it is a whole large number rather than a string of decimal places as is it with mA.

Table 12.5 shows example data from a 0 to 5,000 psig pressure sensor calibrated in 1,000 psi steps entered into data acquisition system using a 24-bit A/D using raw data and stored as a calibration.

TABLE 12.5

Linearisation Table for 24-Bit

A/D Converter

Pressure	Raw
0	1,431,011
1,000	4,308,272
2,000	7,417,172
3,000	10,159,672
4,000	12,535,772
5,000	14,545,472

This response is after the effect of the internal linearisation correcting the transducer is shown in Figure 12.16 along with the actual steps.

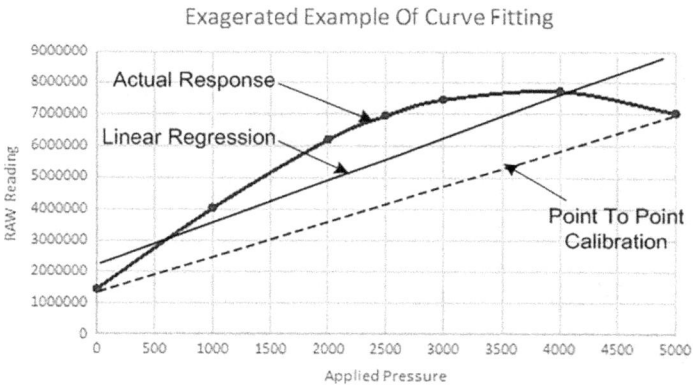

FIGURE 12.16 Calibration Table Response Plot.

The graph in Figure 12.16 shows the plot of the calibration table, with the normal point to point (red) calibrations used in data acquisition systems. Also, the linear regression (blue) is also plotted which is a more accurate representation over the full range. What some of the DAS systems do is take each individual data point and interpolate between these points so that the resultant accuracy is improved.

Figure 12.17 shows the section of the graph with the accuracy of the interpolated pressure as shown joining the dots. These systems go further and actually produce a calibration certificate to show the accuracy.

The two graphs in Figure 12.18 are from a DAS system interpolated calibration certificate. Before the interpolation calibration there was an 11 psi error. After processing using interpolation the error is 0.1 psi. It must be stressed here that this does not affect the resolution, just the accuracy.

The resolution is the smallest part that can be measured whereas the accuracy is the deviation from the action response.

Interpollation Between Two Points

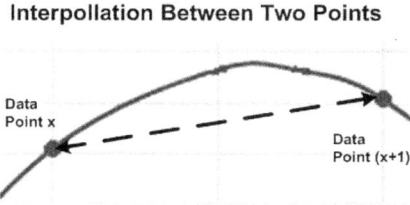

FIGURE 12.17   Interpolated Data.

Example Interpollated Calibration Response Plot

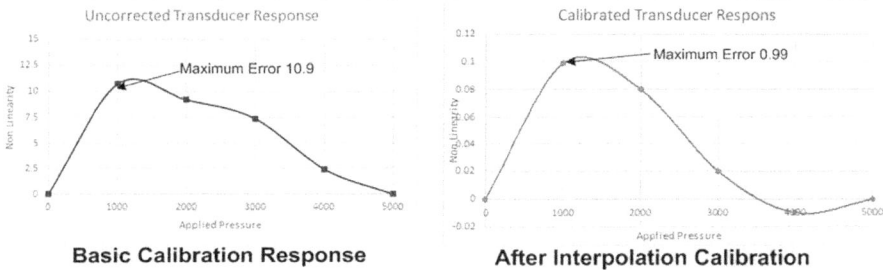

Basic Calibration Response   **After Interpolation Calibration**

FIGURE 12.18   Pressure Response Curves.

In order to understand the true response of a transducer the difference between the applied pressure and calculated pressure should be plotted against applied pressure. This magnifies the errors and discrepancies. If a plot of applied pressure against calculated pressure is used the figures are too large (1,000's of psi) for any detail or errors to be seen as shown in Figures 12.18.

Accuracy can be further enhanced over a set pressure range by adding extra calibration points, decreasing the intervals for calibration points. The increased calibration points, Table 12.6, over a set region will enhance the accuracy of the sensor over that actual operational area, illustrated in the resultant graph in Figure 12.19.

TABLE 12.6
Enhanced Calibration Table

Pressure	Raw
0	1,431,011
1,000	4,308,272
2,000	7,417,172
2,500	8,834,222
2,700	**9,375,394**
2,800	**9,640,484**
2,900	**9,901,910**
3,000	10,159,672
4,000	12,535,772
5,000	14,545,472

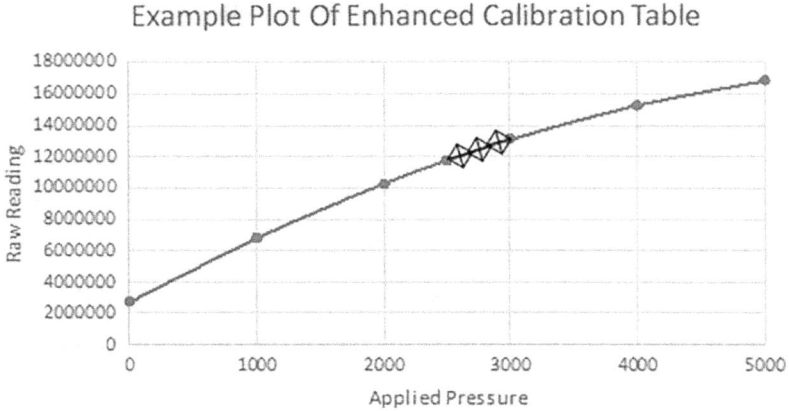

FIGURE 12.19   Plot of Enhanced Calibration Table.

HART Readings

With the smart sensors it is possible to read the pressure directly from the HART processor embedded in the transducer. This negates all errors generated by an A/D converter in a data acquisition system when reading the data.

However, the HART approach again assumes a straight line and the errors should be checked using a deadweight tester.

NOTE: *Wireless sensor systems such as WirelessHART will utilise the readings from the on-board processor with a straight-line calibration.*

12.3   WIRELESS TRANSDUCERS

Wireless transducers use exactly the same mechanisms to measure pressure but instead of using wire as a transmission system they broadcast wirelessly (radio transmission).

There are several different formats for the data to be transferred using wireless techniques and each one has its benefits. The different formats will not be covered in this document because as far as the operation is concerned if they perform their intended function the format is not of concern.

However, from an operational perspective, there are several points to consider.

12.3.1   Transducer Security

Because the data is being transmitted – or broadcasted – it can be considered as an open system, similar to an unsecured wireless router, where a security risk with respect to data intrusion or "hacking" can take place.

There needs to be a level of security built into the system to stop unauthorised users.

12.3.2　Data Transfer Rate

Normally a wireless transducer transmits data at a fixed rate. It is critical to know what this transmission rate is. If the transmission rate is fast, typically one-second readings, then this is acceptable as the data can be filtered and stored at different intervals within the data acquisition system as per the user configuration.

However, if it is a slower system, one-minute readings, then important data can be lost at critical times, such as initial stages of build ups. Also, the logging of important data in case of inadvertent shut-in or with accidental events is limited.

Some systems utilise a variable transmission frequency system where the transmission rate can be changed from the control system. This can have one disadvantage in the time needed to configure the transmission rate change, which is an important variable to consider and often missed by the data acquisition operator.

The main reason why certain systems do not transmit at high frequencies is power. The wireless transducer takes more power when it is transmitting frequently rather than in silent mode. This has an effect on battery life and ultimately the cost of running the system.

Some data transmission systems, such as WirelessHART, transmit a lot more than simple data; their transmission includes the status of the network and other system information. This has a major advantage of a secure system and the information cannot be intercepted or used by unauthorised personnel. The disadvantage to this can be a slow working system and in all cases the compatibility with a data acquisition system should be verified.

12.3.3　Transducer Accuracy and Repeatability

Accuracy and repeatability functions must be independent of the battery voltage or power available. As the battery capacity reduces with time this should have no effect on the readings of the transducer. If the battery condition drops below the power required to maintain accurate readings, then the transducer should stop transmission to indicate an issue.

Some of the more advanced wireless protocols will transmit battery status and alarms.

12.3.4　Batteries

Because the device is wireless it stands to reason that the power source is integral within the transducer. This also applies to down-hole memory gauges. Because of the power capacity compared to the size, most companies use a form of lithium battery. What is ignored is that these batteries under certain conditions are considered hazardous. Therefore, a safety sheet should accompany each battery.

The batteries have an effective "sell by date"; due to the nature of the electro-chemistry within the cell(s), the system slowly self-discharges. So, the first thing that should be checked is the age of the battery.

Despite the battery's age, just by visually inspecting it, we cannot determine its condition. Checking the voltage will not determine the capacity of the battery. The only way to confirm a battery's condition is to physically load the battery in a circuit to draw down the battery. The load should be enough to draw the voltage down and will vary from battery to battery depending upon the battery capacity, electrochem-istry, and manufacturer.

There are battery testers that will do this, but if it is not supplied, a test with a 50–100 ohm high wattage resistor can be used in conjunction with a voltmeter as shown in Figure 12.20.

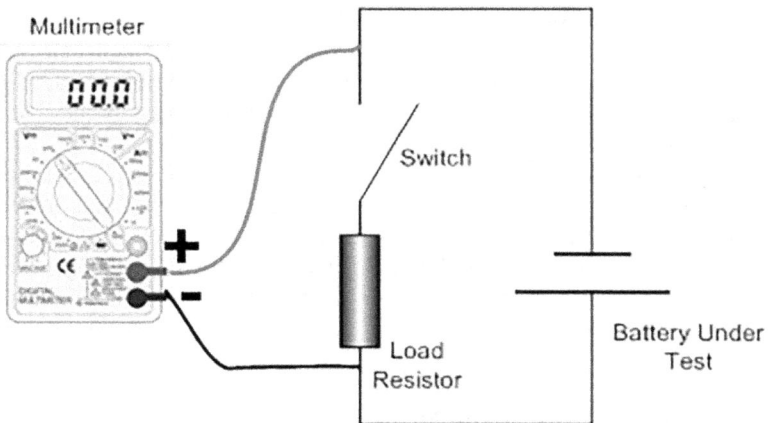

FIGURE 12.20   Lithium Battery Test Circuit.

With the battery connected as shown with the switch open

- Note the open circuit voltage of the battery for one to two minutes
- Close the switch and observe the voltage for approximately a minute and note the final voltage
- Open the switch and observe the voltage recovery response

With a good battery the voltage should go back to – or near to – the open circuit bat-tery within a minute. If the battery is used or defective the recovery response will not return to the open circuit voltage.

The graphs in Figure 12.21 give example potential responses for the batteries. The voltage response curves have been highlighted. It is possible to take readings and plot the results manually.

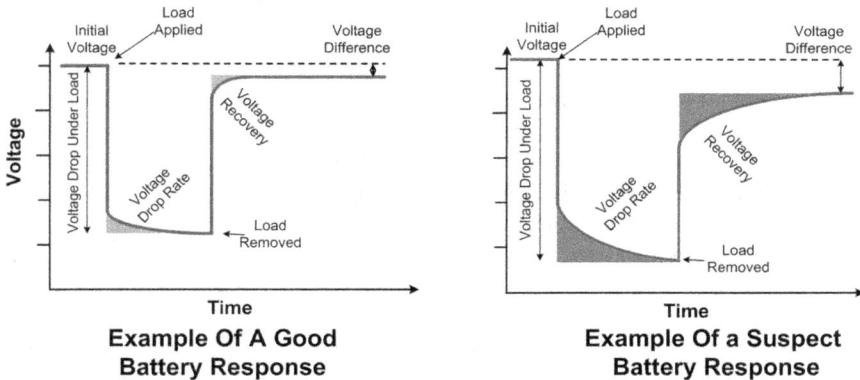

FIGURE 12.21 Battery Response after Loading with a Resistor.

- The bad battery's voltage will drop further and take longer to recover
- The good battery responds quicker and the voltage is not drawn down as far

The batteries should come with an individual usage sheet, used for tracking the use of the battery giving a record of the battery's life expectancy.

Knowing the power consumption of the transducer and the capacity of the battery the amount of hours usage available from the battery can be calculated. This can be a complicated calculation based on the sample rate and transmission power used. However, some of the wireless systems will give a battery life report after the job has finished.

This figure should not be taken as an absolute figure but down-graded by at least 10% especially in the case of memory gauges where temperature cycling is involved.

Battery Changing

Although some systems specify that it is possible to "hot swap" the batteries in a hazardous area this is often not approved by the operating company. From a safety aspect it is better to remove the entire transducer to a safe area where the battery can be tested and verified before replacement of an old battery.

Battery Costs and Spares

There are two ways to look at this as the cost of replacement batteries can be high. On a standard well test the following transducers are required as a minimum.

- Wellhead pressure
- Wellhead temperature
- Annulus pressure
- Separator gas line pressure
- Separator gas line temperature
- Separator differential pressure
- Separator oil line temperature
- Oil flow meter high rate

- Oil flow meter low rate
- Water flow meter

There is a total of ten batteries and the cost can run into a significant sum. Therefore, the battery costs should be factored into the contract rather than considered a consumable. If this type of cost invoice is presented to the company representative without mention in the contract it is unlikely to get approved.

This then leads into the subject of spare batteries. As the life of a battery cannot be exactly predicted then spares should be stocked within the data acquisition department. A minimum stock level of 25–30% spare batteries should be maintained on top of the estimated numbers required for the job(s).

Preference should be given to sensors that use standard alkaline battery that can be purchased locally. The cost factor is then minimised. This usually means that there is a reduction in battery life, but it should be offset against running cost.

Battery Disposal

All lithium-based batteries must be considered explosive and disposed of correctly by certified and approved means, most often returning them to the supplying company.

It may be that the battery is not certified for transport by air; this too should be looked into along with local and international regulations which should be discussed in the TWOP meeting.

12.3.5   WIRELESS NETWORKS

When using an intelligent wireless system that has the facility to utilise a wireless mesh network, depicted in Figure 12.22, the time factor to implement this network should be considered and declared prior to starting any data acquisition.

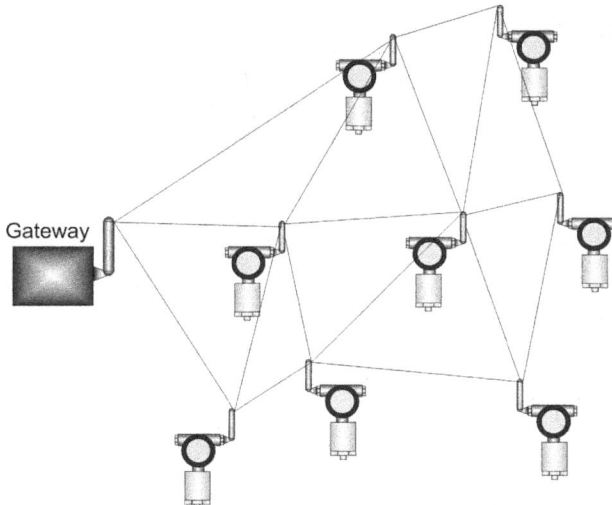

FIGURE 12.22   Wireless Mesh Network.

A wireless mesh network is where each transducer acts as a relay, so the path back to the modem is always maintained. This is a good system, but the following events should be considered:

With this mesh network in place while logging data, if there is a power failure or crash of the wireless gateway/modem, how much time is required to re-establish the network and restart data logging? This is lost data which cannot be recovered. Unless there is a secondary data source, the unacquired data is lost.

12.3.6 WIRELESS FREQUENCY AND POWER

It might be a simple question, but is the frequency at which the wireless transducers operate/transmit with considered legal and free in the country the system is installed in?

Along with the transmitted power they have to be verified and licensed for use in each individual country. Generally, the approach is that it is the same frequency as the cell phones, but this cannot always work as it is being used to transmit data and comes under a different set of rules in many countries.

Many frequencies used by wireless transducers are similar to the mobile phones and the frequency used in a microwave oven. This has the disadvantage of being greatly attenuated by the presence of water, moisture bearing structures, or snow. In this case the base frequency should be shifted to account for the attenuation of water.

12.4 MOUNTING OF TRANSDUCERS

Strictly speaking all pressure measuring transducers should be mounted in a vertical manner with respect to the effect of gravity. In practice this is often not the best option. After calibration issues, the mounting and handling of transducers are the next biggest source of transducer errors.

If the transducer is mounted directly on pipework that is to be processed or assembled then the transducer must be removed before work starts. This particularly applies to any hammering that has to take place as all shocks and mechanical impacts can be transmitted to the transducers and cause damage. This is a very common cause of damage, where the well test operators either are unaware or unwilling to remove sensors before working.

There are several solutions to avoid this issue, by using a remote mounting with the pressure being transmitted via a steel liner or certified hose, to using a coiled stainless-steel liner that acts as a spring-type shock absorber.

Figure 12.23 shows a pressure transducer mounted on a coiled liner that is filled with hydraulic fluid to buffer the well fluid but allow the transmission of pressure. The importance here is to keep any mechanical shocks which can damage or affect the transducer's response.

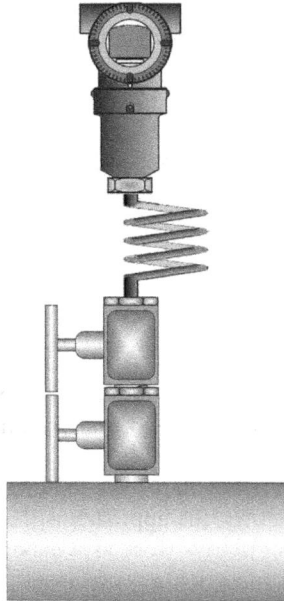

FIGURE 12.23   Shock Absorbing a Pressure Transducer.

NOTE: *Use of double-needle valves is fast becoming a standard.*

Using a liner to remotely mount a transducer has the benefit of avoiding the corrosive effect of the well flow fluids to the transducers, with the hydraulic oil acting as a buffer. This is the identical principle used with the liner on the deadweight tester and surface chart recorders. This method assists if the well fluids are hotter than the maximum operating temperature of the transducer as it thermally isolates the transducer also.

As the hydraulic fluid is not isolated from the well fluid it is necessary to replenish/purge the isolating liner regularly.

12.5   ELECTRICAL SAFETY

In order to utilise electronic and electrical transducers and instruments in a hazardous area the instruments must be certified according to an authorised body that is recognised in the country where the work is being performed.

Due to rules varying from country to country and as to avoid any issues only the basics are covered in this section.

The most widely accepted certification worldwide is now ATEX but other country-specific certifications can be used.

There is a general rule with regard to electric wiring that enters a hazardous area, being that all electric cabling and equipment, no matter what the voltage or current, must include some form of electric safety. This includes computer interfacing, there are no exceptions.

There are two more common systems followed:

i. Intrinsic Safety (IS)

Intrinsic safety systems are also known as IS systems. An intrinsically safe system has a protection technique for the safe use of electrical equipment in a hazardous area by limiting the energy available to the device, disabling its ability to act as a source of ignition.

Intrinsic safety limits both electrical and thermal energy of a system.

In order for intrinsic safety to be valid, all the devices and transducers, must be certified with an IS certified system, without which they become invalid. This is because intrinsic safety is a system certification where all devices and sources must comply with the energy storage limitations imposed by IS.

Intrinsic safety is considered a low-voltage low power system and as such is usually limited to instrumentation.

ii. Explosion Protection

Explosion protection systems are also known as an Exd system. An explosion protection system is a technique for the safe use of higher power devices in a hazardous area. The primary function of Exd is to prevent the propagation of an internal explosion or fire to the hazardous area by containing the explosion within its enclosure.

Exd systems tend to be bulky and heavy and are normally limited to high current devices. As a by-product the use of heavy-duty enclosures help to protect the internal components and connections from humidity, dirt, dust, or water and any accidental spillages.

In order for Exd to be valid, all the housings, connectors, glands, cabling, and interconnecting systems must comply with the rules and regulations of Exd.

iii. Certifications

With the previous two systems the use of a non-certified components or devices will invalidate the entire certification.

All the safety certifications for all devices, whether a single certificate for all transducers from the same manufacturer or for each individually, should be available and stored on location with the calibration and other certifications.

There should also be a permanent detail of the transducer's safety level etched, engraved, or on a permanently fixed metal plate on each individual transducer.

If there are no safety details then the transducer cannot be used.

12.5.1 Solar Power

The use of solar powered instruments is becoming more common. However, this too must comply with the electrical safety rules and regulations. As solar power is normally associated with low power then the IS system certification is usually used with the instrumentation, therefore an intrinsically safe solar system must be used.

There exist solar power systems, including the batteries, that meet the relevant certifications, therefore the use of an off-the-shelf solar system should be avoided.

Solar Power outside a Hazardous Area

It is possible to place a solar power system outside a hazardous area provided:

i. Intrinsically Safe Equipment

The output from the solar power module must pass through an intrinsically safe barrier system/module located on the power module in a safe area. This way the power from the solar power module is limited to the requirements of the instrumentation supplied.

In this scenario the barriers used would be a galvanic type where a common earth is not required.

ii. Explosion-Proof Equipment

The output from the solar power module has to use certified Exd cable from the solar module in the safe area to the equipment in the hazardous area.

Again, all certified fittings must be used in the hazardous area.

12.6 MAINTENANCE AND CARE

All the calibration and "tweaking" in the world will not be of any use if the transducers are not correctly maintained and stored. These are instruments that are calibrated specifically, often by third-party vendors, then just thrown in a box for shipment. What is worse is they are often just thrown in a box before shipping to or between jobs.

This type of handling can lead to the sensors being "knocked" and the calibration shifted or changed. This then leads to the question of why they are calibrated before shipment. If the timing of the test is correct and sufficient time is allocated, a pre-test calibration can be carried out. Pressure transducers should be checked or calibrated with a deadweight tester.

This has an additional benefit in that a post-test calibration can also be carried out to ensure no sensor damage or drift.

FIGURE 12.24   Sensors Not Stored Properly.

The sensors, strictly speaking, should be positioned vertically in a container or partition that will not allow it to easily move or be knocked when being shipped.

Cleaning the sensors can have serious effects. The more inexperienced data acquisition operators can clean the sensors in a manner that renders both the data from the sensor inaccurate and the sensor un-calibratable using a two-point or straight-line fit.

Operators pulling the sensors across the ground – or deck – by the cabling has similar effects.

Sensor Positioning

Damage to sensors is quite often preventable. Within the testing environment damage to sensors can be prevented by correct positioning and due care. The positioning of the sensor on a flowline should be one that ensures it cannot be damaged or knocked.

The classic example is sensors mounted on a flowline with hammer unions – or flange bolts – that need tightening using sledge hammers – the standard practice is that ALL instruments must be removed before these activities take place.

It is common for pressure sensors to be mounted remotely. The use of high-pressure liners and associated fittings is common to move sensors into a safe area. This technique is also used if the process media is at a higher temperature than the working temperature of the device or if corrosive fluids can damage the sensor plates. All these liners should be purged with hydraulic oil. The deadweight tester normally has a built-in pump and can be used to both purge and pressure test liners.

A damaged sensor should not be used, and a lot of damage cannot be identified visually.

Within a pressure sensor there is usually a diaphragm that the pressure acts upon directly. As the majority of pressure sensors below 10,000 psig use a 1/2" NPT female port fitting, it is subject to blockage due to mud, high viscosity fluids, etc.

Cleaning these sensors is relatively easy, however, tools like screwdrivers are often used to clean out the ports, and using metal objects for cleaning results in damage to the diaphragm profile as shown in Figure 12.25.

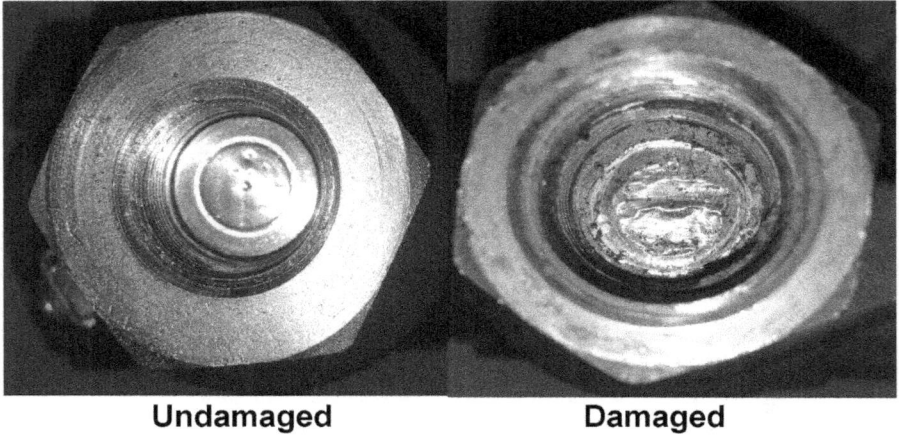

Undamaged   **Damaged**

FIGURE 12.25   Sensor Diaphragms.

The resulting damage **WILL** change the response of the transducer and alter the calibration. This type of damage is often NOT picked up by third-party calibration, for either reason, and it is not picked up using HART calibration methods.

i. Examination is not part of the service
ii. No physical pressure is applied to the sensor – an electronic calibration is executed

Quite often this type of damage will make the transducer's response non-linear and unsuitable for accurate data acquisition systems. Therefore, physical inspection should be part of a third-party calibration and certificate.

If we take a standard surface pressure transducer, the majority of them have a diaphragm in place either acting as a buffer or the pressure measurement plate.

Figure 12.26 shows the position of the diaphragm with no pressure and with pressure applied.

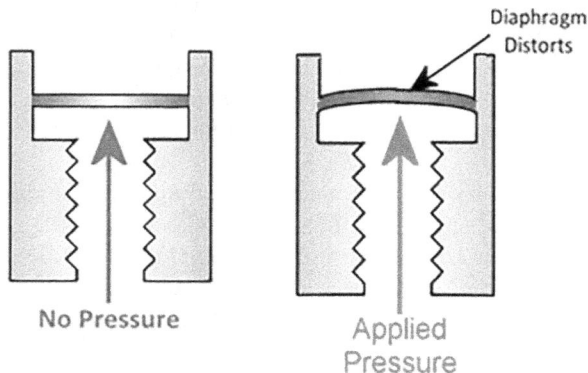

FIGURE 12.26   Diaphragms with and without Pressure.

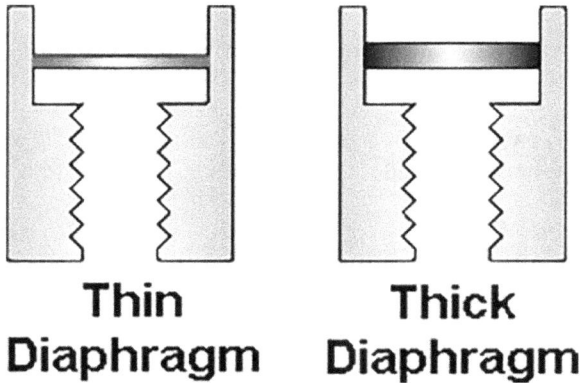

Thin Diaphragm   **Thick Diaphragm**

FIGURE 12.27   Pressure Transducer Diaphragm Thicknesses.

Figure 12.27 shows two identical sensors with differing diaphragm thicknesses.

As with Bourdon tube sensors, the thicker the sensor walls the more pressure is needed to distort the diaphragm in equal measurement. The thicker diaphragm intended for higher pressure ranges will not give the same resolution when working at lower pressure ranges compared to a thinner diaphragm.

Therefore, when a higher range (with a thicker diaphragm) transducer is re-calibrated at a lower full-scale range, the readings are often electronically amplified within the transducer to imitate an equal resolution. This amplification can cause the transducer to have a loss of accuracy or resolution, be inaccurate, be unstable, and be affected by rapid temperature changes.

This is similar to an audio amplifier; it does not just amplify the current but the noise and imperfections as well, which can make the readings appear unstable.

Most often the effects of these amplifications are seen on the well test separator's pressure sensor. The vast majority of well test separators are 1,440 psig; the use of a 10–15,000 psig transducer, working on its full range or having been re-spanned, will not respond as accurately to the small changes in pressure as lower pressure range.

Temperature Effects

The modern pressure transducers utilise temperature sensors within the body of the transducer which are used to compensate for temperature effects. There can be up to three temperature sensors in a single pressure transducer. These are not normally accessible to the user but used in an internal complex algorithm to correct the pressure.

This technique works well, but when the transducer has been re-spanned from a high range to a lower range, the temperature compensation can become an issue. The temperature compensation only works to its true intent if all the temperature transducer sensors are in equilibrium (nearly the same). If the temperature sensors are different then the pressure reading can be in error.

As a test of transducer stability pour cold water (hot if in a colder environment) over the transducer and monitor both the output and the digital display, if applicable. If it changes by a significant amount then the suitability is under question.

The use of the onsite deadweight tester to validate and calibrate sensors is a practice that has fallen into disuse and should be re-instated, to shed light to the operators on all these factors.

12.7   MAIN DATA SOURCES

The main source of data within an acquisition system is supplied from the transducers and sensors that are mounted on the equipment. These can be either a single sensor such as a pressure sensor or multiple sensors mounted in a complete module, such as a Coriolis meter.

Therefore, the data stream transmitted to the acquisitions system can be one of two options.

   i. Raw data
  ii. Calculated or processed data

The calculated/processed data involves an internal microprocessor that reads the data from multiple sensors and calculates the required result(s).

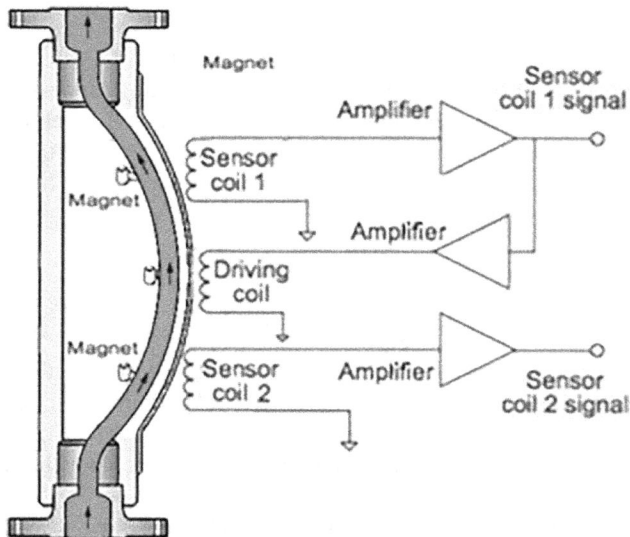

FIGURE 12.28   Basic Coriolis Meter and Electronics.

Nowadays most devices, including simple pressure transducers, also incorporate a microprocessor to manage and correct the data often giving the option of differing output formats.

Taking calculated/processed output from meters such as Coriolis poses an issue if the calculated data has an issue due to inputted parameters (two-phase flow, different SGs, etc.). Unlike dealing with raw data, reverse engineering the processed data to correct it for these issues and errors becomes extremely challenging.

Transmitted data can be acquired from these instruments in different formats, and it is important that the transmission rate of these parameters suits the purpose and is not slower than the other monitored parameters as we want continuity of data with all data taken at the same interval.

In some routine well tests the data acquisition system data storage rate is as low as 15-minute storage increments since that is what the report printout uses. This is a waste of a DAS system as manual readings could be easily used instead. DAS is there to monitor data which should be no slower than one minute as the purpose is to monitor the well response. With the advantages in computer technology and data storage capability, there should be no limit on data storage rates or what data is stored.

Often the transmission format or system, other than 4–20 mA systems, has no safety considerations unless they are cabled using certified Exd armoured systems. This is a safety issue that must be corrected.

12.8 DOWNHOLE SURVEYS

When running any form of pressure gauge in the well, whether a memory gauge or a PLT, it is a good idea to put a deadweight tester on the lubricator with the gauges inside before running in, illustrated in Figure 12.29.

When the well is stable then take a reading from the deadweight tester and the value and time.

Perform the survey as normal; pull the string out of the hole into the lubricator and allow it to stabilise. Do not shut any well valves until finished.

Again, with the instrument inside the lubricator take a reading using the deadweight tester and note the value and time.

Complete rigging down.

When all the data from the down-hole gauge is acquired and available then compare it with the deadweight readings.

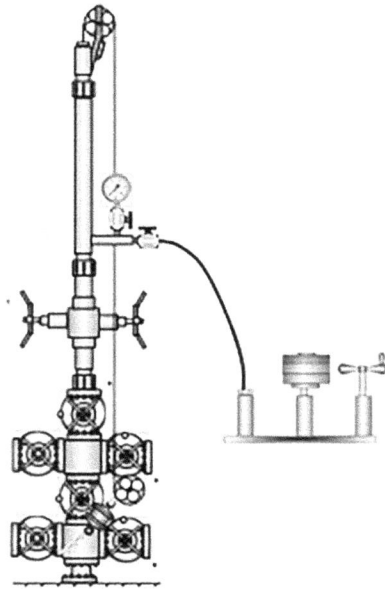

FIGURE 12.29 Illustration of Lubricator with Connected Deadweight Tester.

- DHGauge pressure reading RIH – deadweight tester pressure = Error 1
- DHGauge pressure reading POOH – deadweight tester pressure = Error 2

Errors 1 and 2 should be similar if the down-hole pressure data is valid.

This cannot be carried out using a standard dial gauge or pressure transducer that is potentially affected by hysteresis errors. It must be a deadweight tester.

NOTE: *Downhole pressure is normally reported in psia but deadweight pressure is psig.*

$$psia = psig + 14.73$$

Gauge Resolution

A lot of importance is placed on gauge accuracy but gauge resolution and repeatability are far more important. Repeatability is normally associated with hysteresis but it is critical.

For example: If the pressure at a set point is 4,850 psia running in the hole but 4,800 psia pulling out of the hole with constant pressure, then there is a repeatability issue.

This manifests itself in the basic reservoir engineering principles, the build-up. If the gauges at both surface and downhole do not give a correct pressure then how accurate the pressure is becomes irrelevant.

The resolution of the gauge is the smallest change of pressure that the device can measure sometimes referred to as a Measurand. Although the transducer can be specified to have the required resolution, the effect of the measuring equipment, the data acquisition system, needs to be considered.

Analogue sensors are measured with an A/D converter which is at the heart of the data acquisition system. The resolution of the data acquisition system as a whole is now dependant on the resolution of the A/D converter as much as the transducer's.

12.9   TRANSMITTED DATA

With all the different service companies involved in a well test, it is inevitable that data from different instruments need to be collated into a single database which can then be stored or retransmitted; this is usually the well testing DAS database.

In order to accurately merge data, it is critical that the real-time clocks on the different systems are synchronised, otherwise we face mismatches in data, from different timestamps.

There are several types of protocols involved in transmitting real-time data, and it is essential that the companies involved have systems with the same protocols. Without similar protocols there would be incompatibility between the data transmission and the data reception. It is recommended that a Systems Integration Test (SIT) is carried out that ensures compatibility between different systems, where all the data transmitting and receiving systems are tried and verified. This will include all the interfaces, cabling, and transmission modules necessary to do the job on the location.

Arrangements for SIT require coordination from the operating company and all the system owners. It can be carried out in the facilities of any of the involved service companies. It is important that the SIT is witnessed and signed off by a representative of the operating company.

On the completion of a successful SIT an operating company representative signed report is submitted which includes the list of the equipment used by each service company and their respective serial numbers; these same equipment would then be delivered to the operations and can be verified when on location.

Ensuring that the same equipment used in the SIT is used in operations and recording the serial numbers is to ensure no substitutes are introduced at operations. Any substitute equipment may not have the same facilities, outputs, or computer software versions that could not function or cause technical issues on location.

These sorts of technical details are often lost with logistical personnel, where they assume one box is the same as another. Although the systems are in technical terms primal, they are very delicate. Therefore, to ensure the fewest complications and technical issues, ALL equipment used in the SIT must be sent to the location, no substitutions. This does not mean issues will not be faced; it only makes sure that there are no unaccounted-for issues.

There are now three ways of transmitting data from one service company to another:

   i. Cable
  ii. Wireless
 iii. Fibre optic

12.9.1 CABLED TRANSMISSION

With the exception of the fibre optic system, which is now rarely used, both the cable and wireless systems must comply with the electrical safety regulations, usually the intrinsically safe (IS) regulations.

Many companies would like to argue "it's only instrument or computer data so it does not need any certification; it isn't a three-phase generator cable." Wrong: all electrical cabling that passes through a hazardous area MUST have an electrical safety certificate.

To make life easier there are specialist companies that manufacture interfaces and modules that turn an uncertified signal into a certified signal. This interface usually needs to be mounted in a safe area. Often another interface to convert the IS signal back to conventional signals is needed in the receiving lab to ensure compatibility.

Figure 12.30 shows the basic layout of a cabled system; the potential problem is the length of the IS line; the IS converter modules can have a limited operating distance.

Here we should note the importance of the location layout and that it has to be adhered to; any changes in distances without consultation could result in the system not functioning properly. An extra foot of cable could kill the transmission.

Although the IS cable does not have to be armoured and normal cable can be used, it is often not the case of "just" adding more cable. Again, the SIT will pick any issues up if the cables are laid out to the full expected length, otherwise it is not a valid test.

NOTE: *Intrinsically safe (IS) cables should have a blue sheath or labelling along the length of the cable.*

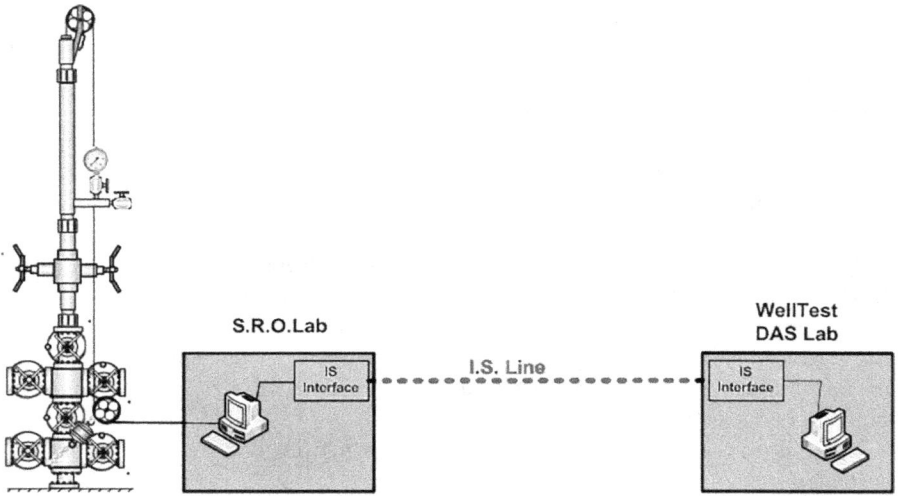

FIGURE 12.30   Data Transmitted Using Cable.

12.9.2   WIRELESS TRANSMISSION

The option to use wireless interfaces, as shown in Figure 12.31, is becoming more popular even though they present their own issues.

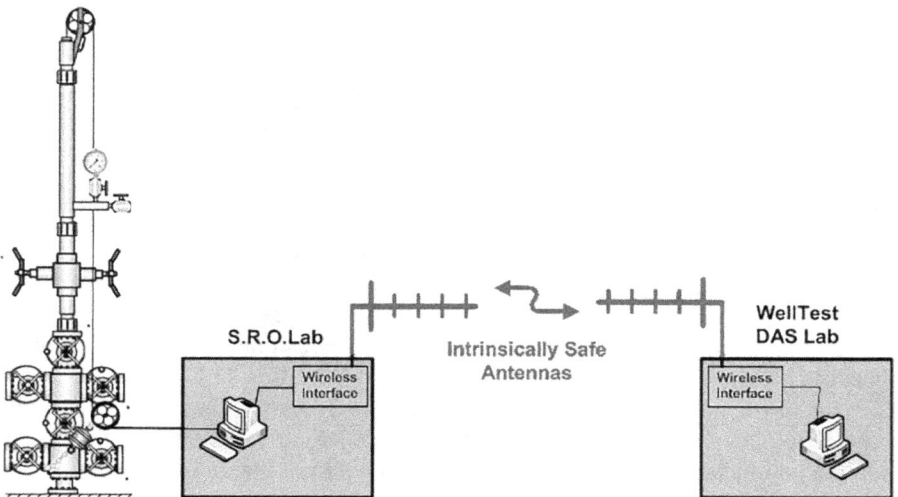

FIGURE 12.31   SRO Data Transmitted Using Wireless.

The working frequency and power of the wireless system must be checked; in many countries and locations there is a limit on the frequency and power used by both the interface and the antenna.

One of the major considerations with this type of communication is that it works better with line of sight, meaning that if an object is put in between the two antennas there can be communication loss. This also applies to any objects suspended on a crane that is moved into the path.

Interference with the line of sight may also include rain or humidity as the signal is attenuated by the water drops which restrict transfer and therefore the loss of data. This is also known as "rain fade" when applied to satellite systems.

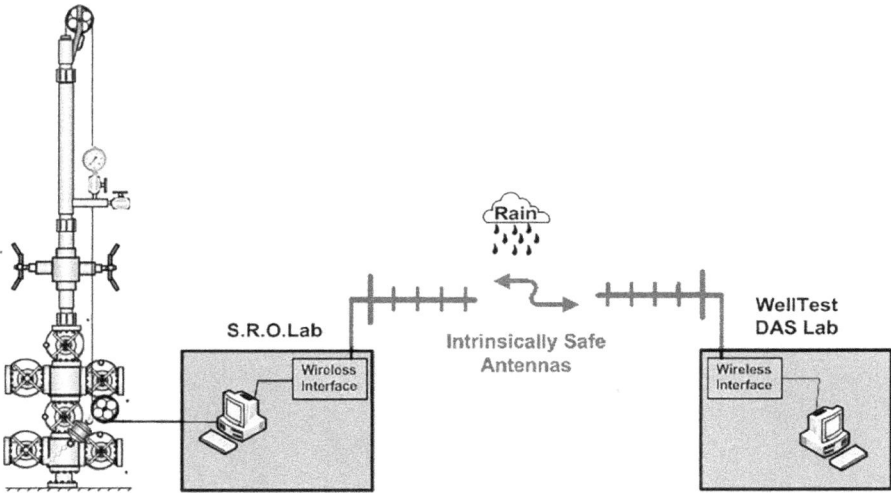

FIGURE 12.32   Interference due to Rain, Humidity, or Snow.

The interfaces do not need to be specific but it is critical that the antennae have an electrical safe certification and are within the permissible limits of the country or location.

12.9.3 DATA TRANSFER

The need for internal clocks on the computers and synchronisation of systems is re-emphasised here to avoid any mismatches with the data points.

With the timing synchronised the rate of data transfer is crucial relative to the data storage interval. The rate of data transmission should be higher than the rate of data storage on the DAS server (receiving system).

Example: If data is being transmitted every five minutes and the data on the DAS server is stored every one minute, there is a gap in the storage data of four (4) minutes. The DAS server would resort to auto-fill the missing data points every minute, using the last valid data received.

i. Network

This type of hardware configurations can be part of a wireless network if TCP/IP protocol is used as a transmission medium.

ii. Antennas

There are two distinct types of antennae used

Yaggi Antenna

The Yaggi Antenna, shown in Figure 12.33, is a directional antenna where the data stream is transmitted in a single direction used mainly to link two systems.

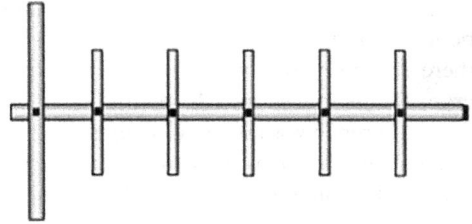

FIGURE 12.33   Yaggi Antenna.

Omni Antenna

The Omni Antenna, shown in Figure 12.34, is a multi-directional antenna used when there is more than one system transmitting often to a single server. This is mainly used in networks.

All the antennas used irrespective of the model should have an electrical safety certificate.

FIGURE 12.34   Omni Antenna.

12.10   REAL-TIME TRANSMISSION

With wells that have a high importance, the use of real-time transmission through the internet is becoming very popular.

Unless there is an existing internet connection available on location then this is not an easy five-minute fix; it takes a lot of organisation and involvement with the communications companies to get a dedicated satellite with the required bandwidth for the allocated time.

What should be considered with any form of real-time transmission is the criticality of the manual readings as all this data is now transmitted in real time or near real time, so any errors, mistakes, or late readings can be easily identified. The well test crew must be aware under all real-time transmissions that the customer is effectively looking over their shoulder and any mistakes, usually corrected by the DAS operator, will be transmitted to the customer.

12.10.1   INTERNET CONNECTION

On locations with existing internet connections, it can be a relatively simple task of transmitting data in real time. IT engineering assistance may be required to set up and for the technical details such as the computer addresses allocated to match the existing network.

Each component or computer will have to have a dedicated computer address or range of addresses referred to as IP addresses. These will have to be entered into all the equipment that connects directly to the internet; generally, this would only be the DAS server.

The issue with connecting to an existing internet network is security, as there are other users on the network. This can be addressed by restricting access to the DAS server or by transmitting encrypted data. If encrypted data is used then a version of the original DAS software, often referred to a as a client station, will have to be installed on the remote computer to decode the data. This type of software is often dongle, hardware key, protected to stop unauthorised use.

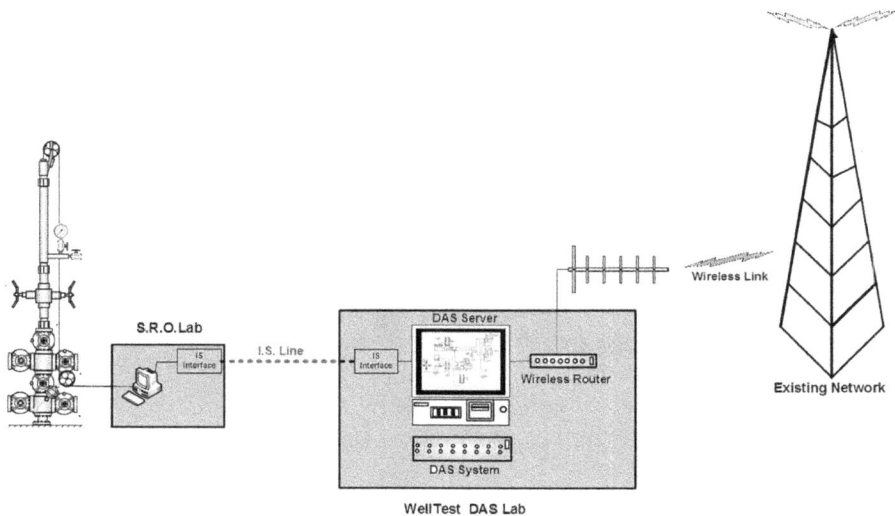

FIGURE 12.35   Schematic Used in Existing Network.

If the Yaggi type antenna is replaced with an Omni type antenna then the setup can be used for a conventional wireless network, instead of connecting to an existing network.

In the option illustrated in Figure 12.32, it is common for the DAS server database to be replicated/duplicated on the remote or customer's client computer. What the replication does is reduce the load on the network so any historical reports or plots do not repetitively download data through the network; this replication of data therefore reduces the network load and bandwidth restrictions. The replicated database will usually be proprietary to the company providing real-time data service and cannot be opened with any software other than authorised programs.

Other options are to restrict the data that can be accessed from the DAS server to keep the load on the network low. Normally a Wi-Fi link will be used to join the existing network, but if cabling is used then the cabling must have an electrical certification if it goes through a hazardous area.

12.10.2 SATELLITE REAL-TIME TRANSMISSION

A lot of the rigs will say that they have a satellite and this can be used for real-time transmission. This is not always the case if the rig satellites are used for other systems such as personnel internet access being used for downloads and consuming all the available bandwidth.

> The term *bandwidth* refers to the transmission capacity of a connection on a network and is defined as a measure of the amount of data that can be transferred from one point on a network to another within a second. Bandwidth is expressed as the data bit rate measured in bits per second (bps).

Therefore, if the data transfer rate from the DAS system is in excess of the available bandwidth, then the transfer will not function correctly.

As an example:

Working offshore on a deep-water appraisal, a real-time monitoring system was installed using the rig satellite so the operating company could monitor operations from its office.

The system worked, but not very well. The only way a continuous signal could be achieved was to turn the rig television satellite system off. This did not sit well with the rig crew.

We managed to continue but suffered "black outs" an hour before shift changes that lasted for about two hours. This was tracked down to the rig crew using the internet to use "FaceTime" to talk to their families. This effectively used all the available bandwidth on the satellite and the real-time monitoring ceased.

The rig then disconnected all the internal networks so the real-time system was the only thing connected to the satellite, and it worked fine from then on, until an unhappy rig crew cut all the leads from the real-time monitoring system and re-plugged the TV and network.

This is not an isolated incident.

This example has been used to highlight the potential problems of using a pre-installed satellite. Given the problems with a shared satellite it is recommended that a dedicated DAS satellite is used for the duration of the test period.

As far as the DAS equipment necessary for real-time transmission it is principally the same equipment used in a wired data transmission system, but in this instance a SIT cannot be carried out to ensure all the system transmission and broadcast capabilities function correctly before shipping to the location.

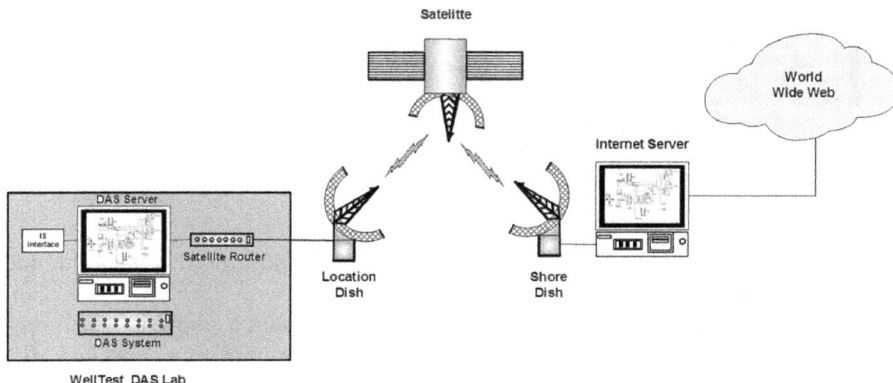

FIGURE 12.36 Typical Satellite System.

The data from the satellite is posted on the World Wide Web (WWW)/internet so all the security issues and procedures should be set in place.

In this type of option, it is common for the DAS server database to be replicated, duplicated, on the internet server computer. What the replication does is reduce the load on the network so any historical reports or plots do not repetitively download data through the network; this replication of data therefore reduces the network load and bandwidth restrictions. The replicated database will usually be proprietary to the company providing real-time data service and cannot be opened with any software other than authorised programs.

Other options are to restrict the data that can be accessed from the DAS server to keep the load on the network low.

Satellite systems' transmissions are subject to the weather conditions at both the transmitting location and receiving station; precipitation will attenuate the signal to the degree that it is no longer transmittable.

It is important that any personnel on the location do not access the data through the satellite; a standard network should be configured on location.

The well testing final reports and their associated accurate data are the most important product of the actual well testing program.

13 Reporting, Data, and Security

Yes, there are equipment, procedures, operations, skill, and expertise involved, and all focus towards a single artefact – *The Report.*

With most trades and professions there is an end product; a shipwright will produce a ship, so what does a well test produce? Nothing; we have either burnt or "disposed" of all the physical material and all we have left is the data.

Several of the procedures are laid down in this document and the non-compliance of company set processes and procedures affect the quality of the data in the final report.

It is the job supervisor's ultimate responsibility to ensure that the customer's report needs are met by checking with the on-site engineers and operating company representatives, if not subject to specifications within the contract. If this report is not presented clearly, concisely, and accurately, the well testers have not done their job and the test is a failure.

Some abbreviations and terminologies may be agreed with the operating company and made standard throughout the well testing campaign to ensure the reference does not lose intended meaning. If unusual, confusing, and obscure abbreviations are used then a glossary of terms should be included in the final report.

What this section is trying to impart is just how important the end of test report is and what an end of test report should consist of. The final report is then passed to the partners in the field/well; this is all they see of the service company.

The golden rule here is that if the report is bad then the entire test is bad.

13.1 MANUAL DATA AND EVENTS

The well test operators must operate a Manual Readings Sheet – or manual entry sheet – where all and every event, reading, and associated times are written down.

If in doubt write it down.

The well testers are being paid by their customer to safely operate their equipment and take readings and observations which then act as a backup to the DAS readings. This is important and mandatory.

All data gathering times must be synchronised with the data acquisition system that generates the final report. This applies also to down-hole instrumentation and surveys.

During the well test the operators MUST be taking manual readings from the deadweight testers, Barton, and other instruments. This should not be optional but

DOI: 10.1201/9781032623689-13

mandatory in all instances as it is a correlation of the data acquisition instrumentation and a backup. Manual data must also be kept while the well is not flowing, recording any activity on the well such as Slickline operations, etc.

The rule that there will be well test personnel stationed outside on the equipment, not in the break room, while pressure is on the equipment SHOULD be enforced, without fail.

This manual report must include all events and factors that affect any parameter of the well in comment form. As an example, wireline rigging up and down should be included along with any openings or disturbances of the well parameters for whatever reason. These readings are not intended to be included in the final report, although they are very often specified by the operating company. Nevertheless, the service company should store them with the final report, scanned if necessary and available if requested by the operating company for confirmation of results.

The manual reading sheets should be kept clean and free from oil and dirt as much as possible. The standard manner of doing this is to utilise a clipboard with a cover. Pencils are used in preference to pens due to limitations of the pen on damp paper or card. With busy sequence of events sometimes the operator may record incomplete or shorthanded notes; using a pencil makes it easier to rewrite these when operations have calmed down or at the end of a shift. The quality and accuracy of these readings are critical to a good test.

All sampling details must also be recorded on the manual data sheet. These are not the formal PVT samples; PVT samples should have their own individual records but samples and results for BS&W, gravity, etc. determination. This then can give the reservoir engineer valuable data on when and where the sample was taken from as the final reports generally just list the sample details on a regular sample interval. In the case of BS&W *both* centrifuge tube readings *must* be on the manual data sheet.

There should be no attempt to squeeze the manual recordings on a single page as it is a requirement that all the data is clearly recorded. Pre-printed forms with the well, client, service company details, and all relevant information are the most common practice and an essential one. With the normal computer equipment and a DAS system it should be possible to customise manual data sheets by adding columns. Removing columns should not happen as this gives the operators opportunity to miss out valuable data.

The important process of manually collecting data and reporting it is vastly underestimated in well testing operations. It is critical that all manually collected data is reported accurately, from the specified location and at the exact time. This should be verified randomly by the well test supervisor who is accountable for missing data.

13.1.1 EXAMPLE OF SHUT-IN RESPONSE

As an example, using the graph in Figure 13.1, the sequence of events reports that the well was shut in at 12:00. However, if we look at the real-time plot in detail it can be seen that this is not the case.

From the expanded section of the plot, Figure 13.2, it can be seen that the well was actually shut in at 12:03 and from the pressure response the drop in pressure is indicative that the separator was by passed before shutting in the well.

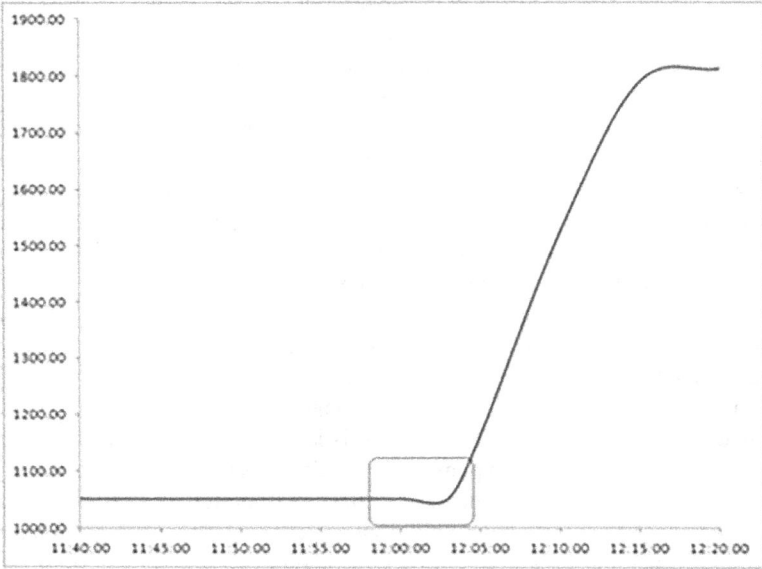

FIGURE 13.1 Bottom Hole Pressure Shut-In Response.

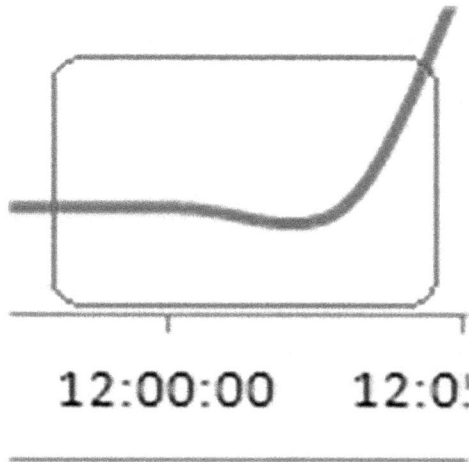

FIGURE 13.2 Expanded Build-Up Section.

This is not then a "clean" shut-in the reservoir engineers are looking for. It is this type of inconsistency that is picked up by the reservoir engineers when carrying out the analysis of the well. Having the sequence of events and data properly recorded would assist the reservoir engineer in understanding the cause of the pressure drop – and/or other consequences – which could be investigated in other ways that do not make sense.

The manual changes and the readings have dramatic changes to flow rates and it is often the case that a critical manual reading has been forgotten or missed until after the event has happened. There has to be an understanding in place between the well test operators and the DAS operators that allows for instant communication of all factors and events.

The main factor here is where the data is recorded; it is no good everyone recording their own information or shouting it over the radio. It simply gets missed. The solution is pure and simple and has been used by well testers for decades but with the advent of DAS systems it has fallen into disuse: the use of the manual readings sheets.

13.2 DATE AND TIME FORMAT

There are two date standards within the oilfield.

- American System   MM/DD/YYYY
- European system   DD/MM/YYYY

In order to avoid confusion, it is suggested that the systems shown be used for either:

- American System   MMM/DD/YYYY
- European system   DD/MMM/YYYY

This is where the month is not a numerical figure but an abbreviation of the true month name.

Example: 01/03/17 can be interpreted as either:

1 March 2017 3 January 2017

Depending on the system used the true date can be confused. Using 01/Mar/2017 clears all confusion. The time standard should be declared in the nomenclature.

13.3 ON-SITE WELL TEST REPORT

As the majority of reports are prepared in Excel or similar format, then it stands to reason that the following personnel should be trained and have certificates in the correct operation and usage of the computers and relevant software:

  i. Data Acquisition Engineers
 ii. Well Test Supervisors
iii. Base Support Staff

All reports should be spelling and grammar checked and corrected where necessary before any printing or submission.

The use of Excel is common but the drawback with Excel is the graphic plots available; these should not be underestimated. Plots are a great visual summary of data and considered the summary off the well tests so they must be as clear as possible; an experienced person can just about detect any sequence of events or errors from different plotted graphs. There are public domain graphic plotting packages available which will overcome this shortfall.

It is normal for a site report – or shift/morning report – to be submitted to the customer throughout the test. This is usually a correct report but the report should state on its header, on each page, that it is an:

On Site Report and Subject to Corrections.

The reason is, decisions are and will be made on the contents of the report so the quality of the data is expected to be high. However, there are possibilities that the reported data can be corrected to account for new or updated information so the on-site report must reflect this.

Example: gas rate is reported but after the flow period had finished the orifice plate size was found to be in error. This makes a significant change in the calculated gas flow, cumulative and GOR data.

The on-site well test reports/daily reports passed on to the client during operations would normally have a similar format as the final well test report, with fewer variables and details reported as agreed with the company representatives.

13.4   FINAL WELL TEST REPORT

As standard the well test supervisor is ultimately responsible for the report. However, with today's digital capabilities the report is usually generated and compiled by the data acquisition engineer.

There are often pre-well test campaign meetings, such as the TWOP, dedicated to data acquisition, daily reporting, and final well test reports which should be adhered to when compiling the final report. These meetings would have discussed and decided on formats, variables recorded, decimal points, units, and other relevant issues.

Often there is carry-over of forms and data from other reports on previous wells/ tests where standard reporting templates have been completed. This often has erroneous data left in the report template and should be avoided; it can cause issues and confuse the reader. New templates must be used for a new well or job in all instances.

This is one of the main reasons why data acquisition engineers *must* be experienced well testers so they know what to report and have the ability to highlight any issues or abnormalities. Too many of the reports are such that the charts and data are just a plot of the entire test and a simple document dump. This is not what is expected or contracted for a well test report.

Upon completion of the report by another engineer, the well test supervisor must then check and correct the report until it reaches a satisfactory standard. Only when they are satisfied should their name be put on the report for issuing to the customer.

If a report is issued without the job supervisor's name, not initials or combination of initials on the front page, then it should be rejected by the customer. Without the supervisor's details, the customer or the partners cannot put forward any questions or queries to anyone relevant in the near or far future.

In the case of multiple supervisors, the supervisor who is principally responsible for producing the report – or last on the job – should have his name on the report. If necessary, a separate sheet can be included in the report listing dates and supervisor attendance.

Any errors or corrections in the report, if rejected or returned by the customer, should be revised, investigated, and remedied out by the final job supervisor.

The final well test report should contain **ALL** the relevant data pertinent to the operation and as a minimum should include the following unless otherwise specified by the customer.

- An introduction
- All the relevant measured data and how it was measured
- A summary of the flow periods
- All manually collected data pertinent to the test
- Any corrections to flow rates
- All relevant plots
- A full list of all events in the correct format
- A list of all samples and measurements taken
- Title and index pages
- All formulas used in ANY calculation
- A list of all data files supplied with the report
- The full names of the people producing the report
- Any company representatives present
- Nomenclature
- Test layout drawing with sensor positioning
- A well completion diagram if possible
- List of all sensors and calibrations

Generally, all these sections are prepared in a proprietary reporting package and exported to a standard format so the report preparation is not as odious as it initially suggests. Also, the forms can be completed while the job is in progress.

The final report format should be in PDF form submitted to the client in printed format but also on a CD or memory stick. Normally the original calculation sheets in excel may be included, scanned copies of manual readings, charts, etc. Many clients also request the raw data files from DAS.

The electronic report should include a file with ALL the DAS data on it, not filtered by time. When the reservoir engineers have a query about a particular response all the data is available with them for clarification without constant need to contact the well test company.

The **Portable Document Format (PDF)** is a format that was developed by Adobe in 1993 to allow documents to be presented using a standard system that is independent of the software package, the hardware used, and the operating system. Originally the PDF format was hard to edit but the software packages available at the moment allow PDF files to be edited.

The service company should physically print and bind the final report and submit it along with any electronic data. The submission of a printed copy of a final report acts as black-and-white proof of what the service company has submitted to their client, where no editing is possible without leaving a trace. The printed report should contain a list of what has been submitted; this includes descriptions of all electronic files.

Often service companies would have one printed out copy stored in their data room/storage as an uncorrupted reference as well. PDFs in turn can be password protected to equally fulfil the same data security and legal obligation.

The original data files should also be submitted as a separate set of data to allow the reservoir engineers to access and use the data. This data should be in ASCII format to allow easy access of the data and ease of importing into reservoir engineering software.

All submissions must be checked, verified, and declared by a recognised anti-virus/anti malware system before submitting it to the customer.

The submitted report, the number of printed and electronic copies submitted to the client, is based on the contractual agreement. Contractually they must be officially received and signed for by the operating company, which is proof of the report being submitted. The original signed receipt form is then filed with the final report in the service company archives.

What has happened with a regular occurrence is the well test was not always analysed immediately but sometime later. The operating company's engineer who the does the analysis will ask for more data or for downhole data from another company to be merged. As the service company now has to assign what is normally a senior technician for this job and take them off their normal duties, then this assignment out of contractual agreement would then be independently charged to the operating company.

It is critical that the operating company specify *exactly* what they require in the report before it is compiled, to avoid any reports and data being revisited. In the worst-case scenario tests might have to be repeated when multiple service companies are involved and data and events are missing. Again, the responsibility is with the well test supervisor to ensure all the data is encompassed in the report.

It is essential that all the data, the submitted report and manual data be permanently stored with the service companies unless instructed by the operating companies.

13.4.1   FLOW RATES AND AVERAGING

The supplied report usually has the flow rates averaged over a set period of time, nominally 15 or 30 minutes. This means that in a gas calculation the following parameters are averaged over the time period:

   i. Separator pressure
  ii. Separator differential
      Gas temperature can be averaged but it tends to change slowly so it is rarely averaged.
      Oil calculation usually has only the raw reading from the flow meter averaged; again, temperature is not usually averaged.

If the calculation sheet is included in the final report, then it is these averaged figures that must be reported. No matter what data sheets are reported the technique of averaging – and the time period used – must be declared on the sheet.

Without the declaration, if the calculations are verified, un-averaged data will yield different results and hence the data can be questioned. This is most likely when the raw data from the DAS is submitted to the client in its minimal sampling rates; the engineer may choose values at any time as a point value rather than average it and find discrepancy.

13.4.2 Sensors and Data Acquisition

The DAS engineer may think the customer knows what data was measured and where it was actually taken from. However, with the possibility of erroneous data being reported it is essential to list where the sensors were mounted and the details of each individual sensor.

The report should include the following sensor information:

- Actual position
- Type of sensor
- Range
- Calibration details
- Abbreviations used for the sensor

The abbreviation WHP does not always mean that the sensor was on the wellhead; it is normally upstream of the choke manifold, so it is important to specify exactly where the sensor was positioned.

The name of the data acquisition system used and its version number

This is necessary as often data files are not always compatible between versions and in the case of any in built calculation errors.

A full listing of the calibrations for all sensors where applicable

This is to show that the sensors have been regularly checked and calibrated. The customer is paying to have calibrated and accurate sensors, prove it.

TABLE 13.1
List of Sensors Used

Abbreviation	Location	Range	Units
WHP	Upstream of Choke Manifold	0–10,000	psig
WHT	Upstream of Choke Manifold	0–350	Fahrenheit
Annulus	Specify Which Annulus Monitored	0–2,000	psig
D/S Choke Press	Downstream of Choke Manifold	0–10,000	psig
D/S Choke Temp	Downstream of Choke Manifold	0–350	Fahrenheit
Sep Static	Separator Gas Line	0–1,500	psig
Sep Oil Meter	Separator Oil Lines (*)	0–35,000	Bbls
Oil Meter Temp	Separator Oil Lines	0–350	Fahrenheit
Sep Water Meter	Separator Water Lines	0–35,000	Bbls
Gas Differential	Separator Gas Line	0–1,000	Inches H_2O
Gas Temp	Separator Gas Line	0–350	Fahrenheit
Sand Detector	Downstream of Choke Manifold	—	Impacts Per Second

(*) This can be 1-, 2-, or 3-inch meters or multiples of meters depending upon the flow rate; all of them should be detailed.

The external variables, as listed in Table 13.2, are those that have been electronically transmitted in real time via computer interfaces such as standard formats ASCII, WITS, WITSML, Modbus, HART, WHART, etc.

TABLE 13.2

Table of Transmitted Sensor Data

External Variables	
Downhole Press	psia
Downhole Temp	Fahrenheit
U/S Prod Press	psia
U/S Prod Temp	Fahrenheit
D/S Prod Press	psia
D/S Prod Temp	Fahrenheit
Annulus Press	psia

All sensors used are certified to either ATEX or CSA intrinsic safety standards

These variables are generally not calculated by the well testing company but by a separate service company and transmitted in real time to the well test server computer which collates all the data. An example of this would be downhole pressure data.

Other external data can be transferred using a memory stick and merge into the well test data base if the DAS system has the capability.

NOTE: *Any external data should be declared before the job starts, preferably at the TWOP, so that any extra channels can be added to the well test data base as required.*

Use of memory sticks to import data should be avoided due to the risks of infection from viruses.

13.4.3 Graphic Plots or Charts

Within the final report there should be a section for the graphic plots. Normally these are at the discretion of the DAS engineer, however a "that will do" attitude has been adopted by DAS several engineers so details like the build-up discrepancy described earlier are missed for lack of reported detail.

What the customer is generally looking for is information pertinent to the well's performance, not just an overview of the entire test. Intervals in the test that should have their own plots – or graphs – that should be included with the final report are:

- Perforation sequence
- Initial flow period
- Initial build-up

- Main flow periods
- Main build-ups
- Surface and subsurface sampling periods
- Any relevant sand production
- Any unforeseen occurrences (safety valve trip etc.)
- Kill or abandonment sequences

A section header detailing all the plots should be included in the final report.

The excuse that is often offered is the work load. Yes, but, the data acquisition system is a multi-tasking system with the majority of them based on the windows operating platforms. This effectively means that the DAS engineer can be working on the reports while the test progresses to the next phase.

For example, once the flow sections have finished the DAS engineer can be working on the report as there is no manual data or corrections to implement during the build-up phases. Also, often the service companies will automatically supply a "report" computer to allow the reports to be completed while on the location, so there is no "great load" at the end.

If the DAS operator is capable then the report can be virtually finalised before leaving the location. If data from downhole memory instruments have to be included then this means that the final report will have to be completed in one go. But this does not mean to say that the report preparation cannot be carried out while on location.

Where possible the main events should be included on the charts. A scan of the Barton and wellhead recorder charts should be included if they are used.

13.4.4 Sequence of Events

Without a concise sequence of events that cover all activity pertinent to the well, then, the well test report is incomplete. As before this relies on the well test crew ensuring all the events (and data) are noted and passed to the DAS operator.

What is not widely adopted is that there is a standard for report comments or events in English. This is known as *third-person past tense*.

Basically, this means that the event is reported as it has already been performed, not as if it is actually being performed.

01/Mar/17:12:15 increase adjustable choke to 32/64". ✗
01/Mar/17:12:15 increased adjustable choke to 32/64". ✓

The correct usage is that the choke has been adjusted, not that the operators are actually in the act of adjusting the choke.

It is also important that all operators use the same terminology and format when inserting events. This makes spelling and syntax easier. This way the customer cannot tell the difference between one operator/shift and another.

Increased Fixed Choke

If we take an event like: "01/Mar/17:12:15 increased fixed choke to 32/64."

This is not representative of true events and actually what should be detailed on the manual reading sheet is:

    i. Well flowing on a 28/64" fixed choke
   ii. Switched to a 28/64" adjustable choke
  iii. Increased adjustable choke to 32/64
  iv. Switched to 32/64" fixed choke

So, if the graphic plot in Figure 13.3 is examined, it is borne out.

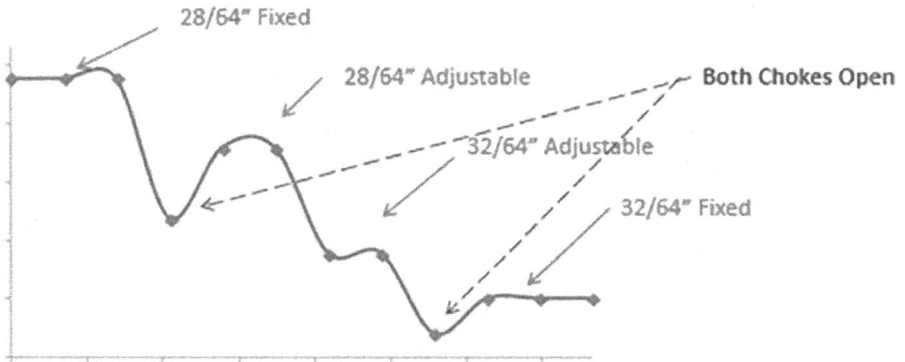

FIGURE 13.3   Response from a Downhole Gauge (BHP) during a Choke Change.

Unless you have the full sequence of operations/events, the reason for these pressure changes can be attributed to other reasons than choke change, which if assumed could result in serious detrimental or costly decisions being made on false assumptions.

One of the more disturbing trends is actually to report the event detail incorrectly an example is as shown:

"01/Mar/17 12:15:00 Bean up to 32/64."

This is not possible and just an expression in common use that should not appear on any reports. The bean is fixed in the choke body and cannot be removed without switching to an adjustable and bleeding off the pressure in the choke body. Again, the full actions and events must be detailed.

13.4.5 Syntax

Taking into account that the reports are normally produced in English, which may not be the DAS engineer's first language, it can lead to misleading and unprofessional comments.

Different styles of writing often shown up at shift changes, where different DAS engineers will use different phrases for the same operation. In order to standardise the report and ensure the use of the correct syntax a simple table can be configured.

This is where all the events in a well test are detailed in an excel table or equivalent, such as the example given Table 13.3.

TABLE 13.3
Event Fill Table

Increased adjustable choke to **X**/64".
Decreased adjustable choke to **X**/64".
Switched to **X**/64" fixed choke.
Switched to **X**/64" adjustable choke.

Table 13.3 shows a small section of the proposed events where the operator picks the relevant event and pastes it into the WT report changing the **X** for the numeric value of the choke.

This has two benefits: standardising the report comments but also ensuring the spelling is correct. This procedure will start off with just a few comments but if coordinated by the base staff, then this will grow with time to cover most aspects of the test. This means that the job supervisors should submit any missing events to the base. This system can be expanded for DAS operators who are not fully fluent.

A new column is added to the left of the original with the event in the DAS operator's native language as shown in Table 13.4. This can be expanded to encompass other languages.

TABLE 13.4
Language Translation Table

Spanish	English
Aumento del choker ajustable a **X**/64"	Increased adjustable choke to **X**/64".
Choke ajustable disminuido a **X**/64"	Decreased adjustable choke to **X**/64".
Choke fijo cambiado a **X**/64".	Switched to **X**/64" fixed choke.
Choke ajustable cambiado a **X**/64".	Switched to **X**/64" adjustable choke.

This enables the operator to verify the phraseology in a report, a report in a language that is not their mother tongue.

13.4.6 Surface Sampling (Non-PVT)

Throughout the well test, manual sampling should occur at least once an hour as a routine procedure. The sample taken every hour is not necessarily the same but can vary from the main three fluids that affect the rate calculations.

These are:

- Gas gravity
- Oil gravity
- BS&W

Often what is not made clear in the well test reports is the trend of these variables, as these figures are normally entered as a comment. Therefore, a separate sample detail sheet should be used for all samples including PVT.

The sampling sheet is of importance to the well test supervisor and DAS engineer compiling the reports as it details times and dates so correlation with the main report becomes easier.

Throughout the length of the test the physical properties of the produced fluids can change therefore it is essential that ALL and ANY samples are accurately recorded:

- Accurate time and date
- Location of sample position
- Type of fluid
- Type of sample container
- Status of sample, pressured or atmospheric
- Serial number of sample and any description of container
- Details of any accompanying forms
- *Any toxic or corrosive fluids should be made clear*

Within the final report a completely separate section should be included to give a full list of all samples as detailed here.

This applies to all samples taken for calculation purposes as there may be a flow rate change that is not obvious from the primary data but is related to a gravity or other manual readings. Therefore, inclusion of this data in the report will help explain any sudden changes.

- Gas   Gravity, H_2S, CO_2 N_2, and any other information.
- Oil   Gravity, BS&W, viscosity, and any other information
- Water   Gravity, salinity, chlorides, and any other information

Table 13.5 is an *example* format of the manual readings sheet. Note that there are two columns for BS&W. The separator BS&W has solids included; this is not always required but if the well has been stimulated it gives an indication of any solids in the separator.

TABLE 13.5
Example Manual Readings Sheet Header

Date and Time	Choke 64ths	Wellhead BS&W		Sand Detector	Gravities		CO_2	H_2S	N_2	Separator BS&W		Chlorides	Zinc Concentration	Shrinkage
		Water	Solids	Impacts per sec.	Gas	Oil	%	ppm	%	Water	Solids	ppm	mg/l	%

Wellhead BS&W

This is used as an *indicator* that the well has cleaned up and can be put through the separator. Also used in separator two-phase flow rate calculations.

Separator BS&W

This is the BS&W from the *separator oil line* used in three-phase oil flow rate calculations.

13.4.7 Supplied Data

If any data is given to the customer prior to the final report, it is important that it is labelled as "*Field Data* subject to correction."

The final report should include all the pertinent information that applies to the test performed, but this is a snapshot or a reduced set of data.

Therefore, it is essential that the customer has access to the full set of data. A full set, every reading, must be included as an ASCII file in the accompanying disc along with other data sets. It is recommended that data compression, zipped files, are not used as some company's computer systems will not accept them.

The physical file size issue must be addressed by the service company and the oil company defining the method utilised for file transfer. Anti-virus systems must be up to date and assurance should be provided that the data has been scanned prior to submission.

Often it is weeks or months before the company petroleum engineers/technologists actually examine or evaluate the data and they will ask for more data or sections to be expanded. If a full set of all the data has been supplied then it can be referred to rather than having to re-visit a job that has been archived by the service companies.

13.4.8 Names

There are two lots of names that should be on a well test report:

The name of the authors

This is important because after the report has been submitted the company may want to ask questions or query the data.

Whether the report is good or bad we need to know who wrote it. If it is multiple authors, then all names should be included; if this is impractical then the name of the most senior engineer and the person preparing the report.

Under no circumstances must just initials be used. This is a cop out and indicates there is something amiss.

Be proud of the reports.

The name of company representatives

This is not as critical as the name of the authors. However, if the company representatives are named then when the report is submitted there is someone who has first-hand information available for questions.

13.4.9 Nomenclature

Within a well test report some common abbreviations are acceptable; however all abbreviations should be explained. Within the safety departments it is documented that **TWAs** are the source of several accidents and near misses.

TWA = three-word abbreviations

$$SSV = \begin{array}{|l|} \hline \text{Sub Surface Valve} \\ \hline \text{Surface Safety Valve} \\ \hline \text{Sub Sea Valve} \\ \hline \text{Safe Sampling Volume} \\ \hline \end{array}$$

It is essential that all abbreviations used are documented and explained as they can mean multiple things. When abbreviations are used which are uncommon or exceed four letters then these must be included in the nomenclature.

However, with the power of today's document processing systems the need for abbreviations is effectively eliminated. It is just as easy for the engineer to run a search and replace on abbreviations to the full name which will both enhance the report and avoid any confusion.

13.4.10 Report Appearance

The final appearance of the report is a matter for the individual companies but what should be stressed is that once the report format has been set then it is final.

The individual operators who compile a report *are not* at liberty to change the appearance of the report, the colours, the font, etc. It is a company specific fixed format usually decided at meetings with the client before the job commenced.

Adding and removing of data columns can also take place if required depending on customer requirements but not to the detriment of the report or data.

13.4.11 Report Conclusions

It may at first glance seem a lot of work to carry out the recommendations detailed earlier, but most of the forms can be created as ready-to-use templates.

As previously stated, the well test report is the final result and should be professionally presented.

13.5 VIRUS AND MALWARE

In this day of computer viruses and malware it is critical that the data acquisition systems have a professional, active (automatically scans any device plugged into the computer), combined anti-virus and anti-malware program installed. No free ware anti-virus software should be used. The company is paying for a computerised data acquisition system which should include professional protection. This should apply to all service company computers on site.

Internet or games should not be configured on the data acquisition computer; these are the prime sources for all infections. All emailing, transmission, and other services should be from an independent reporting computer; the data acquisition computer should not be used for anything but its sole function.

The other method of reducing infection is to restrict the use of memory sticks/ flash drives/pen drives authorised to be plugged into the computer. Quite often the driller or company representative will ask for the day's data on their portable drive.

Scanning

Any device being written to by the computer must be scanned for viruses before and after writing. This is a bit of a chore but in this day of multiple infections it is essential. If there is only a single data file on the memory stick then this scanning is virtually instantaneous; multiple files and large files take longer.

Preferably a single and dedicated external storage should be used between computers; having more than one becomes risky on a number of security and virus levels. This has to be enforced by the operating company.

By reducing the number of devices that can plug into the computer it will reduce the potential of security breach and infection.

Any digital data should be accompanied by a certificate, soft copy, saying that the device has been scanned clean. Draconian measures, yes, but it is one of the only ways that we can keep the data safe.

The virus definitions should be up to date at all times if possible.

Security

All memory sticks must only contain a single copy of the data/report, older and non-essential files to be deleted.

Personal memory sticks with music etc. on them must not be used.

Strictly speaking, according to the confidentiality clauses in the contract, the drilling company should have to get the company employee's permission to have this data which should be limited to the bare minimum rather than a disc dump. This is critical in "tight holes"

This applies to all other service companies.

13.5.1 Digital Security

With the continuing development of mobile devices, the risk of data leakage is increased as every person on the rig site will now have some form of smart device.

It is easy to say that in a remote or offshore environment the access to the internet is limited; however, the operating and drilling companies will have their own links.

The security then is reliant on the radio operators and users of the company networks not to divulge the passwords. This, however, does not stop hacking.

The better way, already employed by some companies, is to limit the use of the smart devices to inside the accommodation only. Some companies have gone further and limited the type of phone to a basic phone with no camera, etc. on their locations. The level of internet security is then down to the operating company but the days of "tight hole" are now not applicable.

The simple fact that a standard mobile phone cannot be used in a hazardous area also bears this out, although there are certified mobile phones for hazardous operations available.

Another area of data leak is the passing of information from one company to another, while it is necessary for some services but not others. Just because a form calls for the flow rate it is often not necessary, especially in drillers' morning reports.

All data movement must be sanctioned by the company, and in secure or tight locations then the data transfer must be individually authorised and recorded. This applies especially to any information, digital, verbal, or otherwise, that leaves the location.

Glossary of Well Testing Terms

2-Phase Flow: This is where there are only two phases of the well stream can be separated by using a separator, usually gas and a fluid.

2-Phase Separator: A separator normally only used in production as it is designed to separate gas from liquid. This also includes water knockout pots.

3-Phase Flow: This is where all three phases of the well stream are separated into their individual components by a separator; oil, water, and gas and individually measured.

3-Phase Separator: The main separator used in well testing; designed to separate oil, water, and gas. The three (3)-phase separator can be configured and used as a two (2)-phase separator.

A

A (Ampere or Amp): The SI unit of electrical current, named after the French physicist André-Marie Ampère.

Most instruments use mA (milliamperes), a thousandth of an ampere.

Absolute Pressure: Pressure that is referenced to absolute zero expressed as psia, pressure absolute.

Normally gauge pressure is the standard; this is the pressure displayed on a dial gauge, which is absolute pressure plus the atmospheric pressure and is expressed as psig, pressure gauge.

Within the oilfield Psig = psia + 14.73

Acid: A chemical composite that is usually a liquid containing hydrogen atoms that can react with other substances to form salts. Depending upon the acid and strength, an acid can burn or dissolve other substances.

Within the oilfield, several acids are used, all of which should be considered hazardous and should be accompanied with a material safety sheet.

Acidizing: Acidizing is normally meant to be the action of pumping an acid into the formation with the purpose of increasing the well permeability by dissolving/removing any materials that inhibit the flow of the fluid from the reservoir.

Adjustable Choke: An adjustable choke is basically a mechanical device with a cone that enters a female profile, allowing restriction of flow. The adjustable choke size can be changed by rotation either clockwise, closed, or anticlockwise, open of the choke, which in turn controls the well head pressure and flow rates.

Normally found on the choke manifold along with a fixed choke option. Used primarily in clean-ups before switching to the fixed choke.

Air Pollution: Covers a multiple of options but it is essentially referring to any particles or elements, smoke, dust, etc., in the air in amounts that can be harmful to safety.

AISI: The abbreviation for the American Iron and Steel Institute.

Alkaline: Having a pH greater than 7.

Alkalinity: Alkalinity is a measurement of dissolved alkaline substances in water (higher than 7.0 pH)

Alloy: A metal composed of two or more different elements.

Alternating Current (AC): An alternating current is a flow of electrons (current) that alternates its direction of flow at a set frequency. Main electricity uses AC and varies between 50 and 60 Hz. In Europe the frequency is 50 Hz and in the USA 60 Hz.

Ampere (A or Amp): The SI unit of electrical current. Named after the French physicist André-Marie Ampère.

Most instruments use mA (milliamperes) a thousandth of an ampere.

Amplifier: A device to amplify or increase signals.

Amplitude: Is a measure of the intensity, loudness, power, strength, or volume level of a signal.

When referring to an electrical circuit operating on alternating current, the amplitude is measured as the voltage (V), stated as +V and V from the zero, or mean, depending on the direction of the current.

Annulus: The space between the tubing string and its surrounding casing.

ANSI: The American National Standards Institute.

Antifoam (or Defoamer): A substance usually injected at the choke manifold to prevent foam formation in the fluid by decreasing the surface tension.

API: The acronym for the American Petroleum Institute.

API Gravity: The gravity standard implemented by the American Petroleum Association (API) for measuring the density of oilfield hydrocarbon liquids. Normally expressed in degrees API.

ASME: American Society of Mechanical Engineers.

ASTM: American Society for Testing and Materials.

Atmospheric Pressure: The pressure applied by the weight of the atmosphere at sea level which is accepted as 14.73 pounds per square inch (psi) although this varies according to height.

B

Back Pressure: The pressure ensuing from the restriction of a fluid flow by a valve or other restriction.

Back Pressure Valve: A valve designed to control flow rates, normally on a three-phase separator, so that the pressure upstream of the valve remains constant.

Backpressure: The pressure opposing the flow through the back pressure valve.

Backpressure Valve: The valve installed on the separator gas line that controls the pressure in the separator, mounted after the Daniel Box.

Baffles: Plates or diversions that are built into a vessel, separator or tank that changes the direction of liquid flow entering the vessel.

Barrel (Bbl): The standard unit of volume used in oilfield that is equivalent to 42 gallons, abbreviated to Bbl.

One barrel (1 Bbl) is equivalent to 42 gallons (158.97 litres).

Barton: A chart recorder used on the separator gas line that records gas line pressure and differential pressure.

Basic Sediment and Water (BS&W or BS&W): As a part of the crude oil flow BS&W is a measurement of the water and other material in the liquid stream. BS&W is measured using a centrifuge.
B = Basic, S = Sediment, W = Water

Bbl (Barrel): The standard unit of volume used in oilfield that is equivalent to 42 gallons, abbreviated to Bbl.
One barrel (1 Bbl) is equivalent to 42 gallons (158.97 litres).

Bbl/D (BPD or B/D): Barrels per day.
BOPD for barrels of oil per day.
BWPD for barrels of water per day.
BLPD for barrels of liquid per day.

Bean: A fixed orifice or choke installed in either a choke manifold or similar pressure reduction area, such as a heater, and is used to regulate the flow of fluid through it from a well.

Berm: The mound or wall of earth or sand that surrounds a flare or burning pit.

BHP: Bottom Hole Pressure. Normally expressed in Psia or Bara.

BHT: Bottom Hole Temperature normally expressed in Deg F or Deg C.

Bleed Off/Bleed Down: The action of slowly draining off liquid or gas normally through a needle valve or similar valve system.

Bolting: The action of using threaded studs, bolts, screws and nuts to assemble pressure vessels and connect piping to valves and manifolds.

Bonnet: The top part of a gate valve that seals around and houses the valve stem mechanism.

Borehole: Another name for the drilled well sometimes referred to as Wellbore.

Bottle Brush: A long-handled brush used for cleaning sight glass internals.

Bottom-Hole: The deepest of the well or Borehole.

Bottom-Hole Assembly (BHA): In drilling, it is referring to the lower part of the drill string; comprised of the drill bit, centralizers, drill collars, etc.
In a completed well, this is the lower part of the completion, the mule shoe, perforated joint, packer, and other profiles and valves.

Bottom-Hole Pressure: The bottom-hole pressure is the pressure at the producing zone rather than the total depth, normally measured in psia or bara

Bottoms Up: When the fluid that was at the bottom of the well has reached the surface. Important in cleanups and surface sampling.

Brass: A metallic alloy of zinc and copper

Brine: Within the oilfield Brine is water that has significant amounts of salt/s dissolved in it.

Bring In a Well: After a well has been drilled, completed, perforated this is the act of flowing the well.
Sometimes used after a stimulation programme.

Bronze: A metallic alloy of tin and copper used in the place of Brass for non-sparking hammers.

BS&W or BS&W (Basic Sediment and Water): As a part of the crude oil flow BS&W is a measurement of the water and other material in the liquid stream. BS&W is measured using a centrifuge
B = Basic, S = Sediment, W = Water

Bubble-Point Pressure: Often referred to as Saturation Pressure; the pressure conditions (at certain temperature) at which first bubbles of entrained gas are released from the crude oil.

Buddy System: A system of working safely where one person actually carries out the task and another watches and is on standby to prevent any accidents or issues. Used predominantly when working with H_2S

Build-Up: The pressure increase at both surface and Downhole when the well is Shut In.

Burner: A customised burner specifically designed to burn crude oil and gas

C

Calibration: Comparison, verification, and adjustment of an instrument to a standard of certified accuracy and precision.

Capacity: The amount of a substance (fluid) that can be safely put or flowed through in a vessel.

Carbon Dioxide (CO_2): A colourless and odourless gas that is a part of breathable air.

When dissolved in the well water it becomes carbonic acid, which is a mild acid.

CO_2 and its derived carbonic acid are corrosive to some of the seal elastomers, O-rings, these elastomers have to be certified for their resistance to CO_2.

Cased Hole: A well, normally producing, in which casing strings have been installed.

Casing: A steel pipe used in wells to prevent the walls of the hole caving in and to seal off hydrocarbon and water-bearing formations. Casing sizes vary from 4 to 20 inches.

Casing Pressure: The surface pressure of the well casing cavity.

cc or Cubic Centimetres: The metric system (SI) units for measuring the volume of a vessel.

1 cc is equivalent to a millilitre and commonly used interchangeably.

Cellar: A hole or excavation dug around the proposed well to allow the casing head to be mounted.

Centipoise (cp): The commonly used unit of viscosity equal to 0.01 poise, which is used as the poise is a large unwieldy unit.

1 poise equals 1 g per metre-second. The centipoise is 1 g per centimetre-second.

Centrifugal Force: The force which tends to force outwards on a body spun around a centre point.

Centrifuge: A device for separating component parts of a crude oil by rotating the samples at high speed to determine the BS&W of the fluid.

Often a heater is associated with the centrifuge to reduce the viscosity before centrifuging.

Centripetal Force: The opposite of centrifugal force in that centripetal force propels the material inward.

Check Valve: A valve that only permits flow in one direction.

Chicksan: A form of swivel jointed pipework that can be configured to bend into different positions. May be used by the rig but now banned in well testing.

Choke: A type of orifice used in the choke manifold for limiting the surface pressures and flow rates.

Within well testing there are two types of chokes.
i. Adjustable choke
ii. Fixed choke

Choke Bean: The replaceable or interchangeable fixed choke profile used in choke manifolds to control the flow.

Choke Manifold: The configuration of valves, chokes (adjustable and fixed), and pipeline used to control the well's flow and pressure. Normally four gate valves with one adjustable and one fixed choke, with instrumentation, sampling and bleed needle valves.

Christmas Tree: The valves, chokes, adapters, and fittings assembled at the surface of a completed well that enables the flow of the well fluids.

The size of the Christmas Tree will depend upon the well pressure and the tubing size.

Clamp Connection: The method of joining two pressure bearing pieces using two bolted pre-formed clamps using screwed, flange, or hammer union type joints.

Classification: A classification or zone is an area where flammable materials (gasses and vapours) can exist in flammable amounts in the atmosphere.

Generally divided into several classes or divisions depending on the gasses present.

Unclassified Locations: An area or location that does not have a classification.

Classified Area: Any area or location that has a classification associated with it.

Clean Out/Clean Up: The operation of removing sand, scale, and other associated materials from the well or wellbore usually by flowing the well.

Closed-In Bottom-Hole Pressure: Usually referred as shut-in bottom-hole pressure.

This is the pressure measured by subsurface pressure transducers (Surface Readout or Memory Gauge) at or close to the producing zone with well shut in at the surface.

Coalescing: The process small dispersed water-drops in combining into larger volumes which are separated easier under gravity.

Coflexip: A proprietary brand of high pressure hoses that are generally resistant to attack by wellbore gasses and chemicals. The tendency is to refer to all armoured hoses as "Coflexip" but this is not always the case.

Coils: This also may be referred to as a tube bundle.

A coil is, as its name suggests, a tube that folds back on itself. Usually used in indirect heaters where the coil is in a heated bath, either steam or water, indirectly heating the fluid inside the coil.

Combustible Liquid: Any liquid that has a flashpoint at or above 100 °F/37.8 °C. Normally split into classes according to the temperature of combustion.

Combustion: The chemical process of rapid oxidation that is associated with the emission of heat and light in the form of a flame.

Commingled: The act of fluids of differing composition mixed together to make a single liquid stream.

Company Man: The operating company's representative on the location responsible for ensuring that the test is carried out properly and safely.

Company Representative: The operating company's representative on the location responsible for ensuring that the test is carried out properly and safely.

Competent: Usually applied to a person who has the experience, knowledge, ability, and skill to do something correctly.

Competency: A system of well test training and assessment based on tasks and questions and answers.

Compression: The action of compressing.

Compressor: A machine used to compress air, usually diesel fuelled.

Condensate: A liquid hydrocarbon that is in a gaseous state when in the reservoir but condenses to become liquid in the surface equipment.

Contaminant: An undesirable constituent or substance that affects the properties of a fluid or gas.

In the case of gasses, the contaminants are usually H_2S, CO_2, and N_2.

Contamination: The occurrence of a foreign material that has the potential to produce a detrimental effect on the fluid or other material.

Corrective Action: Any actions or measures taken to rectify/correct an unplanned event and to prevent its re-occurrence.

Corrosion: The chemical reaction of a material or the eating away of the metal by a reactive atmosphere or fluid.

Corrosion-Erosion: A combination of a material first being corroded and then undergoing an erosive process due to flow.

Couplant: A gel-like substance that has the ability of allowing ultrasonic waves to pass through unimpeded from a transducer to the material under test.

Coupon: A metal strip, or tube, with a known thickness that is installed in a pipeline which is exposed to the erosive flow. The degree of erosion measured on the coupon is; used to assess the amount of erosion taking place throughout the installation.

Critical Flow: The term Critical Flow is defined as the velocity through a choke, usually fixed, which is equal to or greater than the speed of sound.

Under critical flow conditions the system upstream of a choke becomes independent of conditions downstream of the choke and therefore any changes downstream will not have any effect on the upstream parameters.

Crossover: An adapter from one pipe format to another, either pipe size or thread.

CSA: The acronym for the Canadian Standards Association.

Cubic Centimetre (cc): The SI unit for measurement of volume.

Cu Ft: An abbreviation of cubic feet (foot)

D

d: The internal diameter of a pipe or vessel.

Daily Report: A summary of the activity and the results from the well test submitted to the operating company.

This report is subject to correction and must not be considered the same as the Final Report or End of Job Report.

Daniel Box: The short name for a Daniel Senior Orifice Plate Carrier that holds the orifice plate in the separator's gas line.

DAS: The acronym for Data Acquisition System.

Data Header (Flow-Line Header): A short piece of flow line, usually with hammer unions for installation, that has multiple tapings, usually ½" NPT, which allows instruments to monitor the conditions in the flow line and controlled access to the fluid flow for sampling as well as fluid injection as required.

Dead Oil: Crude oil that contains no dissolved gas when sampled.

Dead Weight Tester (DWT): A hydraulic instrument that uses calibrated weights to measure or apply a precise pressure. Normally in increments of 1 psi.

Dead Well: A well that does not flow.

Decibel (db): The unit for measuring the volume or level of noise and sound.

Defect: A fault within or on a piece of equipment.

Deg F (°F): Degrees Fahrenheit temperature.

Dehydration: The removal, loss, and evaporation of water from a material.

Demulsifier: The liquid chemical used in the process for the separation of crude oil-water emulsions into its constituent parts.

Used as an additive in BS&W measurements.

Density: The weight or mass of a substance per unit volume.

Density is calculated by dividing the measured mass of the object by its actual volume.

Design Pressure: The maximum allowable working pressure of a pipe or vessel at its design temperature.

Detergent: A liquid used in cleaning that has the properties of lowering surface tension, wetting, and dispersion.

Differential: The difference between two measurements.

Differential Pressure: In general, it refers to the difference in measurements of pressure between one reference point and another.

Most commonly used to measure the difference between two gas pressure tappings either side of an orifice plate on a separator gas line.

Direct Current (DC): An electric current flowing in one direction.

Dispersant: A chemical, usually a liquid, used to break down or separate the concentration of liquids or material in liquids.

Disposal: The planned removal of unwanted material.

Disposal Well: Usually, an unused well that has waste fluids pumped into it for safe disposal.

Dissolved Solids: Material dissolved in a fluid, usually water that can be either organic or inorganic in substance.

Distillation: The method of boiling a liquid into steam and then condensing the vapour back into a liquid.

Diverter: A diverter is as valve/series of valves that are opened/closed in order to redirect a flow of a liquid or gas.

DnV: The acronym for the Det Norske Veritas. A certification body.

Documentation: Recorded information pertaining to the well or well test.

Drill-Stem Test (DST): A test where the well is flowed at various rates to prove its viability before it is fully completed.

Drill String: A combination of the drill, collars, centralisers, and bit used in drilling a well.

Driller: A person employed by the rig owner/operator who oversees the operation of the rig, its personnel, and the actual drilling of the well.

Drilling Rig: Either a land or marine mobile group of equipment designed for drilling the well on either land or water/sea environments.

Dry Gas: A natural gas that is produced from a well that has no liquid hydrocarbons in the flow

It can also be a gas that has been processed (dehydrated) to remove any liquids.

Dry Hole: A well that has been drilled and has failed to discover any hydrocarbons or useful fluids.

E

Earthing (Electrical): Electrical earthing is achieved by connecting the metal body of the equipment to an earth point.

Earth Point: A point with a direct physical connection to the earth.

Elastomer: A group of materials that will return to their original profile or shape after being deformed by pressure or mechanical deformations. Mainly used in O-Ring and pressure seals.

Elbow: A piece of pipework that causes the flow to be diverted through an angle of less than 180 degrees. Most elbows are either 45 or 90 degrees and have a bore to suit the application.

Electric Line (E-Line): A braded cable of varying sizes that has one or more electrical conductors at the centre that is used to lower electronic instruments into the wellbore.

Electrical Classification of Areas (or Zones): A series of area classified according to the potential presence/release of hydrocarbons and the level of electrical certification required to safely operate electrical equipment in the area.

Electrical Enclosure (Junction Box): A case, box, or housing used for the mounting of electrical cabling, switches and associated apparatus that protects from atmospheric conditions and prevents personnel from accidentally accessing live electricity.

The enclosure can be from a simple plastic box to a fully armoured box used in hazardous areas where the box acts as a barrier to flammable atmospheres.

Electrochemical: A chemical action that is associated with flow of electricity through the chemical compounds.

Electrolyte: A liquid or gel that has the ability to allow the flow of electricity.

Electromagnet: A ferromagnetic material, such as an iron bar, that becomes a magnet when a current flows through a coil wrapped around the bar.

Elevation: The measured height above sea level.

Emergency Shutdown System (ESD): A series of manual pull buttons that are positioned about the location that, when pulled, causes a hydraulic shut in valve to close and shut the well in.

EM: The acronym for Electromagnetic Inspection used in wall thickness inspection.

Emission Standard: A measurement of the maximum level pollutant that is permitted to be emitted from a source.

Emulsifier: A liquid that assists in enabling the mixing of two liquids that normally do not mix to form an emulsion.

Emulsion: A mixture of crude oil and formation water that will not normally separate and need treatment, demulsifier injection/centrifuge before the two liquids will separate.

Enclosure: A structure used to provide environmental protection for equipment.

Enclosure, Explosion-Proof: An electrical box or enclosure, usually manufactured of metal, whose main purpose is to prevent the ignition of a flammable atmosphere. The enclosure will also have the capability of withstanding a fire or explosion.

Enclosure, Purged: An enclosure that has a continuous clean air flow at a positive pressure to reduce the ingress of any flammable atmospheres into the enclosure.

Entrained Gas: Natural gas that is suspended/dissolved in a liquid.

Entrained Liquids: Vapour-like liquid drops that are suspended in a gas.

Environment: The surroundings and conditions in which a living organism lives and survives.

EP or ExD: The Explosion Proof marking, normally on equipment or junction boxes.

EPA: The acronym for the American Federal Environmental Protection Agency responsible for establishing and enforcing environmental standards.

Equipment: A combination of different components that is assembled and manufactured to perform a specific purpose.

ESD: A series of manual pull buttons that are positioned about the location that, when pulled, causes a hydraulic shut in valve to close and shut the well in.

Excess Temperature: A temperature that exceeds the rated designed working temperature.

Exploratory Well: An exploratory well is usually a single well drilled in order to verify the presence of producible hydrocarbons.

Explosive Limits: The explosive limits are the limits or concentration of a gas, vapour, or dust, which details the concentrations, in air, at which an explosion or fire can exist when in the presence of an ignition source.

There are defined upper and lower explosive limits which are different for every gas, vapour, or dust.

External Thread: A thread normally cut on the outside of pipework.

Extra Heavy Oil: Crude oil with an API gravity <10.

F

Fb: Orifice Plate Factor used in gas calculations.

oF or Deg F: Degrees Fahrenheit temperature.

Fail Safe: A system in the event of failure or breakdown of a specific piece of equipment that will be automatically activated to ensure continual safety of the process.

Fail-Safe Device: A device which will automatically activate and make the process safe.

Failure: Impaired or restricted performance and operation of a system that prevents the design function being achieved.

Fatigue: The failure of a metal under repeated loading and stress during normal operations.

Fatigue Failure: The continual and cumulative use of equipment that results in repeated stress.

Female Connection: A connection with the threads on the inside.

Field: A group of wells drilled in the same reservoir individually completed.

Field Repair: Disassembly, reassembly, and functional testing of the well test equipment or instruments.

This type of repair does not include manufacturing operations.

Any field repairs carried out or attempted must be recorded and the functionally witnessed which may include pressure testing.

Filtration: The method of separating solids from a liquid.

Fire: Combustion of a flammable material resulting in light flame and heat.

Fitness-For-Purpose: Designed and manufactured specifically for a job or purpose.

Fittings: Smaller pipework, nipples, and needle valves used primarily for instrumentation.

Flame Arrestor: A device that is normally placed on an exhaust or vent that prevents the propagation of a flame from a potentially flammable atmosphere.

Flammable: A substance that is capable of igniting very easily or having a fast rate of flame spreading

Flange: A collar-like edge mounted at the end of a pipe with holes to allow studding to be used to securely join flanges together.

There is a seal ring (gasket) placed between the two flanges that have a recessed profile, in both flanges, where the gasket sits.

Flange, Blind: A type of flange with no bore that is used to seat the pipework

Flange, Threaded: A flange having a sealing face on one side, a female thread on the other and regular bolt holes.

Used for joining pipes to a flange but in most places not allowed for use with hydrocarbons.

Flash Gas Liberation: Flash gas is where the pressure is lowered rapidly allowing uncontrolled gas to escape from solution.

Flash Point: The lowermost temperature where the vapour pressure of the fluid becomes an ignitable mixture with air.

Flaw: A gap or damage in a product or assembly which does not necessarily call for rejection of the product.

Flow Line: A pipe used to flow the well fluids through to various pieces of equipment necessary in oil and gas production.

The flow line can have multiple sizes and multiple types of connections.

Flow-Line Header (Data Header): A short piece of flow line, usually with hammer unions for installation, that has multiple tapings, usually ½″ NPT, which allows instruments to monitor the conditions in the flow line and controlled access to the fluid flow for sampling as well as fluid injection as required.

Flow-Samples: A sample of the fluid flow that can be taken either at atmospheric pressure or at the pressure in the flow line.

Flow samples can be taken from differing locations on the well testing package depending upon purpose.

Flowing Bottom hole Pressure: The pressure taken using Downhole pressure recording instruments at a depth near producing interval/s while the well is flowing. Different flow rates will result in differing bottom hole pressures.

Flowing Pressure: The pressure measured at the surface, usually the well head or choke manifold.

Flowing Well: A completed well that flows/produces without any aid or artificial means.

Fluid: A fluid is generally referred to as a substance that flows. Therefore liquids, gasses, and vapours are all considered as fluids.

Fluid Flow: The condition of movement of a fluid.

Within well testing Turbulent Flow and Laminar Flow are the two most common types of flow.

Fluid Level: Usually only applies to liquids. The measurement from the bottom of the fluid containment vessel to the top or surface of the liquid.

Fm: Frequency Modulation as used in radio waves and broadcasting.

FM: The acronym for the Factory Mutual Research Corporation and American certifying association.

Foam: An oil foam is a crude oil that contains dispersed gas bubbles.

Formation Pressure: The measured pressure of a shut in well. Measured at the reservoir face.

Foxboro: A chart recorder similar to the Barton Recorder usually used to record well head pressure and temperature only.

Fracture: A large crack or fissure in the formation that is either natural or generated by the process of hydraulic fracturing.

Fracturing (Fracing): The process of applying a hydraulic pressure to the reservoir formation in order to break or induce a fracture crack in the formation to permit easier movement of reservoir fluids.

Free Gas: The gas that is present and produced from a gas cap on an oil reservoir. This is different from gas that is dissolved in the reservoir liquids.

Frequency: The rate that a number of times a fluctuation repeats in a unit of time, usually a second.

Originally referred as cycles/second but the SI unit of Hertz is the unit of frequency.

Friction: Friction is the resistance to a body or mass to movement when the two surfaces are in physical contact. Heat is also generated when friction occurs.

G

g: The acceleration due to gravity expressed as 9.8 m/sec^2 in the SI system of units.

Gas Contaminants: Gasses occurring in a natural gas stream that are not hydrocarbons. H_2S, N_2, CO_2, etc.

Gauge Pressure: The measurement of a pressure that includes the pressure exerted by the atmosphere.

In the case of normal oilfield units, pounds per square inch (psi) is expressed as psig to indicate gauge pressure. Where the atmospheric pressure is not included in the pressure, this is known as Atmospheric Pressure and expressed as psia. Psig = psia + atmospheric pressure (14.73 psi)

Gaging: The method of using a calibrated measuring tape (line) to measure the liquid level in a non-pressurised tank.

Gain: To increase the output from an instrument by increasing the "amplification" of the input signal.

Gamma Rays: Part of the electromagnetic spectrum; gamma rays are a high-energy, shortwave length electromagnetic radiation that is emitted by a nucleus of a radioactive atom. Gamma rays are the strongest and more penetrating than the other two radiation types with energies of 0.010 to 10 MeV.

Gas: A fluid-like material that is in a state in which it will freely expand fill and fit to the shape and confinements of the vessel it is in. Gasses have no fixed shape, unlike solids, and possess no fixed volume unlike a liquid. A gas's volume is totally dependent on the size of its containing vessel and its pressure. Gasses are compressible while most liquids are not.

Gas Breakout: Crude oils and associated fluids that contain dissolved or entrained gas in solution will release the gas when the pressure in the vessel is reduced or as the temperature of the fluid increases.

Gas Cap: The part of the oil reservoir that is gas alone.

Gas Detection System: A system of individual gas detectors linked to a central controller that monitors the presence of gasses, both combustible and toxic. When a sensor reaches an alarm condition, the controller instigates audio visual alarms and if a critical condition can instigate a shutdown.

Gas Hydrates: These are ice-like objects that are formed by combination of gas molecules encapsulated in frozen formation water.

Gas-Liquid Ratio (GLR): The amount of standard cubic feet of natural gas that is produced from a standard stock tank barrel of crude oil.

Gas-Oil Ratio (GOR): The amount of cubic feet of gas at base conditions that is produced to an individual barrel of oil. Units are cu ft/barrel.

Gas Reaction Tube: A glass tube full of reactive chemicals that change colour in the presence of a specific compound or element. Used to detect non-hydrocarbon gasses, or contaminants, in the gas flow.

Gas Regulator: A system for controlling the pressure of the gas that flows in a flowline or vessel.

Gas Well: A completed well that predominately produces gas.

Gate: The sliding component of a gate valve that facilitates the seal.

Gate Valve: A high pressure valve that uses a sliding gate to open or close the valve usually used on choke manifolds or flow heads.

Glycol: A form of antifreeze injected into the well flow to prevent hydrate formation.

GOR (Gas-Oil Ratio): The amount of cubic feet of gas at base conditions that is produced to an individual barrel of oil. Units are cu ft/barrel.

Gradient: The pressure exerted by a fluid for each foot of fluid depth in the well. Usually in psi/ft or bar/m.

Fresh water exerts a gradient pressure of 0.433 psi/ft.

Gradient, Temperature: The measured temperature that exists with the change in depth. Usually in Deg F/ft or Deg C/m.

Gravel Pack: Gravel that is put around a slotted liner across the producing zone to stop or reduce sand production from the well.

Gravity, Specific: The density of a material that is defined as the ratio of the volume of substance in question to the volume of an equal volume of a standard known fluid.

With liquid and solids, the standard is fresh water.

With natural gas, or other gases, the standard is air.

Grind Out: A drilling term for the processes involved in BS&W.

Groundwater: Naturally occurring water present in the reservoir known as the aquifer.

H

Hammer Unions (Weco): A quick union allowing pipework to be connected without rotation of the pipe and sealed using hammers to tighten the union.

Often referred to as Weco unions.

Hand: A generic term for just about anybody who works in the field within the in the oil industry.

Green hand – A trainee:

Hard Hat: A certified moulded plastic hard hat that is mandatory for all personnel who work in the field and that protects the head from injury. Plastic hard hats have a "use by date" embossed into the material after which it is not certified safe.

HAZOP: An acronym for hazard and operability.

Hazardous Substance: Any substance that has the capability to cause illness, harm, or death.

Header: A small piece of pipework that enables samples of fluid to be taken from the flow.

Heat Exchanger: Generally, a sinusoidal tube fitted into a body that is filled with steam or hot water. The crude oil is flowed through the tube and is heated by the surrounding steam/water. This aids in separation of the crude oil and reduces its viscosity.

Heater: As heat exchanger.

Hertz (Hz): The unit of frequency defined as one cycle per second.

High Liquid Level: A high liquid level is where the liquid in a vessel is above its normal highest operating level.

HPHT: A well that exhibits a high pressure and high temperature.

High Pressure (HP): i.A definition of a well that exhibits a higher pressure than normal at the surface

ii.The pressure in a vessel/pipeline that exceeds its maximum allowable working pressure

High Temperature (HT): i. A definition of a well that exhibits a high temperature at the surface

ii. The temperature in a vessel/pipeline that exceeds its maximum allowable working temperature

Hoisting: The act of lifting a piece of equipment.

Hoisting Equipment: Equipment either manually or powered used to vertically lift/raise equipment, materials, or supplies. This can be a crane or hoisting block and tackle.

Hole: The drilled well or wellbore, normally completed.

Homogeneous: Material that is of a uniform or similar composition throughout. Usually applied to a reservoir.

Horsepower (HP): The unit of power equal to 550 foot-pounds per second.

Hot Work: The act of carrying out work that can involve an explosion/fire risk and usually requires a hot work permit.

HP: See Horsepower

HPHT: An acronym for High Pressure High Temperature.

HR: The acronym for Human Resources a description given to a company's personnel department.

HSE: The acronym for Health Safety & Environment. Originally the British government department but now applies to all safety departments.

HVL: The abbreviation for a Highly Volatile Liquid.

Hydrate: A compound or mix of a hydrocarbon and water (ice crystals) that can be formed in gas streams due to a temperature reduction resulting in an expansion of the gas present when the pressure is reduced. A hydrate can look like normal ice and has the capability to plug or reduce flow.

Hydraulic: Refers to the operation of equipment by application of fluid pressure or flow.

Hydrocarbon: The group of compounds made up of only carbon and hydrogen in multiple combinations that are found in crude oils and gasses.

Hydrogen Sulphide (H_2S): The most common of the toxic and flammable gasses found in oil and gas wells. The gas has a distinguishing smell of rotten eggs in low amounts and in higher amounts can kill the sense of smell. The presence of H_2S is corrosive to materials, tubular, wellhead, and subsequent processing equipment; all these equipment must be H_2S certified.

H_2S gas is toxic even in low concentrations and can result in death.

Hydrometer: A glass instrument that floats in the liquid to determine the liquids specific gravity or density. Temperature must be taken for correction to 60 °F.

Hydrostatic Head: The pressure that is applied by a column of fluid. Usually used in the wellbore for the bottom hole pressure.

Hydrostatic Pressure: i. Pressure exerted uniformly by a pressurised liquid, normally a water.

ii. Pressure at a set depth in the wellbore as exerted by the fluid column.

Hydrostatic Test (Hydro-Test): Pressure testing using water, or water with antifreeze, in order to determine its suitability and to find any leaks. Applicable to vessels, pipework, and tubing.

Hz: The abbreviation for Hertz (cycles per second)

I

I: The abbreviation used for current in electronics. Applied to direct current systems.

IADC: The abbreviation of International Association of Drilling Contractors.

ID (id): The abbreviation for inside diameter. Usually applied to a pipe or vessel.

IEEE: The abbreviation for the Institute of Electrical and Electronics Engineers.

Incineration: The burning action in which materials (solid, liquid, and gas) are burned and converted into gasses and non-combustible residues.

Inhibit: To stop or prevent an action, whether a mechanical or chemical action.

Inhibition: To reduce the degree of a reaction normally applied to corrosion or its affect.

Inhibitor (Corrosion): A chemical that is normally injected into a process that reduces or stops a chemical reaction. Usually implemented as injection at the choke manifold for the following:
  i. Hydrate formation
  ii. Corrosion
  iii. De-foaming

Injection Pump: The pump that is used to inject inhibitors and associated liquids into the well flow.

Inlet: The actual opening of a vessel through which fluids enter.

Inspection: The physical process of examination of equipment and goods for defects and wear that can affect performance.

Interface: The boundary between two liquids or surfaces.

Internal Thread: The thread profile on the inside of pipework or coupling.

Interval: Usually, the producing or perforated section of a reservoir.

Intrinsically Safe System (IS): An electrical safety system that functions by limiting the amount of energy that could cause a spark or ignition.

> **NOTE:** All equipment connected to an IS system must be similarly certified.

Intrusive Sand Detector: An erosion system where the detector protrudes into the inside of the pipework and is subject to the effects of erosion.

J

Joint: Usually refers to a section of pipe or tubing normally in drilling and completion but can apply to well testing pipe work.

K

Katz Tables: A set of tables dealing with the properties of natural gas.
  Often referred to as Standing-Katz tables after the originating engineers.

Kill: To physically stop a well from producing or being active in any form so that the well can be worked over (serviced) or to remove the tubing and surface facilities.

Knockout: A type of vessel that is designed to remove water. Also, to remove water from the flow of a fluid as it passes through it.
  Sometimes referred to as a knockout pot.

KVA: An electrical abbreviation short for Kilo Volt-Ampere.

KW: Abbreviation of Kilowatt. A measure of 1,000 watts of electrical power.

L

Laboratory: In well testing the laboratory is a location where safe area testing can be carried out, the well tester's office and usually the location of the DAS system.
  i. Laboratories placed in a safe area do not need certification.
  ii. Laboratories placed in a hazardous area must have a certified and tested pressurised/purged air system taken from a safe location.

Laminar Flow: This type of flow is also referred to as parallel, streamline, or viscous flow.

In well testing it is where all the components of a fluid stream flow in the same direction parallel to each other and parallel to the walls of the pipe or vessel.

Leak: An accidental release of a liquid and/or gaseous hydrocarbons into the surrounding atmosphere.

Lease: The area that is defined for which a permission to drill and produce oil and gas. Also, legal document giving permission to drill and produce.

Lease Operator: The company that is responsible for the operation/maintenance of all equipment and facilities on the location. This also includes responsibility for the safe running of any wells on the lease.

LEL: An abbreviation for Lower Explosive Limit.

Liner: Pipework installed in the wellbore and cemented into place.

Liquid: Matter that freely flows and assumes the shape of the container it is placed in and retains its volume. A liquid normally can be both seen and sensed.

Unlike gas it does not expand infinitely.

Liquid Meter: An instrument to determine the volume of a liquid that flows through it. Normally measured in barrels or square meters.

Live Oil: Crude oil with gas entrained in it.

Location: The area in which an oil or gas well has been drilled, often referred to as the well site.

Log: Originally referred to as paper strip chart. Now usually refers to any data that has been recorded.

Logged: Data that has been stored usually by computerised systems.

Logging: Now refers to the act of recording data usually by a computerised system.

LOPA: The acronym for Layer of Protection Analysis.

Low Flow: Fluid flow that is below the expected minimum designed operating design or parameter.

Low Liquid Level: Where the level of a liquid in a piece of processing component is below its operating design or parameter.

Low Pressure: i. Part or equipment designed for low pressure use only.

ii. Where the pressure in a piece of processing component is below its operating design or parameter

Low Temperature: The temperature in a vessel/pipeline that is below its minimum allowable working temperature.

Lower Explosive Limit (LEL): Often referred to as LEL. The lowest concentration of explosive or combustible gases that when mixed with air that can be ignited, at ambient conditions, when in presence of an ignition source.

Lubricator: A section of line pipe assembly, often of different sizes, that has quick unions to mount on a wellhead/flow head to facilitate Slickline/E-line operations under pressure, in allowing wire with tools to pass into the wellbore.

The quick unions are to allow quick and easy connection and removal

M

Make-Up: To assemble or connect equipment.

Malfunction: A problem or issue with equipment that causes it to malfunction or operate improperly.

Manifold: A collection of valves and fittings that have the purpose of isolating pressure or flow as well as diverting/combining flow from/to other equipment.

Manual Readings.: Readings taken by the well test crew at regular interval independent to the DAS sensors.

Used for data checking and a backup.

Manual Readings Sheet: A record of ALL the manual readings and events during the well testing operation. Normally mandatory.

Manufacturer: The company that has fabricated the equipment or component parts.

Mass: Mass can have many meanings but in an engineering context it is the weight of a substance considering its specific gravity and particle size.

Mass Flow Meters: A specialised meter that is capable of measuring fluid volume and fluid density.

Master Valve: The principal valve in a Christmas tree or flow head that is used to isolate the well by blocking the path of the well flow.

Maximum Allowable Operating Pressure (MAWP): The highest or maximum operating pressure that can be applied to any piece of equipment.

MCF: The oilfield abbreviation for 1,000 cubic feet which is usually applied to natural gas.

MCFD: The oilfield abbreviation for 1,000 cubic feet of gas per day.

Measured Depth: A depth in the wellbore that is referenced to a set point at the surface.

Rotary Kelly Bushing on a rig or swab or maser valve on a production well.

Measuring Tank: A non-pressurised tank with a calibrated scale that enables the volume of the tank to be measured.

Meniscus: When liquid is in a column, the top of the liquid sets into a concave form which is defined as the meniscus. As seen in a liquid thermometer.

Mesh: Mesh size is defined as the average distance between the two parallel wires on a mesh measured from the centre of the wires.

A 300-mesh sieve has 300 openings per inch.

Metal Loss: Metal loss is the loss of material in sections of pipework or equipment that have been exposed to a corrosive fluid. (H_2S etc.)

Meter Factor: A calibration performed on flow meters against a standard to correct any volumetric errors.

Carried out using the fluid flowed through the separator.

Methane (CH4): Usually the main component of natural gas. Methane is a colourless and flammable hydrocarbon gas that has the chemical composition CH4. Methane is often abbreviated as CH4.

Micron (µ): A unit of length or size in the SI units a millionth part of a meter.

1,000 microns = 1 millimetre; 25,400 microns = 1 inch

Microns have the symbol µ pronounced Mu

Mil: A unit of length one (1)-thousandth of an inch (0.001 in)

MilliDarcy: The basic unit of permeability in reservoirs, a thousandth of a Darcy (0.001 Darcy)

Millilitre (mL): Part of the SI units for the measure of liquid volume. The standard unit of volume is the Litre (L) A millilitre is a thousandth of a litre (0.001 litres). The millilitre can be used interchangeably with cubic centimetres (cc), which is also a unit of volume.

Mist Extractor: Part of the internals of a three-phase separator that helps in the removal of moisture and condensable hydrocarbons in the gas chamber of a separator.

MMCF: The oilfield abbreviation for 1,000,000 cubic feet used exclusively to measure gas volumes.

The more common representative of gas volume is MMSCF, where SCF is the abbreviation for standard cubic feet (gas corrected to standard conditions).

Gas flow is usually MMSCF/D.

MMSCFD: The standard oilfield unit of gas flow 1,000,000 standard cubic feet of gas per day.

Molecule: Defined as the smallest part of any substance that can exist and exhibit all the properties of its original substance.

Molecular Weight: This is the summation of all the individual atomic weights of all the atoms that constitute the molecule.

Morning Report.: A summary of the activity and the results from the well test submitted to the operating company.

This report is subject to correction and must not be considered the same as the Final Report or End of Job Report.

MPI: An abbreviation for Magnetic Particle Inspection.

MSCF: An oilfield term for a thousand (1,000) standard cubic feet of gas.

MSCF (MCF): An oilfield term for thousands (1,000's) of standard cubic feet of gas which is used to measure the gas flow or volume from a separator or another instrument.

MSCF/B (MCF/B): An oilfield term for thousands (1,000's) of cubic feet per barrel. Which is a measure of the amount of gas entrained in a barrel of crude oil.

MSCFD: A standard oilfield measurement of natural gas, thousands (1,000's) of standard cubic feet of gas per day.

MUD: A complex liquid mix of chemicals and that is circulated round the well-bore during drilling and work over operations as a lubricant and to contain the reservoir pressure.

MWP: The acronym for Maximum Working Pressure.

N

NACE: National Association of Corrosion Engineers

Natural Gas: A naturally occurring gas which is a mixture of different hydrocarbons and other non-hydrocarbon gasses that exist in different concentrations dependent upon the reservoir and depth.

NDT: An abbreviation for Non-Destructive Testing.

Needle Valve: A standard valve in the oilfield, a small, usually half inch (1/2″) valve having conical seal assembly, hence the name needle valve. A needle valve is limited in its size, therefore it is mainly used on instrumentation and where control of small quantities of fluid is required as in surface sampling.

Nipple: Normally a small pipe fitting that is threaded on both ends using an NPT thread.

Non-Conformance: Deviation from a specified process or requirements.

Non-Destructive Testing (NDT): A technique of mechanical inspections for flaws and defects in material that does not damage the equipment undergoing test.

Non-Intrusive Sand Detector: A sand detector that senses sand particles in the flow stream impacting on the internal side of the pipe that cause an acoustic profile that is measured by a sensor clamped to the outside of the pipe.

Non-Sparking Hammer: A hammer that is used in hazardous areas made of metal that does not produce a spark when used.

Normally Closed Valve: A hydraulic, pneumatic, or electrically activated valve that closes upon the loss of pressure or power rendering the process safe. The oil control valve on a separator are of this type of valve.

Normally Open Valve: A hydraulic, pneumatic, or electrically activated valve that opens upon the loss of pressure or power rendering the process safe. The gas control valve on a separator are of this type of valve.

O

Offshore: The term offshore relates to oil exploration and production that is carried out below the bed of the sea, ocean, or natural body of water.

Operator: Generally, refers to a company that is responsible for the process of drilling and running an oil field.

Oil-Water Interface: Due to oil having a lower specific gravity than water it will automatically migrate to the top of the water body. The interface between the two liquids which is the bottom of the oil body and the top of the water body is referred to as the oil water interface.

Orifice Meter: The most common form of gas metering on a separator where gas flows across a known opening (orifice), there is a pressure drop (differential) created. By measuring the differential pressure and other variables associated with the orifice plate, the gas flow rates can be calculated.

Orifice Plate: A metal plate, usually stainless steel, with a precision drilled hole in the centre that is installed in the separator gas line to determine the gas flow. Orifice plates are manufactured in differing sizes to suit the gas flow rates.

OSHA: The Occupational Safety and Health Administration.

P

Packer: A packer is an expandable plug that is installed in a wellbore, normally near the bottom, that isolates the tubing and the annulus.

Paraffin: A heavy wax like hydrocarbon that tends to build up on the walls of the production tubing and if left untreated can restrict the well flow.

Parts Per Million (ppm): Parts per million is a measuring unit used to express a measurement of small amounts of a substance with respect to its surrounding materials. Primary use is in the measurement of H_2S gas component.

Particle: A small piece of material that is often a single crystal or sand of a regular proportion.

Particle Size: As its name implies, it is the physical size of the measured particle. Particles are measured as a diameter in microns.

PDA: An acronym for Personal Digital Assistant is basically palm-sized computer/phone that can connect to the web and email.

Percentage Water: Generally, the amount of water entrained or dissolved in crude oil, measured as a percentage of the volume. Used in BS&W readings along with solids.

Percentage: The standard oilfield measurement to express the volume of a material that is suspended/contained in another material.

Perforating: The process of punching holes into a body usually by an explosive type device. Perforations are carried out into the formation directly or through the casing.

Perforation Depth: The depth or range of depths of the perforations in the wellbore. Usually measured in feet.

Permeability: The rock or formation property to inter-connect a porous medium so that fluids flow from the reservoir to the wellbore. Permeability is measured in Darcy's (D) but this is such a large and unworkable unit the milliDarcy (mD) is standard in the oilfield.

Permit to Work (Work Permit): A permit issued to the company to enable hazardous operations to be carried out by the safety representative or operations. On a rig this is usually the rig Toolpusher.

The permit must be signed and approved before work commences. A copy is held with the issuer and signed off when the work is completed or a shift change.

Petroleum: Hydrocarbon fluids that are produced from underground reservoirs.

pH: Usually accepted as a measurement of acidity or alkalinity, pH is an abbreviation for Potential Hydrogen ion.

Pipe (Pipeline): In the oilfield a pipe is a long hollow piece of metal to conduct the flow of fluids

Pit: A large hole or pit that is usually dug by the drilling operator to store liquid materials used in the drilling operations.

Other pits can be dug for the storage of water and other chemicals.

Pilot Tube: A liquid flow measuring instrument that works on differential pressure.

Poise: The unit of viscosity in the SI units.

Pollution: Where discarded or emitted unwanted matter is discharged into the environment and causes undesirable effects and consequences.

Porosity: A reservoir engineering term porosity is the void space or volume in a formation that can contain fluids.

Porosity is usually measured in percentage of voids per volume.

Portable Well Test Unit: This is usually a mobile single trailer unit, although it can be more, with all the necessary equipment to carry out a temporary well test.

Positive Choke: Often referred to as a Fixed Choke. Unlike an adjustable choke the positive choke has a fixed orifice size normally lined with tungsten carbide for protection against erosion.

Positive-Displacement Meter (PD Meter): A meter used in measuring the flow of a fluid by utilizing chambers of a known volume. Often used as a calibration or meter check instrument.

PPE: The acronym for Personal Protective Equipment.

Pressure: Pressure is defined as force that is applied over a surface, force/area. The normal units of oilfield pressure are Psig, Psia, and Bar.

Pressure Drawdown: Pressure reduction created by reducing the bottom hole pressure at the wellbore to drive fluids from the reservoir into the wellbore.

Pressure Gauge: The main instrument for measuring pressure.

Pressure Gradient: The change in pressure from one depth to another in the wellbore measured in psi/ft.

Pressure Relief Valve.: A safety valve which vents the internal pressure if it exceeds the maximum set limit. Mounted in pairs on separators or other pressure bearing equipment (mainly vessels) for safety.

Pressure Maintenance: The process of assisting to maintain a reservoir pressure by injecting a fluid, normally treated water or gas.

Pressure Sensor: An instrument that detects pressure by the deformation of a material.

Produced Gas: The actual gas volume that is produced from the wellbore.

Production: The fluids that are produced from an oil or gas well or the department responsible for the flowing of a well through the surface processing equipment to export and storage.

Proppant: Generally, a sand-like substance used in the fracturing process that serves to prop open the fractures after the process of hydraulic fracturing has forced a crack open in the formation rock.

Psi: The abbreviation for pounds per square inch. Psi = Psig

Psia: The abbreviation for pounds per square inch atmospheric. This does include atmospheric pressure. Generally, Psia = Psig + 14.73

Psig: The abbreviation for pounds per square inch gauge. This does not include atmospheric pressure.

PTFE: The acronym for polytetrafluoroethylene, marketed under the name Teflon. Used in a tape form to aid in sealing NPT threads.

Pump: i.A device used to move fluids from one vessel to another vessel in a different location.

    ii. A device used to increase the pressure in a vessel or pipeline, used in pressure testing.

Pumping Unit: Usually part of a rig's equipment, a high-pressure unit for pumping fluids, mud, and cement into the wellbore. Often used for pressure testing surface equipment.

Pup Joint: A piece of pipe/pipeline or tubing that is shorter than the standard length. Used in spacing out in the wellbore and at the surface facilities.

Q

Qualified Person: A person who has the qualifications, training, experience, and ability to perform an operation.

Quality Assurance (QA): Actions that are designed to ensure that all materials and actions are carried out correctly to prevent any malfunctions in the equipment.

Quick Union: A type of union that does not need specialist tools to open and close it. The union has a coarse thread with O-Ring seals to ensure a pressure seal. Used on Slickline Lubricators, Sand Filters, etc.

R

Ranarex: A version of a gas gravitometer used for measuring natural gas gravity.

Rankine: A unit of temperature similar to the Fahrenheit scale

Deg R = Deg F + 459.69

Rated Working Pressure: The maximum pressure that a piece of equipment is designed to work under. Working pressure is not the same as test pressure.

RAW: RAW readings are the unprocessed data sets from complex electronic devices and transducers and is shortened from Raw Readings.

Records: Generally, information and data that has been recorded, usually on electronic media, that pertain to the well and any well test performed.

Relief Valve: A safety valve which vents the internal pressure if it exceeds the maximum set limit. Usually mounted in pairs on separators, and other equipment (mainly vessels) for over pressure safety.

Reservoir: i.In reservoir engineering a reservoir is a subsurface permeable rock formation that holds hydrocarbon deposits.

ii. Storage of a liquid used in hydraulics that automatically keeps the system full of liquid.

Reservoir Fluid: The fluid naturally occurring in the reservoir.

Reservoir Pressure: Generally, the average pressure that is in a hydrocarbon or water-bearing formation or reservoir.

Retention Time: The length of time that any of the fluids are physically retained in the separator, or other vessel.

Reynolds Number: In well testing a dimensionless number that is used in the calculation of gas flow rates.

Rigging Up: Assembly of all the well test, or other service company associated equipment, on the location. Usually followed by a pressure test to verify the integrity of the rig up.

Ringlemann Scale: A measure of smoke density.

RP: Abbreviation for Recommended Practice, which is an API guideline on how an operation should be carried out.

S

Safety Device: A separate instrument to either monitor the parameters of a piece of equipment or a device that shuts down or vents any excess pressure or fluid.

Safety Factor: The ratio of the calculated or tested working parameter to the maximum permitted working parameter.

Safety Hat: A form of certified head protection made of a plastic. Metal safety hats are banned.

Safety Valve: A safety valve which vents the internal pressure if it exceeds the maximum set limit. Usually mounted in pairs on separators, and other equipment (mainly vessels) for over pressure safety.

Salinity: The amount of salt dissolved in water, measured in ppm or as mg/l.

Salt: In oilfield terms the general description of "salt" is applied to multiple chemical compounds but primarily to sodium chloride, NaCl–

Sand: A small granular substance remnant of rocks that have disintegrated and eroded over time, mainly quartz grains but other rock fragments can also be present.

Sandstone: A form of sedimentary rock that is mainly made of the mineral quartz.

Saturated Liquid: A liquid at a given temperature contains as much of a dissolved material as it possible, without any material dropout.

Saturated Solution: A solution is said to be saturated if it contains, at a given temperature, as much of a dissolved material as it can without any material dropout.

SCADA: The acronym for Supervisory Control and Data Acquisition.

Scale: In oilfield terms, scale is a deposit formed in place from a reaction with either, chemicals, temperature, or pressure variations, that are in contact or immersed in water. Scale often is formed on the inside of the production tubing.

SCF: The abbreviation for Standard Cubic Foot of fluid that has been corrected to standard conditions of 14.73 psia and 60 °F.

SCF/STB: The abbreviation for Standard cubic feet per stock tank barrel.

Sensor: An instrument that measures the operating conditions of the equipment it is mounted on. Can also be used to trigger a shutdown or safety alarm.

Separator: The main vessel in well testing, the separator separates the fluids from the well and separates them into their main components oil, water, and gas.

Separator Gas-Oil Ratio: A calculated value from the ratio of the separator gas rate divided by the separator oil rate.

SG or sg: The standard abbreviation for Specific Gravity.

Shake Out: Alternative name for a BS&W.

As a part of the crude oil flow BS&W is a measurement of the Water and other material in the liquid stream. BS&W is measured using a centrifuge. B = Basic, S = Sediment, W = Water

Shrinkage: The reduction of the measured volume of a crude oil as a result of dissolved gas escaping from the liquid due to the pressure being reduced.

Shut In: The action of closing the surface valves on the wellhead so it ceases production.

Shut-In Bottom-Hole Pressure: The pressure that is at the bottom or perforations of the wellbore when the well is Shut-in.

Shut-In Pressure: The pressure that is at the surface of the wellbore at the well head or choke manifold when the well is Shut-in.

Slickline: A method of running instruments or tools into the wellbore while under pressure using a smooth (slick), single-strand, high-strength, steel wire.

Solvent: A normally volatile liquid used as a cutting agent or to soften or dissolve a substance. Always check the safety sheets.

SOP: An abbreviation for Standard Operating Procedures.

Sour Crude: The term for oil that has a toxic gas entrained in it, H_2S.

Sour Gas: The usual definition for a natural gas containing Hydrogen Sulphide (H_2S)

Spark Arrestor: A device placed on the exhaust of an engine (diesel) to prevent the discharge of any sparks that may result in a fire or explosion.

Specific Gravity: Defined as the ratio of the weight of a material when compared to an equal weight of an equal volume of a standard material.
Water is the standard for liquids and air is the standard for gases.

SSCSV: The abbreviation for a Sub Surface Controlled Safety Valve.

SSSV: The abbreviation for Subsurface Safety Valve

Stage Separation: Where one separator is not sufficient to separate the crude oil, the flow is passed through two (2) or more separators.

Standard: A name that is given to the data that details all the Specifications, Operations, and Recommended Practices that set out how a process is to be followed.

Standardisation: The exercise of ensuring that all trainings, routines equipment, and instruments are all conforming to the same agreed practice.

STB: The abbreviation for Stock Tank Barrels.

STB/D –: The abbreviation for Stock Tank Barrels per Day.

Stock Tank: In well testing a stock tank is a tank for the temporary storage of produced liquids at atmospheric pressure and for use in determining a combined meter factor for liquid flowmeters.

Subsurface Equipment: A general term for any equipment that is installed in the wellbore to perform operations below the wellhead.

Sulphur Dioxide: A toxic gas that is a by-product of burning Hydrogen Sulphide (H_2S).

Supervisor: A well test operator who has gained sufficient experience and qualifications to supervise and control all aspects of a well test.

Surface Pressure: The pressure measured at the wellhead or choke manifold.

Surge Tank: In well testing a surge tank is a tank for the temporary storage of produced liquids at the separator pressure - and for use in determining a meter factor for liquid meters.

Suspended Solids (SS): These are small solid particles that resist separation by conventional means.

Swabbing: The operation of assisting the well to flow normally using Slickline.

Swab Valve: The top valve in a standard well head used mainly for slickline and e-line operations as well as any coiled tubing or swabbing.

T

Tap or Tapping: A threaded port into a flow line or vessel where a needle valve can be installed. Gauges or other instruments are installed after a needle type valve as a safety factor so they can be removed while pressure still applied.

TD: The abbreviation for the Total depth which is the maximum depth that the hole has been drilled.

Teflon: The commercial name for polytetrafluoroethylene (PTFE). Used in a tape form to aid in sealing in NPT threads.

Temperature: The physical quantity that defines hot and cold as degrees Fahrenheit (°F) or metric units as degrees Celsius (°C).

Test Pressure (TP): The pressure that is to be applied to a piece of equipment.

Texsteam: A high-pressure low-volume pump used primarily in chemical injection.

Thermowell: A sealed piece of pipe that is inserted into a pipeline to facilitate the entry of a thermometer.

Third Party Inspector: A specialist that is an independent contractor who is brought in to determine whether the equipment complies with the specification. This can be from material to pressure testing.

Thread Half: The part of a hammer union that holds the elastomer seal and has the wing nut hammered on to it.

Three-Phase Flow: This is where all three phases of the well stream are separated into their individual components by a separator; oil, water, and gas and individually measured.

Three-Phase Separator: The main separator used in well testing; designed to separate oil, water, and gas. The three (3)-phase separator can be configured and used as a two (2)-phase separator.

Tool Box Talk: Generally, an informal safety meeting carried out before performing a potentially hazardous operation. Must be documented and all attendees signed as present.

Toolpusher: The manager in charge of all aspects of operation on a drilling rig.

Transducer: Another name for a sensor but a transducer is an instrument that converts one parameter to another. As an example, a pressure transducer converts pressure to an electrical current.

Tubing: In wellbore drilling the tubing is the pipe that reaches from the top (surface) of the well to the producing zone (perforations) and is the medium for transporting fluid to the surface.

Tungsten Carbide: Used predominantly in well testing as the liner for fixed chokes and the cone for adjustable chokes. Tungsten carbide a very hard substance made from tungsten and carbon at high temperatures, very resistant to erosion.

Two Phase Flow: This is where there are only two phases of the well stream can be separated by using a separator, usually gas and a fluid.

Two Phase Separator: A separator normally only used in production as it is designed to separate gas from liquid. This also includes water knockout pots.

TWOP: The acronym for Test the Well On Paper. A process where all the involved companies and key personnel meet to discuss a forthcoming test.

U

Union: A device to connect or couple that negates the need to rotate the pipe.

V

Vacuum: Where the pressure in a vessel or pipe is below the atmospheric pressure.

Valve: A piece of equipment that is designed to open or close a line that contains pressure; shutting off flow and isolating pressure.

Valve (Gate): A gate valve in well testing is a multi-turn valve that lifts/lowers a gate into position to effect either opening or closing. A gate valve only has two positions open or closed any attempt to partially open the valve for flow control will result in permanent damage to the seat and failure to hold pressure.

Valve, Master: The lowest valve in the Christmas tree/flow head used to completely isolate the pressure and flow from the wellbore.

Valve, Wing: An output valve located on the cross-piece valves, used for isolation and to shut off the flow from the wellbore.

Velocity: The speed of flow in a pipeline or wellbore.

Vent: An uncontrolled opening in a vessel to allow gas to escape and prevent pressure build up.

Vertical Depth: The total depth of a wellbore measured vertically.

Viscosity: Viscosity is defined as the resistance of a fluid to a change in shape, or movement of adjacent portions relative to each another. Viscosity denotes opposition to flow.

Volatile Liquid: A flammable liquid that has a flash point below 100 °F (37.8 °C).

Voltage (V): An electromotive force or potential difference measured in volts and standard symbol V, with two different types' direct current (DC) and alternating current (AC)

Vortex: A whirlpool like whirling fluid that forms as liquid exits a smaller outlet. As observed in wash basins.

W

Weir Plate: A weir is considered as a barrier, which a liquid flows over. In a separator the water level is kept below the level height of the weir plate so only oil flows over the weir plate.

Weld Joint: Two parts that are joined together by welding, normally pipework.

Well: The actual drilled hole from the surface to the total depth drilled to facilitate the flow of hydrocarbons from the reservoir.

Well Depth: A measurement of the well depth that is referenced from a point on the rig floor, usually the rotary Kelly bushing on a rig or from the ground on a land location as a reference point.

Well Test: An action carried out by a specialist company to flow the well through specialist equipment to determine the pressures and flow rates that the well is capable of producing.

Well Tester: An individual who is part of a crew that operates the well test equipment.

Wellhead: Generally, an assembly constructed of casing and tubing headers, valves, chokes, and fittings used to control and regulate the flow from a well. Often referred to as a Christmas Tree.

Wet Gas: Produced natural gas that has substantial amounts of liquids in its stream. The liquids can be hydrocarbons or water.

Wing Nut: The part of a hammer union that is tightened using a hammer. Tightens on to the thread half.

Wireline: An earlier name of Slickline now referred to as Slickline.

A method of running instrument or tools into the wellbore while under pressure using a smooth (slick), single-strand, high-strength, steel wire.

Wireline is braded cable with one or more individual wires at the centre.

Working Pressure: The pressure that a piece of equipment or pipeline that can be applied to during its normal operation cycle.

Work Permit (Permit to Work): A permit issued to the company to enable hazardous operations to be carried out by the safety representative or operations. On a rig this is usually the rig Toolpusher.

The permit must be signed and approved before work commences. A copy is held with the issuer and signed off when the work is completed or a shift change.

X

Y

Z

Zone: An area of safety dependent upon the possibility of hydrocarbon release and exposure.

Index

For Product Safety Concerns and Information please contact our EU
representative GPSR@taylorandfrancis.com
Taylor & Francis Verlag GmbH, Kaufingerstraße 24, 80331 München, Germany

9 781032 623658